THE ANTHROPOLOGY OF THE FETUS

Fertility, Reproduction and Sexuality

GENERAL EDITORS:

Soraya Tremayne, Founding Director, Fertility and Reproduction Studies Group, and Research Associate, Institute of Social and Cultural Anthropology, University of Oxford.

Marcia C. Inhorn, William K. Lanman, Jr. Professor of Anthropology and International Affairs, Yale University.

Philip Kreager, Director, Fertility and Reproduction Studies Group, and Research Associate, Institute of Social and Cultural Anthropology and Institute of Human Sciences, University of Oxford.

For a full volume listing please see back matter.

THE ANTHROPOLOGY OF THE FETUS
BIOLOGY, CULTURE, AND SOCIETY

Edited by
Sallie Han, Tracy K. Betsinger, and Amy B. Scott

berghahn
NEW YORK • OXFORD
www.berghahnbooks.com

First published in 2018 by
Berghahn Books
www.berghahnbooks.com

© 2018, 2019 Sallie Han, Tracy K. Betsinger, and Amy B. Scott
First paperback edition published in 2019

All rights reserved. Except for the quotation of short passages for the purposes of criticism and review, no part of this book may be reproduced in any form or by any means, electronic or mechanical, including photocopying, recording, or any information storage and retrieval system now known or to be invented, without written permission of the publisher.

Library of Congress Cataloging-in-Publication Data

Names: Han, Sallie, editor. | Betsinger, Tracy K., editor. | Scott, Amy B., editor.
Title: The anthropology of the fetus / edited by Sallie Han, Tracy K. Betsinger, and Amy B. Scott.
Description: New York : Berghahn Books, [2018] | Series: Fertility, reproduction and sexuality : Social and cultural perspectives ; 37 | Includes bibliographical references and index.
Identifiers: LCCN 2017037761 (print) | LCCN 2017042428 (ebook) | ISBN 9781785336928 (ebook) | ISBN 9781785336911 (hardback : alk. paper)
Subjects: LCSH: Physical anthropology. | Human biology. | Fetus—Social aspects.
Classification: LCC GN60 (ebook) | LCC GN60 .A56 2017 (print) | DDC 306—dc23
LC record available at https://lccn.loc.gov/2017037761

British Library Cataloguing in Publication Data

A catalogue record for this book is available from the British Library.

ISBN 978-1-78533-691-1 hardback
ISBN 978-1-78920-501-5 paperback
ISBN 978-1-78533-692-8 ebook

To my children, Sabrina and Sam. — SH

To my own little fetuses, Anderson and Garrett. — TKB

To Declan, who defied the odds. — ABS

Contents

List of Illustrations	ix
Acknowledgments	xi
Foreword. How/Shall We Consider the Fetus? *Rayna Rapp*	xii
Introduction. Conceiving the Anthropology of the Fetus *Sallie Han, Tracy K. Betsinger, and Amy B. Scott*	1

PART I: The Fetus in Biosocial Perspective

Chapter 1. The Borderless Fetus: Temporal Complexity of the Lived Fetal Experience *Julienne Rutherford*	15
Chapter 2. The Biology of the Fetal Period: Interpreting Life from Fetal Skeletal Remains *Kathleen Ann Satterlee Blake*	34
Chapter 3. Pregnant with Ideas: Concepts of the Fetus in the Twenty-First-Century United States *Sallie Han*	59

PART II: Finding Fetuses in the Past: Archaeology and Bioarchaeology

Chapter 4. The Bioarchaeology of Fetuses *Siân E. Halcrow, Nancy Tayles, and Gail E. Elliott*	83
Chapter 5. Fetal Paleopathology: An Impossible Discipline? *Mary E. Lewis*	112

Chapter 6. The Neolithic Infant Cemetery at Gebel Ramlah
in Egypt's Western Desert   132
Jacek Kabaciński, Agnieszka Czekaj-Zastawny, and Joel D. Irish

Chapter 7. Excavating Identity: Burial Context and
Fetal Identity in Postmedieval Poland   146
Amy B. Scott and Tracy K. Betsinger

PART III: The Once and Future Fetus: Sociocultural Anthropology

Chapter 8. Waiting: The Redemption of Frozen Embryos
through Embryo Adoption and Stem Cell Research in
the United States   171
Risa D. Cromer

Chapter 9. Deploying the Fetus: Constructing Pregnancy
and Abortion in Morocco   200
Jessica Marie Newman

Chapter 10. Beyond Life Itself: The Embedded Fetuses of
Russian Orthodox Anti-Abortion Activism   227
Sonja Luehrmann

Chapter 11. The "Sound" of Life: Or, How Should We Hear
a Fetal "Voice"?   252
Rebecca Howes-Mischel

Conclusion   276
Tracy K. Betsinger, Amy B. Scott, and Sallie Han

Glossary   279

Index   284

Illustrations

Figures

1.1	Spatiotemporal borderlessness of fetal experience.	17
1.2	Implantation of the anthropoid primate blastocyst.	23
2.1	Features of pelvis.	39
2.2	Right pubic bone.	41
4.1	Second trimester fetal burial from the site of Non Ban Jak, Thailand.	87
4.2	Full-term neonate buried alongside an adult female from Khok Phanom Di.	90
5.1	Congenital defects in the thorax of a thirty-nine-week-old perinate.	119
5.2	Possible aneuploidy in a thirty-eight- to forty-week-old perinate.	123
6.1	Infant cemetery of Gebel Ramlah—distribution of burials.	137
6.2	Infant cemetery of Gebel Ramlah. Burial 20—skeleton in clearly visible grave pit.	138
7.1	Example of a copper coin burial.	154
7.2	Drawsko 1 site burial map.	156
8.1	Faux frozen embryo cryotank.	178
9.1	Embroidery work picturing a fetus, umbilical cord, placenta, and uterus.	207
10.1	Poster commissioned by the Russian Orthodox diocese of Nizhnii Novgorod, 2012.	239

10.2 Icon of the Holy Innocent Infants of Bethlehem used by
 the Saint Petersburg Life Center. 243

11.1 Public health mural in Oaxaca. 263

Tables

2.1 Time Period Definitions for In Utero and Birth
 Time Frames 36

7.1 Distribution of Burial Treatments across Age Categories 155

Acknowledgments

Many hands go into the making of a book. The editors are so grateful to the authors of these chapters for their brilliant contributions as well as for their effort, energy, and patience. We thank our series editors, the anonymous reviewers who commented on the manuscript, and Berghahn Books, especially Marion Berghahn, for their enthusiasm and support for this project. The editors would also like to thank the Harrison McCain Foundation and SUNY Oneonta for their assistance with publication costs. This has been a richly rewarding experience for us, and we encourage other scholars and researchers to seek collaborations with each other across subdisciplines and disciplines.

Foreword
How/Shall We Consider the Fetus?

Rayna Rapp

The fetus, a fetus, and the differential life chances of fetuses everywhere constitute a perfect storm of what the feminist theorist Donna Haraway would call materialsemiotic objects. Liminal in the most profound sense, fetuses serve as lightning rods for any ontology you'd care to imagine, providing our meaning-making species with a continually self-reproducing nature-culture, a biosocial or material-vitalistic entity to which every generation must necessarily address itself. Fetuses have been protected through taboo, amulets, and secrecy; through medicalization; through religious acceptance and denial etched into bioarchaeological findings exhumed from cemeteries. As the cascade of chapters in this volume shows us, fetuses represent and are represented in a dizzying array of contexts. Sometimes, they index "suspended pregnancies" in Moroccan folk practices to protect their own legitimacy, or in pagan-turned-Christian negotiated postmedieval Polish burials. Under the sign of both scientific scrutiny and misrecognition, tiny fetal osteo-traces are easily mistaken for rodent bones. North American feminist theory had a heyday analyzing the distorted sonogram images of fetuses popularized by the Right to Life Movement's propaganda in the 1980s, and Lynn Morgan long ago taught us how unstable local meanings of these "maybe babies" might be in her "Imagining the Unborn in the Ecuadorian Andes." Her *Icons of Life* is a key anthropological landmark that helps us to map the local meanderings of this domain. There is nothing standard about this biological universal as an imagined and material object. Indeed, it is open to the ascription of a surplus of meanings as various stakeholders with interests in gender and generational, institutional, and religious relations all imagine, image, and sometimes contest the status of these creatures-becoming-us.

The book you are about to read offers rich scholarly interpretation and empirical research on fetuses, insisting—rightfully—that anthropology has many superb tools for their consideration. Yet, until this volume, our field has never addressed the liminality of this consequential core to our own origins across our many research practices. Rather, from "man the hunter" through "woman the gatherer," our multiple disciplinary methodologies have taken live-born infants as the ground zero for studying the demography, sociology, and cosmology of what it means to be human. But surely, "we," in all our breathtaking chronological and cultural-geographic diversity, begin in a fetal state on whose survival or destruction every population depends. Might this be yet another androcentric layer to the human story that takes collective feminist practice to partially deconstruct?

As Julienne Rutherford points out in chapter 1, the fetus never floats free; its Janus-faced evolution depends on the virtual scientific terra incognita of the placenta and an ever-adapting maternal body for its very development. Fetuses and their women-made afterbirths-before-births are literally intertwined, interdependent—there is no living "fetus" independent of placental and maternal support. Recent fetomaternal and bioarchaeological research is beginning to reveal philosopher of science Georges Canguilhem's choreography of the normal and the pathological. This interdependency encodes health and illness, developmental pathways, flourishing, pathology, stages of vulnerability, survival, and much more. Fetuses provide exciting research material in which the species' past, its interdigitation with our environments, and health potential and futurity may all be read. And increasingly, fetal health is understood to hold potential clues to understanding the well-being and suffering of not only themselves but also their mothers and their communities.

Yet, anthropological researchers took a long time before beginning to offer rich reports of this problematic yet vital subject-object, a true ontological and evolutionary entity. Whatever we might opine about the intellectual genealogy of fetuses as work objects, we surely can say this: with the onslaught of new genomic tools for diagnosis (like the noninvasive prenatal tests that screen for certain disabilities in the first trimester of pregnancy) and dating (comparative bone mineralization across the maturation of a pregnancy in bioarchaeology), or CRISPR/Cas9 (the gene-editing tool whose protocols for human use are currently under ethical debate in many countries), the fetus and its centrality to any story of human origins are now front and center in our life sciences, our bioengineering market economies, and—of course—in our reproductive politics. The potent mash-up between differential reproductive rights/consciousness in public culture and rapid developments in technological platforms may yield both better access to maternal/child/population health information and a muffling of women's interests in insuring the centrality of their own expertise in describing pregnancy and birth in all its biosocial power-laden diversity. How/are maternal/communitarian/scientific stakes in fetuses blurred, put into conversation or competition? Fetal conception, development, and birth/death constitute only the beginning of new science-as-culture narratives.

This collection proffers an intellectual invitation: please read it carefully, for it opens up the opportunity to incorporate sensitivity to fetus-as-research-subject-and-object into our future anthropological work across a wide range of subfields and methodologies. However we position these seeds of ourselves as past, present, and future, our editors have given us innovative resources for thinking biosocially about *The Anthropology of the Fetus* and our collective scholarship will be greatly enriched by their work.

Rayna Rapp is professor of anthropology, New York University. Her research, writing, and teaching focus on the politics of reproduction; medical anthropology and science studies; gender, kinship, and disability. The author of *Testing Women, Testing the Fetus: The Social Impact of Amniocentesis in America,* and editor/ co-editor of *Toward an Anthropology of Women, Conceiving the New World Order, Promissory Notes: Women in the Transition to Socialism,* and *Articulating Hidden Histories,* she is currently completing *Disability Worlds: Crippling the New Normal in 21st Century America* with Faye Ginsburg.

Introduction

CONCEIVING THE ANTHROPOLOGY OF THE FETUS

Sallie Han, Tracy K. Betsinger, and Amy B. Scott

To say that the human fetus is a topic of especially vital interest today is an understatement. From studies in epigenetics suggesting links between fetal biology and adult health, to the sentiments and emotions that ultrasound "baby pictures" can arouse in expectant parents, to the conflicts surrounding abortion care and embryonic stem cell research, fetuses (and embryos) figure significantly in the sciences, culture and society, and politics. The fetus matters in so many dimensions of our experiences and expectations because it is both materially and metaphorically a product of the past, a marker of the present, and an embodiment of the future. Fetuses are the fragile bones discovered in prehistoric and historic graves; the specimens that have been collected, preserved, and studied toward an understanding of human growth and development; and the tissue that is used today in medical and scientific research on health and disease. Fetuses and embryos are what we all once were as biological individuals. They also are the signs or representations of our ideas and ideals held in common and in contest. Though they might be most familiar to readers, particularly in North America, as what anthropologist Lynn Morgan (2009) calls "icons of life," this is in fact a rather restricted notion of fetuses as conceived within the particular context of the culture and politics of reproduction in the late twentieth-century United States. In contrast, scholars of religion Vanessa Sasson and Jane Marie Law (2009) remind us that the fetus has been imaged and imagined historically and cross-culturally as more broadly a symbol of "inclusivity, emergence, liminality, and transformation" (3).

A thorough and thoughtful examination of the fetus requires perspectives and approaches that can attest to its complex nature and culture. For this, anthropology is especially well suited. Our discipline's methodological and theoretical frameworks enable us to approach fetuses and embryos as always biological *and* cultural *and* social. Indeed, we use the terms "fetus" and "embryo" advisedly here. While the terms have been defined as different stages of development or "becoming," they also frequently have been understood as different entities or "beings." As anthropologists, we are able to acknowledge and account for the contemporary, historical, and prehistorical ideas, practices, and processes by which fetuses and embryos are conceived and constructed as material and metaphorical bodies. Yet, when we turn toward the literature, we find not an anthropology of fetuses but various anthropologies of fetuses that have yet to come into conversation with one another.

From the vantage point of biological anthropology, the human fetus is a body of interest within the broader context of primate biology (Clancy et al. 2013). It provides evidence of the health of populations, especially of the biological consequences of social conditions and constraints, in particular maternal stressors. Recent biological research on fetal growth and development in connection with adult biology and health—what is called the developmental origins paradigm—also can be translated into changes in future practice and policy (see Rutherford, chapter 1). In archaeology and bioarchaeology, the study of fetuses has been included within a consideration of infants and children for the insights they might offer on the cultural practices and social ideas of the peoples of the past (Lewis 2007; Scott 1999). Cultural anthropology has focused on the social uses and cultural meanings of fetal images, including in medical anthropologist Janelle Taylor's *The Public Life of the Fetal Sonogram* (2008) and Lisa Mitchell's *Baby's First Picture: Fetal Ultrasound and the Politics of Fetal Subjects* (2001). Two recent books taking social and cultural approaches to the study of fetuses include medical anthropologist Lynn Morgan's *Icons of Life: A Cultural History of Human Embryos* (2009) and historian Sara Dubow's prize-winning account, *Ourselves Unborn: A History of the Fetus in Modern America* (2011), which present detailed historical accounts that emphasize that scientific knowledge of fetal biology is itself the product of cultural and social processes.

The present volume aims to begin a discussion about the human fetus that reaches across the fields of anthropology and includes other disciplines. This book presents the recent and continuing work of anthropologists working in sites from North Africa to Europe to Asia to North America and concerned with the human fetus as an entity of biological, cultural, and social significance. While each chapter is grounded in the particular concerns of specialists in archaeology, biological anthropology, cultural anthropology, and linguistic anthropology, taken as a whole, they present a perspective on the human fetus that is biosocial/biocultural, historical, and cross-cultural—in a word, holistic. Readers will find that they are already familiar with some of the material, but some of it will be new to them, especially when coming from outside their fields of specialization. For example, bioarchaeologists likely are aware of the topics covered in chapters 4 and 5, which might be unfamiliar to cultural anthropologists. Speaking from our own experiences as a cultural anthropologist and two bioarchaeologists collaborating on this volume, we encourage readers to step out of their comfort zones and read "across" the discipline. The reward will be not only to discover the work of anthropologists in other subfields but also to connect it to (and it integrate into) their own research.

In anthropology, holism is frequently upheld as an ideal, yet it is a challenge in practice. The goal of this volume is to provide readers with a multifaceted understanding of fetuses, how they are conceptualized, and how they matter as objects and subjects of study, doing so using a four-fields of anthropology approach. The chapters are organized to explore and examine the themes of biology, culture, and society and of past, present, and future. Part I includes chapters that introduce the biological, sociocultural, and archaeological significance of fetuses. The following two sections address fetuses in the past and fetuses in the present and future. Because the book is intended as a resource for scholars both outside and inside anthropology, the authors have attempted to write in clear and concise language that is accessible to readers regardless of their particular specialization, taking care to describe and explain the methods and theories that guide our practice as archaeologists, biological anthropologists, and cultural anthropologists. In addition, a glossary of key terms and concepts appears at the end of the book. In all of the chapters, the authors address a common set of questions:

1) What is a fetus? How is it defined and conceptualized in a particular field of study?
2) What methodological approaches are used—and challenged—in studying fetuses?
3) What does a study of fetuses in a given field contribute not only to scholarship in other fields but also to public concerns such as reproductive policies and practices?

The chapters here represent a range of responses to these questions, which reflect a range of concerns. How the fetus is defined is shaped by the particular modes of inquiry and practice in any given field of study. By laying bare these varied concepts, we can arrive at a more complete and nuanced understanding not only of fetuses but also of the methods, approaches, and perspectives that we might bring to their study.

For biological anthropologist Julienne Rutherford, the fetus is a biological entity with labile boundaries. In her chapter, "The Borderless Fetus: Temporal Complexity of the Lived Fetal Experience," Rutherford notes the fetus is an individual with its own genome, but that genome is the collaborative output of two other individuals, which in turn exponentializes into past generations. She also describes how the watery world in which a fetus develops has a temporal signature that reaches into the past and extends beyond gestation. In addition, the fetus as an entity does not exist without its placenta, an extrasomatic organ that must be conceptually incorporated with the fetus as the biological bridge between generations. According to Rutherford—who has conducted research with marmoset monkeys and vervet monkeys in addition to humans—a biological view of the fetal experience restricted to the time and space of the fetus's body alone is inadequate to fully situate individuals, communities, and species within the intergenerational ecologies they create and inhabit. Framing the fetus as both the fruit of previous generations and the seed from which future generations grow thus gives rise to a biology of life history that is Moëbian rather than linear. In short, Rutherford suggests the need for an understanding of the fetus that is both more expansive and inclusive.

In bioarchaeology, the fetus represents both a biological entity that can inform about past health and lifestyle but also a sociocultural being that can shed light on past cultural practices. While small and often incomplete, fetal skeletal remains can aid anthropologists in interpreting the circumstances within an archaeological population or a forensic setting, argues biological anthropologist Kathleen

Blake. In her chapter, "The Biology of the Fetal Period: Interpreting Life from Fetal Skeletal Remains," Blake describes both how to look and what we learn from looking at fetal remains. As the authors of the other chapters also suggest, Blake maintains it is a misconception that fetal remains do not survive well. Rather, the absence of fetal remains from an archaeological site is likely due to cultural burial practices. From the perspective of biologists, the fetal stage can be defined as a time of development and growth from the embryonic period until birth. During this phase, processes can be influenced by internal and external factors, including the overall health of the mother, genetic disorders, retardation of growth and development, and hormonal influences. By studying fetal remains, we can infer important information about the health and well-being of the mother and the cultural practices, disease prevalence rates, and other patterns within the community. While traits like biological sex cannot be determined with consistent accuracy, the assessment of population variants and trends might enable us to see patterns associated with sexes. A consideration of fetal remains thus might contribute to correcting our interpretations about identity, burial patterns, and gender analyses. Additionally, it can assist forensic researchers in differentiating between naturally occurring conditions and pathology.

Taking the question of what is a fetus in another direction, cultural anthropologist Sallie Han considers the quandary of what to call "it" in the first place. To refer to a fetus, a baby, or a child is to refer not only to it in its material existence but also to the social relations that surround it. In her chapter, "Pregnant with Ideas: Concepts of the Fetus in the Twenty-First-Century United States," Han suggests that to define a fetus is also to describe what is a pregnancy and what is a pregnant woman. She traces shifts in the characterization of the fetus in the United States over the past thirty years, with fetuses characterized as vulnerable (and pregnancies as conflicted and tentative) during the 1980s and then, with the ritual and routine use of imaging technologies in prenatal medical care, imagined as lively and requiring the prenatal parenting and developmental stimulus of "belly talk" during the 1990s. While an understanding of the fetus and of pregnancy as bare facts of biological life is taken for granted in the United States today, Han reminds us that this itself is an effect of particular historical and social processes. At other times and in other places, where and when fetuses do appear, they are not necessarily ascribed with the same moral, political, or scientific and medical importance and meaning.

Finding Fetuses in the Past

The work of archaeologists and bioarchaeologists particularly illustrates the importance and necessity of considering the questions of what is a fetus and the related questions about how to study it that all scholars must address. Siân Halcrow, Nancy Tayles, and Gail Elliott discuss the methodological approaches—and challenges—of undertaking a bioarchaeology of fetuses, which they situate in a broader field of study on children or subadults that has emerged in archaeology and bioarchaeology during the past two decades. Surveying the literature in bioarchaeology, they consider a range of concerns, from the uses of different terms (such as infants, newborns, neonates, and perinates) to the exclusion of infants, especially newborns or neonates (by communities themselves). Because it is, in fact, very rare to find remains in utero in an archaeological context (e.g., enclosed within the skeletal remains of the mother), the authors note that bioarchaeologists are effectively using preterm and low birth weight, full-term babies from bioarchaeological samples as proxies for fetuses. Nevertheless, Halcrow, Tayles, and Elliott argue that with the development of a robust bioarchaeology of fetuses, there is much to be gained in terms of investigations of infant care, including their feeding and weaning; diet, growth, development, and mortality; patterns of health and disease and of biocultural change; and larger cultural practices and ideas.

Examining perinatal remains from past contexts in order to identify skeletal pathology presents a number of challenges, which bioarchaeologist Mary Lewis reviews in her chapter, "Fetal Paleopathology: An Impossible Discipline?" This chapter is especially recommended for specialists in archaeology and bioarchaeology and for other scholars interested in becoming familiar with the methods and analysis of fetal skeletal remains. The chapter reviews how skeletal features—specifically, pathological lesions—can be recognized and used to identify a cause of death that provides insight into the conditions in which individuals might have lived and died. Because the majority of perinates likely died of infectious or congenital conditions, Lewis contends it is critical to develop criteria in order to distinguish pathological lesions from those resulting from the normal growth process. Also, while fetal remains recovered from the pelvic cavities of female graves hint at obstetric hazards, individual perinatal burials have the potential to tell us much about the health of the fertile maternal population, as well as the environmental factors that affect the survival of newborns.

It is frequently claimed that a focus on the fetus is a development of the modern world. According to this logic, the social "value" of children is connected historically with changing conditions that eventually lead to better health—for example, reductions in the risks that childbearing and childbirth pose to women and improvements in pregnancy outcomes—and a culture of expectation that children will be born living, survive, and even thrive into adulthood. Another claim is that the use of modern imaging technologies has cultivated an affective view of the fetus, as when sonograms are seen as occasions for expectant parents to see and "bond" with their expected children—or when ultrasound scans are made mandatory for women seeking abortion care (see Howes-Mischel, chapter 11). However, whether children were less valued in the past can be disputed, based on ancient archaeological evidence. Jacek Kabaciński, Agnieszka Czekaj-Zastawny, and Joel Irish ("The Neolithic Infant Cemetery at Gebel Ramlah in Egypt's Western Desert") describe their research on what appears to be the oldest known cemetery set aside specifically for infants. Among the remains—which have been dated between 4700 and 4350 BCE—are those surmised to have belonged to perinates. The authors contend that the existence of the cemetery is evidence of the status ascribed to infants and possibly perinates, which appear to have been not only treated with respect (in terms of burial) but also considered rightful members of the group. It suggests inclusivity, regardless of age, which they hypothesize might be an element of the complex cultural package brought by late Neolithic desert societies to the Nile Valley, when they were forced to move there because of extremely unfavorable climatic conditions. The authors also suggest a connection to the social developments of local Nile Valley groups, which led to the emergence of the Egyptian state. For Kabaciński, Czekaj-Zastawny, and Irish, this examination of an ancient cemetery for infants provides insight not only into the historical and cross-cultural diversity of ideas and practices surrounding children but also into a prehistory of significance.

What fetuses are, significantly, are cultural artifacts from which we can infer various insights into the practices and ideas of the individuals and communities that imagine, bear, care for, and preserve or dispose of them. This is true in our understanding of the past as well as the present. For archaeologists and bioarchaeologists today, the treatment of the dead represents evidence of life in the past. The meaning of a life (and a death) is made; it becomes ascribed through the deliberate efforts of which we see traces in the mortuary contexts uncovered by archaeologists and bioarchaeologists. Whether—and how—the bodies of individuals, young and old, are treated at death reflects ideas about who the dead were or more particularly what they meant to the living, as Amy Scott and Tracy Betsinger demonstrate in their chapter, "Excavating Identity: Burial Context and Fetal Identity in Postmedieval Poland." Although the skeletal remains of fetuses have often been excluded from archaeological analyses because of their poor preservation and/or misidentification, Scott and Betsinger assert the burial treatment of fetuses provides a unique opportunity to investigate what they call fetal identity. Scott and Betsinger discuss the skeletal remains of individuals, ranging in age from six months in utero to four years, who were recovered from a Polish cemetery dating to the seventeenth century. Based on the authors' examination of various aspects of mortuary context—including coffin use, grave goods, and position within the cemetery—they found no significant differences in the treatment of individuals, suggesting that fetuses were ascribed identity comparable to that of older children. This, Scott and Betsinger suggest, might be related to what they call "potentiality," or a shared perception about what the individuals would have contributed to the community had they survived.

The Once and Future Fetus

For anthropologists and other researchers and scholars, what fetuses are, significantly, are objects of study. Cultural anthropologists are especially concerned, however, with what fetuses are for the individuals and communities that become interested and invested in them. Ethnographic research enables us to document and detail the cultural ideas and social practices surrounding fetuses and embryos, which are both material and metaphorical, and ascribed with private, public, moral, and political significance.

The uncertainty surrounding embryos "left over" after in vitro fertilization (IVF) illustrates all of the above, as Risa Cromer describes in her chapter, "Waiting: The Redemption of Frozen Embryos through Embryo Adoption and Stem Cell Research in the United States." In 1998, two coinciding events in the United States thrust the growing supply of unused frozen embryos into public controversy—the establishment of the first human embryonic stem cell line *and* the creation of an adoption program for leftover embryos. What could these putatively opposing solutions for "saving" the remaining IVF embryos have in common? Cromer conducted a twenty-two-month ethnographic study at two primary field sites in California: a Christian embryo adoption program and a university's stem cell and regenerative medicine institute. Based on her fieldwork, Cromer argues that the remaining frozen embryos themselves are not inherently valuable or controversial, precious or burdensome. Rather, significant efforts at framing, classifying, and otherwise defining what these embryos are transform them into preborn persons, frozen assets, and excess waste; simultaneously, the givers of embryos become parents and sacrificial donors while the recipients of embryos become bearers of responsibility and arbiters of value. Indeed, Cromer finds that not all embryos are considered equal, at either the embryo adoption program or the stem cell research institute. Some embryos are deemed "hot commodities" while others are considered to have "special needs" and, thus, difficult to repurpose so are left "waiting." These "waiting" embryos illuminate notions of personhood and potential.

Ethnographic examinations of the ideas and practices surrounding fetuses across cultures are especially informative, as demonstrated in Jessica Newman's chapter on the fetus as presented and represented in Moroccan media and activism and Islamic jurisprudential texts. Her chapter ("Deploying the Fetus: Constructing Pregnancy and Abortion in Morocco") considers how fetuses figure in local discourses on sexuality and morality, and explores the relationships between the legal, medical, and religious conceptualizations of the fetus in Morocco. The Moroccan penal code outlawing abortion after forty days of gestation (except in cases of grave threat to the mother's health) is firmly rooted in biomedical understandings of conception and gestation. Yet, Sunni *fiqh* (religious jurisprudence) describes the stages of fetal development in rather vague terms, which make space for other more flexible and fluid understandings of pregnancy. In addition, long-standing medical and spiritual practices concerning contraception, pregnancy, and abortion complicate and inform knowledge about the fetus as a potential citizen, subject, and member of the Muslim faith. In sum, Newman's account suggests that fetuses in Morocco are the products of competing systems of knowledge.

Religion figures also in Sonja Luehrmann's "Beyond Life Itself: The Embedded Fetuses of Russian Orthodox Anti-abortion Activism." In English-language scholarship on the fetus, the ascription of personhood has been a central concern. Notably, in North American public debate, fetuses are often able to obtain the status of a personal agent by embodying biological life at its barest. In contrast, however, Luehrmann encountered various theological reservations against ascribing individual personhood to unbaptized fetuses during her ethnographic research among Russian Orthodox Christian anti-abortion activists. She found that assigning value to fetuses and asserting their humanity occurs through a process of embedding them in human collectives, such as families, the church, and the nation. As a result, she writes, ritual commemorations of past abortions do not turn the aborted fetus into a named individual that iconically represents life itself but rather represent it as a protosocial being whose membership in threatened human collectives was thwarted—and it is exactly this protosocial quality that makes fetuses effective participants in Russia's politics of reproduction today. In a setting where conservative activists argue that the fabric of the social is itself threatened, fetuses represent the weakest but also most crucial link between a collective's troubled present and its potential futures.

A focus of scholarship on the fetus has been on its visual presence. In "The 'Sound' of Life: Or, How Should We Hear a Fetal 'Voice'?" Rebecca Howes-Mischel turns our attention to its materiality as a body not just to be seen but also to be heard. Combining ethnographic and rhetorical methodologies, her chapter analyzes the cultural constructions of fetal materiality, juxtaposing two instances in which a fetus's audible heartbeat is used to make claims about its "self-evident" presence—one in Ohio during legislative hearings on a bill to restrict access to abortion care and the other in Oaxaca, Mexico, during a routine encounter between an obstetrician and her pregnant patient. As diagnostic technologies (in this case, a fetal Doppler) are used to make social claims about how to recognize fetal presence and how to respond to them, they rely on entangled cultural assumptions about the heart as the biological locus of both energetic and social life and the immediacy and intimacy of sound as a form of public sensing. In addition, they reiterate expectations about forms of "proof" offered by technological mediation that displace women's sensed and bodily relationships with their fetuses as also authoritative. This contrast between the politicized and the ordinary illustrates some of their shared presumptions, through which fetal bodies are made socially recognizable. Ultimately, this analysis highlights how reproductive politics increasingly rely on the enrollment of diagnostic technologies to make social and affective claims about the public sensing of biological materiality.

In sum, the work featured in this volume presents the directions that anthropologists across the fields have been pursuing already and suggests the rich possibilities of conceiving an anthropology of the fetus.

Sallie Han is Professor of Anthropology at State University of New York at Oneonta, and past chair of the Council on Anthropology and Reproduction. She is the author of *Pregnancy in Practice: Expectation and Experience in the Contemporary US* (Berghahn Books, 2013).

Tracy K. Betsinger is Associate Professor of Anthropology at State University of New York at Oneonta. She conducts bioarchaeological studies of health and mortuary patterns with medieval/post-medieval European populations and prehistoric populations from the Southeastern United States.

Amy B. Scott is Assistant Professor of Anthropology at the University of New Brunswick. Her research interests include biochemical analyses of health and stress, skeletal growth and development, and mortuary burial patterns in medieval and post-medieval Europe and 18th century Atlantic Canada.

References

Clancy, Kathryn, Katie Hinde, and Julienne Rutherford. 2013. *Building Babies: Primate Development in Proximate and Ultimate Development.* New York: Springer.

Dubow, Sara. 2011. *Ourselves Unborn: A History of the Fetus in Modern America.* New York: Oxford University Press

Lewis, Mary E., 2007. *The Bioarchaeology of Children: Perspectives from Biological and Forensic Anthropology.* Cambridge: Cambridge University Press.

Mitchell, Lisa. 2001. *Baby's First Picture: Ultrasound and the Politics of Fetal Subjects.* Toronto: University of Toronto Press.

Morgan, Lynn. 2009. *Icons of Life: A Cultural History of Human Embryos.* Berkeley: University of California Press.

Sasson, Vanessa, and Jane Marie Law, eds. 2009. *Imagining the Fetus: The Unborn in Myth, Religion, and Culture.* New York: Oxford University Press.

Scott, Eleanor. 1999. *The Archaeology of Infancy and Infant Death*. BAR International Series 819. Oxford: Archaeopress.

Taylor, Janelle. 2008. *The Public Life of the Fetal Sonogram: Technology, Consumption, and the Politics of Reproduction*. New Brunswick, NJ: Rutgers University Press.

Part I

THE FETUS IN BIOSOCIAL PERSPECTIVE

Chapter 1

THE BORDERLESS FETUS
TEMPORAL COMPLEXITY OF THE LIVED FETAL EXPERIENCE

Julienne Rutherford

> The completely self-contained "individual" is a myth that needs to be replaced with a more flexible description.
> (Margulis and Sagan 2002: 19)

The idea that the experience and environment of pregnancy constitute an event that produces a singular individual called the "fetus" presumes fetus as discrete entity: a fetus has a distinct body, a specific temporality (i.e., bounded by gestation), and a distinct genetic identity. In essence, something fundamental and bounded must be assumed about the identity of the individual. However, in this chapter I argue that a fetus is not an easily definable entity with clear boundaries. Certainly, a fetus can largely be defined by its existence situated within a specific time and place, but these tangible attributes obscure the spatiotemporal complexity of the fetal experience. This is not merely a semantic or conceptual puzzle but a biological reality, whose borders extend past the flesh and bone of an individual. This borderlessness is in large part due to the lived experience within the womb and the role of the placenta as the interlocutor between "mother" and "fetus"—genetically and somatically overlapping yet distinct entities existing simultaneously in overlapping yet distinct ecologies, in different life history phases.

To critique the concept of the fetus as a distinct biological individual, a brief discussion of the meaning of "individual" is warranted. There are multiple philosophical approaches to this discussion (and indeed entire literatures that are beyond the scope of this chapter, e.g., the sociolegal implications of fetal "personhood"). However, my focus is on biological concepts of individuality, given my broader focus herein on the biological experience of the fetus. This discussion is not exhaustive and is mammal-centric, but it is provided as an organizing starting point for the argument that follows. The biological concept of the individual is labile across varying, somewhat arbitrary, levels of organization (e.g., genetic distinction, physical separation from "others," etc.) (Pepper and Heron 2008: 623, table 1; Wilson and Barker 2014). The conventional concept is that of a single organismal entity that is both genetically and physically distinct from others (e.g., one monkey) (Benson 1989). This entity can reproduce itself either alone via cloning or budding, or through sexual reproduction. The entity interacts biologically and socially with other individuals and with(in) an environment but has a distinct and isolated three-dimensional form that is spatiotemporally limited. This is an organism-centered view of biological identity (cf. Gould 1980: 129) that allows for a hierarchical framework as follows (Wilson and Barker 2014):

(1) An organism (e.g., a monkey)
(2) A part of an organism (e.g., a placenta)
(3) Groups made of organisms (e.g., a family, a troop)

Given this simple framework, what is the fetus? It seems to meet the definition of an organism, but it also resides within and is physically attached to another organism, its mother. As is the case for most vertebrates, the fetus is genetically similar and dissimilar to its mother, owing half of its DNA to its father. Further, the residence the fetus inhabits is shaped by maternal biological processes and events that in turn are shaped both in the moment and through the life course by socioeconomic and psychosocial inputs (Rutherford 2009; fig. 1.1). Thus, the conceptual borders that frame the definition of an "individual" are potentially hampering our view of what it means to be a fetus in a biological sense.

To develop the concept of a biologically borderless fetus connected to multiple individuals and life history stages, I use three interrelated frames: (1) genetic complexity, or how genetic inheritance and epigenetic modifications link us to past and future generations; (2) experimental connectivity, or how the historic experiences of our mothers shape who we are as fetuses and adults; and (3) placental synchronicity, or how the placenta is a part of not only our fetal bodies but also our preconception histories and our adult futures.

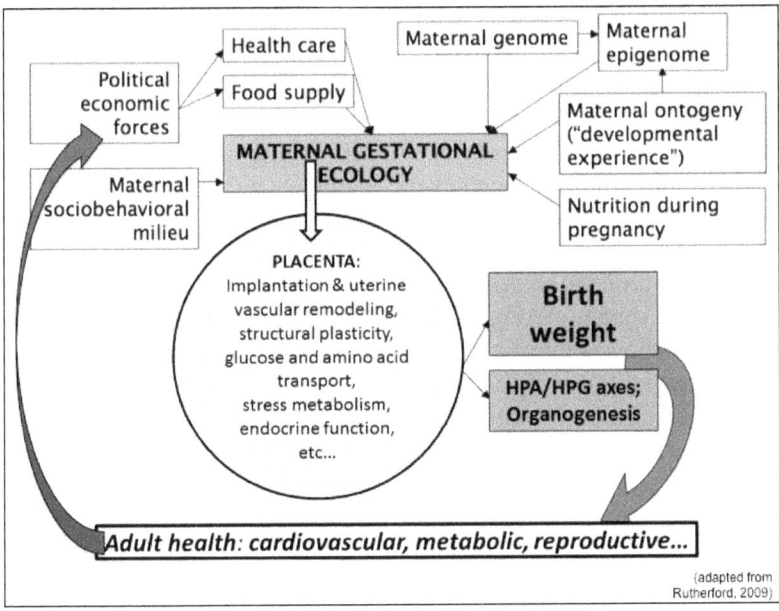

FIGURE 1.1. Spatiotemporal borderlessness of fetal experience.

Frame 1: Genetic Complexity

Who we are is shaped in some part by our genetic composition. I liken this to a musical composition (for a literal interpretation of genes as music, see Søegaard and Gahrn 2009), in which the notes on the page are intended to be played but are subject to transient circumstances that change how they are played: the timing, the duration, the omission, the improvisational noodling (Shank and Simon 2000). That composition is written most immediately by two parents, but of course their contributions are shaped by the genetic composition of their two parents, their four grandparents, their eight great-grandparents, and so on. We have incorporated basic knowledge of inheritance into our understanding of life to the extent that we don't always grasp the weight, enormity, and complexity of its importance. In the sense that we are all individual genetic melting pots, the borders between individuals and bodies dissolve in meaningful ways. We contain strands of DNA that have persisted for millennia through our immediate families and tribes, as well as our species and genera and those other families and orders and classes and phyla and kingdoms. This has great bearing on our conceptualization of the biological fetus.

One of the main reasons this connection is so important to understand is because the genome is mutable within the life course in ways we are only now beginning to appreciate. Epigenetics is the phenomenon of mechanisms acting beyond the genes (Waddington 1942). Whereas the DNA sequence inherited at conception ("genome") remains fixed across the life cycle, the "epigenome" comprises molecular processes that determine how genes are expressed in specific environmental contexts and which can be permanently modified by early life environmental experience. DNA is a tightly folded series of nucleotide sequences ("genes") that form the chromosomes. To initiate gene expression within the cell nucleus, a specific sequence of nucleotides must be briefly unwound. Gene expression can be silenced by applying chemical locks that keep that sequence from being unwound. One prominent epigenetic mechanism involves the addition of methyl groups to regions of DNA that promote gene expression (Berger 2007). The addition of these groups to these regions (i.e., "hypermethylation") locks down that sequence and, in effect, silences that gene. The formation of eggs and sperm as well as early embryogenesis are when most existing locks are erased and, in a sense, developmental trajectories reprogrammed. What is critical to understand is that this reprogramming is subject to experiences like maternal nutrition (Lillycrop et al. 2005), rearing behavior (Weaver et al. 2004), stress (Murphy and Hollingsworth 2013), and even larger societal pressures such as racism and the experience of discrimination (Sullivan 2013) and war (Rodney and Mulligan 2014)—essentially, the lived contemporaneous and historical experience of the mother that is itself a product of complex environmental interconnectivity.

This interconnectivity may span multiple generations. Several studies in humans and animal models demonstrate maternal and grandmaternal effects on offspring growth that are not explained by genetics alone. In humans, mothers who experienced famine conditions when they themselves were fetuses gave birth to small-for-gestational-age babies, with placental size varying as a function of the time of famine exposure (Lumey 1998). Further, the female offspring later gave birth to small babies, independent of adult maternal weight (Lumey 1992). Similar patterns have been demonstrated in nonhuman primates (Price and Coe 2000; Price et al. 1999) and rodents (Drake et al. 2005; Hoet and Hanson 1999). Evidence for the transmission of environmentally triggered epigenetic changes to offspring comes from a variety of animal models, for instance, showing that maternal protein restriction in a pregnant mouse predicts blood sugar in her great-grandoffspring (Benyshek et al. 2006). Similarly, studies have shown that exposing several generations of mice to maternal undernutrition yields decreased birth weights, which are restored to control range only after three consecutive generations of normal diet (Stewart et al. 1980). Epigenetic modifications in response to varying developmental environments likely underpin the large disparities in health that have come to define the lived experiences of many people, how inequality becomes embodied as biology (Gravlee 2009; Thayer and Kuzawa 2011.)

A fetus possesses DNA sequences that overall are distinct from but born of its mother and father. But this genetic code and the developmental and regulatory processes it shapes are forged in an ecological setting that is a function of experiences and processes that both precede the fertilization of an egg and extend beyond gestation, a phenomenon I explore below.

Frame 2: Experiential Connectivity

This understanding of genetic complexity—an alterable, flexible genetic code subject to mechanisms that respond to lived experience—is key to the frame of experiential connectivity. The fetus develops in the context of a complex gestational ecology that is informed by proximate maternal physiology, to be sure, but we must recognize that the boundaries of those "merely biological" processes expand outward in time and space to encompass an almost limitless range of genetic, epigenetic, political economic, dietary, familial, and additional factors occurring not only during that pregnancy but over the course of that woman's life (Rutherford 2009).

In light of this temporally and spatially interconnected developmental context, the developmental origins of health and disease (DOHaD, aka the Barker hypothesis, fetal programming, developmental programming, developmental origins) paradigm has emerged to conceptualize links between early life characteristics—features such as maternal health during pregnancy, gestational age at birth, birth weight and size, number and sex of in utero siblings, early postnatal and juvenile growth and experience—and adult health and function. Robust findings across human populations and nonhuman animal species including primates indicate that we carry with us our fetal experience. Low birth weight, which is viewed as a reflection of some kind of intrauterine restriction or perturbation, has been linked to adult cardiovascular parameters (Adair et al. 2001; Barker 1995), metabolic syndrome (Armitage et al. 2004), inflammatory response (McDade et al. 2014), mental illness (Abel et al. 2010), and other personally and societally detrimental health outcomes, including reproductive function in females. In a profound biological sense, we do not leave behind our fetal identities even as we mature into adulthood, and reaching back into the past, our fetal identities are shaped by factors that precede our conception and even our mother's conception.

My work on the early life factors that shape reproductive function in the common marmoset monkey illustrates this point. This species of marmoset monkey, like the other marmosets and tamarins, produces litters of variable size, typically observed as twins in the wild but ranging up to quadruplets and even quintuplets in captivity. Recent evidence suggests that triplet pregnancies occur across various marmoset and tamarin species in the wild as well (Bales et al. 2001; Dixson et al. 1992; Savage et al. 2009). Twins and triplets equally contribute to nearly 98 percent of all births in the marmoset colony at the Southwest National Primate Research Center, where my research team's work is conducted. Our previous research has shown that triplet fetuses experience a nutrient- and growth-restricted environment relative to that experienced by twin fetuses, with increased perinatal mortality, lower birth weight, and accelerated postnatal growth (reviewed in Rutherford 2012). Low birth weight triplets are more likely than low birth weight twins to grow into exceptionally large adults (Tardif and Bales 2004), a phenotype seen in many humans who have experienced intrauterine nutrient restriction. This is consistent with the concept of "mismatch" in which the prenatal mechanisms that allow a fetus to adjust its growth to survive in restricted intrauterine environments are offset by differential developments of organ systems that are ill-prepared for postnatal environments of relative plenty (Godfrey et al. 2007). All of these outcomes have important implications for adult health.

Because of the importance of maternal ecology and history in shaping the developmental experience of fetuses, we asked whether adult reproductive function differs between adult twin and triplet marmoset females. Life history theory suggests there are tradeoffs between different biological functions and different life history phases. By extension, an investment in merely reaching thresholds of growth and development in the face of an impoverished intrauterine environment may mean that not all body systems develop optimally, particularly if they are not essential for survival. This could play out as maximal investment in the development of the brain, which is central to global functioning, with reduced investment in the nonessential-for-survival reproductive system. Thus, greater detriments experienced during fetal life may well affect adult reproductive functioning. Links between reduced birth weight and impaired adult reproductive function have been observed in many animals, including primates such as macaques (Price and Coe 2000) and humans (Lumey 1992, 1998).

The range of litter size in the marmoset monkey offers a type of "natural experiment" in variable intrauterine environments and long-term consequences for reproduction. Our team has taken advantage of this experiment to demonstrate startling variation in adult reproductive function as a result of developmental experiences (Rutherford et al. 2014). Extensive demographic records of the marmoset colony allow us to determine that twins and triplets in fact produce roughly the same number of fetuses, averaging around 9 spread over about 3.5 litters. However, adult triplet females lose those fetuses to spontaneous abortion and stillbirth at a much higher rate, nearly three times as great, regardless of current adult characteristics such as weight. As mentioned, because triplets tend to be born at lower birth weights, this disparity could be a residual effect of low birth weight. However, when matched to twins in terms of birth weight, triplet females still are losing a significantly larger proportion of fetuses as adult. This suggests that important developmental differences exist depending on the number of fetuses sharing intrauterine resources. It also means that growth does not equal size: the fetal body itself is shaped by complex temporal and developmental processes that do not always manifest in low birth weights (Jansson and Powell 2007; Sibley et al. 2005).

Female marmoset fetuses experience another burden in utero. Marmosets are not monozygotic (produced from a single ovum), so mixed-sex litters are common. Across all females, regardless of litter size, exposure to a brother in utero led to a threefold increase in fetal loss when they were adults. This "brother effect" is possibly due to the production of testicular androgens that masculinize the female reproductive axis, although the mechanisms of this process are currently not well understood. What is clear is that the presence of a brother in utero does play a role in the much higher fetal loss rate experienced by triplet females, but even when triplets from all female litters are compared to twins from all female litters, triplets still suffer greater loss as adults, suggesting that the complex intrauterine environment experienced by triplets reflects baseline disruption that is further exacerbated by exposure to males.

More deeply, the example from the marmosets, as well as from a broader literature, means that who we were as fetuses has a direct impact on the fetuses we produce. This link suggests a sort of relativity of life history stages: though "just" a fetus, developmental processes set into motion trajectories that will affect not only the lived adult experience but also the development of the next generation of fetuses. This necessarily means that generations cycle forward and backward in time simultaneously because of this overlap in life history stages within an individual's life. This is the significance of fetal experiential connectivity.

Frame 3: Placental Synchronicity

The placenta has long been appreciated across cultures as anchoring the living back to the womb, the geographical place of birth, and previous generations throughout the life span (Buckley 2006; Schneiderman 1998). For example, the mode of disposing of the placenta after the child's birth is thought to predict a child's disposition toward home and travel in Visayan traditions in the Philippines (Demetrio 1969). Likewise, from a biological standpoint, the placenta is traditionally viewed as the direct physical interface between two individuals, the mother and fetus, in linear fashion. However, as I have argued throughout, the application of a temporally bounded linear model of the individuality of the fetus may obscure the complexity and temporality of developmental processes. A crucial piece of the gestational ecology puzzle is the placenta, through which historic, economic, and physiological experience are synchronized and conveyed to the fetus. The placenta is the mediator of maternal nutrition, metabolism, and stress that links to fetal growth rate. Amino acid metabolism by the placenta is critical for fetal growth, as amino acids are required for protein synthesis and accretion in the fetus (Regnault et al. 2002). Amino acids are actively transported from maternal to fetal circulation by transporters located in the syncytiotrophoblast, the layer of placental cells that in the human fetus separates its circulation from its mothers (Cetin 2001; Jansson 2001). Amino acid transport function is reduced in intrauterine growth restriction (Cetin 2003), hypoxia (Nelson et al. 2003), and maternal smoking (Sastry 1991), and increased in macrosomia related to gestational diabetes (Jansson et al. 2006), conditions that all have bearing on fetal development and longer-term health outcomes. Maternal condition during gestation thus influences fetal growth through effects on amino acid transport at the maternal-fetal interface. Placental metabolic dysfunction in response to maternal condition may be the primary determinant, as opposed to being a consequence, of fetal growth disruption (Jansson et al. 2006). Insofar as current maternal nutritional state is at least partially a function of life course nutritional history and experiential connectivity, placental metabolic function may be one of the links between a woman's early life events and her offspring's birth outcome via nutrient transport function.

The placenta can be viewed as an extrasomatic fetal organ. It arises from the same fertilized egg as the fetus proper. By day 6 of human development, the fertilized egg has divided scores of times and has developed into a hollow ball of cells—the blastocyst—that will implant into the uterine wall (fig. 2.2). The blastocyst is comprised of an outer cell mass, essentially the outer wall of the ball, and an inner cell mass, a clump of cells attached to the wall of cells. The outer cell mass is formed by tissue called the trophectoderm, some of the body's earliest differentiated cells, that will become the placenta. This early differentiated trophectoderm precedes the fetus

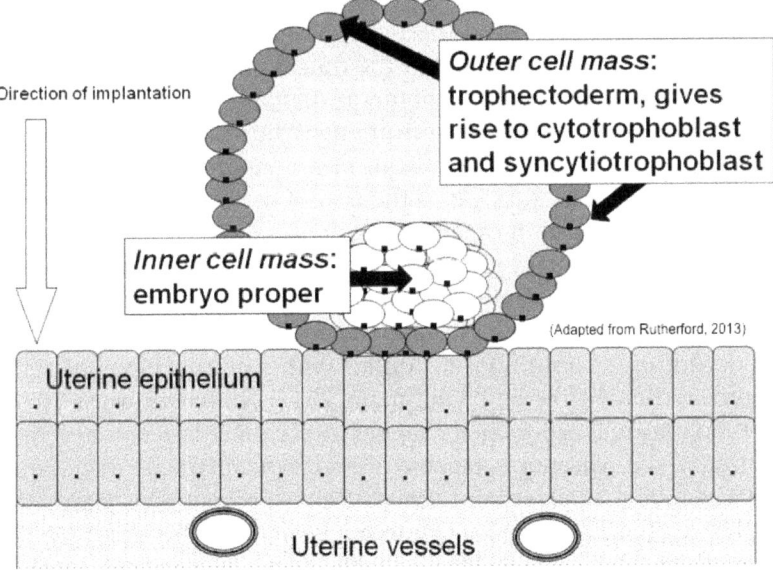

FIGURE 1.2. Implantation of the anthropoid primate blastocyst.

developmentally, differentiating and invading the uterine wall before the inner cell mass exhibits a body plan (i.e., the establishment of a "head" vs. "tail" end, "right" vs. "left" side). There is literally no fetus without a placenta. It is the fetus's most important organ, and it engages in an intimate transactional dance with the mother, coming into direct contact with her blood to withdraw nutrients and oxygen and to deposit wastes.

Gases, nutrients, and wastes are exchanged between mother and fetus through the placenta by passive diffusion, facilitated diffusion, active transport, endocytosis, and exocytosis (Murphy et al. 2006; Redmer et al. 2004). However, the placenta does not act merely as a conduit of maternal resources or as a simple filter but also has its own metabolic needs to meet (Hay 1991a, 1991b; Meschia et al. 1980). For example, the placenta consumes up to 70 percent of the glucose uptake by the uterus, significantly affecting the amount available for fetal growth (Hay 1991b). Glucose is the primary substrate for fetal metabolism and growth, particularly for fetal brain growth (Battaglia and Meschia 1986; Hahn et al. 1995; Murphy et al. 2006). The placenta is capable of modifying the balance of nutrient availability through endogenous metabolic processes such as glycolysis, oxidative phosphorylation, and amino acid interconversion, thus further altering the quantity and quality of fetal nutrition (Cetin 2001; Cetin et al. 2001; Hay 1991b). Thomas Jansson and Theresa Powell (2006) have framed the placenta as a nutrient sensor, communicating to the fetus via molecular and physiological processes information about the nutritional status of its mother, thus providing the fetus a calibration standard for growth in accordance with the availability of nutrients. Thus, the developing fetus responds and adjusts to its world, in a manner simultaneously to its advantage (e.g., survival to the end of gestation) and detriment (e.g., diminished adult capacity) (Rutherford 2009).

In normal pregnancy, the primate placenta grows in size but also in microscopic complexity such that the treelike villi containing fetal capillaries become smaller in diameter and the lining tissue called the syncytiotrophoblast that is in contact with maternal blood becomes thinner (Benirschke and Kaufmann 2012; Kulhanek et al. 1974). Beyond normal age-related changes in placental function and morphology, the placenta is capable of enormous plasticity in response to a changing or adverse intrauterine environment. This plasticity is exhibited both by morphological and functional changes. Placental growth and function respond to fluctuations in maternal nutrition and lifestyle. Evidence from experimental studies show that the response of placental growth to maternal nutrient restriction is sensitive to the timing, duration, and severity of the restriction, and that this variation in the placental growth response is reflected in changes in the fetal/placental weight ratio (Fowden et al. 2006; Myatt 2006). This is exemplified by studies of the Dutch "hunger winter" of World War II (Lumey 1998; Smith 1947; Stein et al. 1995; Stein and Susser 1975). From October 1944 to May 1945, German occupation of the Dutch town of Rotterdam restricted food supplies, lowering daily intake from 1,600 calories before occupation to 1,300 calories afterward. This was further restricted during the latter period of occupation to only 500 to 600 calories per day. Once the German forces surrendered in May 1945, the famine ended abruptly with provisions from Allied forces. Demographic records of newborn and placental weights from births occurring during the famine period have allowed researchers to determine the effect of the timing, duration, and severity of maternal nutritional restriction on fetal and placental growth. Maternal restriction during the third trimester, but not during the first or second, led to significant reductions in placental weight (Stein and Susser 1975). Caloric restriction during the first trimester with subsequent restoration of energy balance yielded no reduction in fetal weight, but placental weight was increased. Third-trimester restriction yielded significant decreases in both fetal and placental weights, but depression of placental growth was even greater than that of the fetal pattern (Lumey 1998). This variation in the ratio of fetal to placental weight suggests that the placenta is capable of responding to restriction of maternal resources during early gestation when placental growth and differentiation is approaching its peak velocity by engaging in compensatory growth pathways.

More recent observations of a Saudi population found that while mean birth weights did not differ from European weights, placentas weighed much less and were therefore more efficient. Further, Saudi mothers who restricted food intake during Ramadan during their second and third trimesters gave birth to babies who were not different in birth weight than the babies of nonrestricted women, but their placentas were smaller and more efficient (Alwasel et al. 2010, 2011). These and other studies suggest that conditions that potentially alter the availability of maternal resources to fetal and placental growth and development will yield measurable differences in the functional morphometrics of placental tissue components. In this way, a direct link between the quality of the intrauterine environment and placental structure is forged. Placental solutions to the problems of metabolic shortages may be adequate to support growth to a minimum threshold required for survival to term, a threshold that might not otherwise be met. Mechanisms of growth at the microscopic level may critically increase the efficiency of the placenta in its support of fetal growth.

Given that placental structure and function play such critical roles in determining fetal size and organogenesis, placental characteristics are, not surprisingly, strongly associated with adult health outcomes. There are associations between placental size and coronary disease (Forsén et al. 1997; Martyn et al. 1996), elevated blood pressure (Barker et al. 1990; Eriksson et al. 2000), and diabetes risk (Eriksson et al. 2000; Phipps et al. 1993). More recently, meetings in 2009 and 2011 at the Centre for Trophoblast Research at the University of Cambridge have generated anthologized discussions of the role of the placenta in human developmental programming (Burton et al. 2011; Constancia and Fowden 2012). This is a growing area of investigation as it offers a clearer path to understanding the mechanisms of developmental programming than an emphasis on birth weight alone. Indeed, Keith Godfrey suggests, "optimizing placental structure and function is likely to have lifelong health benefits for the offspring" (2002: S26).

Conclusion: Moving Beyond a Linear Timeline of Fetal Experience

It is conventional (and evidence-based) wisdom that the intrauterine environment of the fetus as it relates to maternal health is an important driver of appropriate development and positive birth outcomes. What is becoming increasingly clear is that this experience is shaped by an accumulation of experience preceding pregnancy and, in turn, shapes experience outside of the womb, even into the next generation of fetal environments. Given the evidence that environmental conditions such as nutritional stress before and during pregnancy affect birth outcomes, epigenetic modifications that specifically alter placental gene expression may play a critical role in the processes that transmit experiences across generations (Maccani and Marsit 2009).

This view of a genetically complex, experientially connected, and placentally synchronized fetus moves us past a temporally bounded definition of the biological individual. What do we gain by moving past that definition? After all, a single fetus is in many obvious ways indeed a discrete biological entity. However, recognizing the context in which discrete features are shaped offers a more fully realized understanding of the intersections of time, space, and society that shape the individual in ways that hold promise for better addressing the biology of "race" and health inequities (Gravlee 2009; Kuzawa and Sweet 2009) and implementing an effective, rather than genomically essentialized, personalized medicine (Juengst et al. 2012). Explicitly situating pregnancy and fetal development in this biological and social complexity, both backward and forward in time, rather than completely constrained by the borders of an individual uterus in an individual woman's body also provides a foundation for interrogating the supposed primacy of the "personal responsibility" any individual woman holds in insuring positive birth outcomes and life course health for her offspring (Savani et al. 2011). Nicole Stephens and colleagues suggest that explanatory models of social and health inequities that combine both the individual ("choice") and structural models are likely to produce "new tools for developing interventions that will reduce social class disparities in health" (2012: 1). Thus, the best individual and thus societal outcomes are likely to come from a blended approach to supporting women during pregnancy. By viewing the fetus as more than an individual bound by a specific timeline and immediate intrauterine milieu, pregnancy and fetal development hold promise for identifying biologically grounded mechanisms for social change and equity for women, as well as the more limited but still important goal of producing healthier fetuses.

Julienne Rutherford is an associate professor of Women, Children, and Family Health Science in the College of Nursing at the University of Illinois at Chicago. Her research centers on the dynamic maternal environment in which the fetus develops and how that time and space shapes lifetime and intergenerational health.

References

Abel, Kathryn M., Susanne Wicks, Ezra S. Susser, Christina Dalman, Marianne G. Pedersen, Preben Bo Mortensen, and Roger T. Webb. 2010. "Birth Weight, Schizophrenia, and Adult Mental Disorder: Is Risk Confined to the Smallest Babies?" *Archives General Psychiatry* 67 (9): 923–930.

Adair, Linda S., Christopher W. Kuzawa, and Judith Borja. 2001. "Maternal Energy Stores and Diet Composition during Pregnancy Program Adolescent Blood Pressure." *Circulation* 104 (9): 1034–1039.

Alwasel, S.H., Z. Abotalib, J.S. Aljarallah, C. Osmond, S.M. Alkharaz, I.M. Alhazza, G. Badr, and D.J.P. Barker. 2010. "Changes in Placental Size during Ramadan." *Placenta* 31 (7): 607–610.

Alwasel, S.H., Z. Abotalib, J.S. Aljarallah, C. Osmond, S.M. Alkharaz, I.M. Alhazza, A. Harrath, K. Thornburg, and D.J.P. Barker. 2011. "Secular Increase in Placental Weight in Saudi Arabia." *Placenta* 32 (5): 391–394.

Armitage, James A., Imran Y. Khan, Paul D. Taylor, Peter W. Nathanielsz, and Lucilla Poston. 2004. "Developmental Programming of the Metabolic Syndrome by Maternal Nutritional Imbalance: How Strong Is the Evidence from Experimental Models in Mammals?" *Journal of Physiology* 561 (2): 355–377.

Bales, Karen, Michelle O'Herron, Andrew J. Baker, and James M. Dietz. 2001. "Sources of Variability in Numbers of Live Births in Wild Golden Lion Tamarins (*Leontopithecus rosalia*)." *American Journal of Primatology* 54 (4): 211–221.

Barker, David J. P. 1995. "Fetal Origins of Coronary Heart Disease." *British Medical Journal* 311 (6998): 171–174.

Barker, D.J.P, A.R. Bull, C. Osmond, and S.J. Simmonds. 1990. "Fetal and Placental Size and Risk of Hypertension in Adult Life." *British Medical Journal* 301 (6746): 259–262.

Battaglia, Frederick, and Giacomo Meschia. 1986. *An Introduction to Fetal Physiology*. San Diego: Academic Press.

Benirschke, Kurt, Graham J. Burton, and Rebecca N. Baergen, editors. 2012. *Pathology of the Human Placenta*, 6th edition. New York: Springer.

Benson, Keith R. 1989. "Biology's 'Phoenix': Historical Perspectives on the Importance of the Organism." *American Zoologist* 29 (3): 1067–1074.

Benyshek, Daniel C., Carol S. Johnston, and John F. Martin. 2006. "Glucose Metabolism Is Altered in the Adequately-Nourished Grand-Offspring (F3 Generation) of Rats Malnourished during Gestation and Perinatal Life." *Diabetologia* 49 (5): 1117–1119.

Berger, Shelley L. 2007. "The Complex Language of Chromatin Regulation during Transcription." *Nature* 447 (7143): 407–412.

Buckley, Sarah J. 2006. "Placenta Rituals and Folklore from Around the World." *Midwifery Today* 80: 58–59.

Burton, Graham J., David J.P. Barker, Ashley Moffett, and Kent Thornburg, eds. 2011. *The Placenta and Human Developmental Programming*. Cambridge: Cambridge University Press.

Cetin, Irene. 2001. "Amino Acid Interconversions in the Fetal-Placental Unit: The Animal Model and Human Studies In Vivo." *Pediatric Research* 49 (2): 148–154.

———. 2003. "Placental transport of amino acids in normal and growth-restricted pregnancies." *European Journal of Obstetrics and Gynecology and Reproductive Biology* 110 (S1): S50–54.

Cetin, I., T. Radaelli, E. Taricco, N. Giovanni, G. Alvino, and G. Pardi. 2001. "The Endocrine and Metabolic Profile of the Growth-Retarded Fetus." *Journal of Pediatric Endocrinology and Metabolism* 14 (S6): S1497–1505.

Constancia, Miguel, and Abigail Fowden, eds. 2012. "Maternal Fetal Resource Allocation." Supplemental issue, *Placenta* 33 (S2).

Demetrio, Francisco. 1969. "Toward a Classification of Bisayan Folk Beliefs and Customs." *Philippines Studies* 17 (1): 3–39.

Dixson, A.F., G. Anzenberger, and M.A.O. Monteriro Da Cruz, I. Patel, A.J. Jeffreys. 1992. "DNA Fingerprinting of Free-Ranging Groups of Common Marmosets in NE Brazil." In *Paternity in Primates: Genetic Tests and Theories Implications of Human DNA Fingerprinting,* ed. R.D. Martin, A.F. Dixson, and E. Wickings, 192–202. Basel: Karger.

Drake, Amanda J., Brian R. Walker, and Jonathan R. Seckl. 2005. "Intergenerational Consequences of fetal Programming by In Utero Exposure to Glucocorticoids in Rats." *American Journal of Physiology: Regulatory, Integrative and Comparative Physiology* 288 (1): R34–R38.

Eriksson, Johan, Tom Forsén, Jaakko Tuomilehto, Clive Osmond, and David Barker. 2000. "Fetal and Childhood Growth and Hypertension in Adult Life." *Hypertension* 36 (5): 790–794.

Forsén Tom, J.G. Eriksson, J. Tuomilehto, K. Terramo, C. Osmond, D.J.P. Barker. 1997. "Mother's Weight in Pregnancy and Coronary Heart Disease in a Cohort of Finnish Men: Follow up Study. *British Medical Journal* 315 (7112): 837–840.

Fowden, Abigail L., James W. Ward, Peter F.B. Wooding, Alison J. Forhead, and Miguel Constancia. 2006. "Programming Placental Nutrient Transport Capacity." *Journal of Physiology* 572 (1): 5–15.

Godfrey, Keith M. 2002. "The Role of the Placenta in Fetal Programming: A Review." *Placenta* 23: S20–S27.

Godfrey, Keith M., Karen A. Lillycrop, Graham C. Burdge, Peter D. Gluckman, and Mark A. Hanson. 2007. "Epigenetic Mechanisms and the Mismatch Concept of the Developmental Origins of Health and Disease." *Pediatric Research* 61 (5): 5R–10R.

Gould, Stephen J. 1980. "Is a New and General Theory of Evolution Emerging?" *Paleobiology* 6 (1): 119–130.

Gravlee, Clarence C. 2009. "How Race Becomes Biology: Embodiment of Social Inequality." *American Journal of Human Biology* 139 (1): 47–57.

Hahn, Tom, Michaele Hartman, Astrid Blaschitz, Gerhard Skofitsch, Renate Graf, Gottfried Dohr, and Gernot Desoye. 1995. "Localisation of the High Affinity Facilitative Glucose Transporter Protein GLUT 1 in the Placenta of Human and Marmoset Monkey (*Callithrix jacchus*) and a Rat at Different Developmental Stages." *Cell and Tissue Research* 280 (1): 49–57.

Hay, William W. 1991a. "The Placenta: Not Just a Conduit of Maternal Fuel." *Diaebetes* 40 (S2): 44–50.

———. 1991b. "The Role of Placental-Fetal Interaction in Fetal Nutrition." *Seminars in Perinatology* 15 (6): 424–433.

Hoet, J.J., and Mark A. Hanson. 1999. "Intrauterine Nutrition: Its Importance during Critical Periods for Cardiovascular and Endocrine Development." *Journal of Physiology* 514 (3): 617–627.

Jansson, Thomas. 2001. "Amino Acid Transporters in the Human Placenta." *Pediatric Research* 49: 141–147.

Jansson, Thomas, and Theresa L. Powell. 2006. "Human Placental Transport in Altered Fetal Growth: Does the Placenta Function as a Nutrient Sensor? A Review." *Placenta* 27: S91–S97.

Jansson, Thomas, and Theresa L. Powell. 2007. "Role of the Placenta in Fetal Programming: Underlying Mechanisms and Potential Interventional Approaches." *Clinical Science* 113 (1): 1–13.

Jansson, T., I. Cetin, T.L. Powell, G. Desoye, T. Radaelli, A. Ericsson, and C.P. Sibley. 2006. "Placental Transport and Metabolism in Fetal Overgrowth: A Workshop Report." *Placenta* 27: S109–S113.

Juengst, Eric T., Richard A. Settersten, Jennifer R. Fishman, and Michelle L. McGowan. 2012. "After the Revolution? Ethical and Social Challenges in 'Personalized Genomic Medicine.'" *Personalized Medicine* 9 (4): 429–439.

Kulhanek, Janet F., Giacomo Meschia, Edgar L. Makowski, and Frederick C. Battaglia. 1974. "Changes in DNA Content and Urea Permeability of the Sheep Placenta." *American Journal of Physiology-Legacy Content* 226 (5): 1257–1263.

Kuzawa, Christopher W., and Elizabeth Sweet. 2009. "Epigenetics and the Embodiment of Race: Developmental Origins of US Racial Disparities in Cardiovascular Health." *American Journal of Human Biology* 21 (1): 2–15

Lillycrop, Karen A., Emma S. Phillips, Alan A. Jackson, Mark A. Hanson, and Graham C. Burdge. 2005. "Dietary Protein Restriction of Pregnant Rats Induces and Folic Acid Supplementation Prevents Epigenetic Modification of Hepatic Gene Expression in the Offspring." *The Journal of Nutrition* 135 (6): 1382–1386.

Lumey, L.H. 1992. "Decreased Birthweights in Infants after Maternal In Utero Exposure to the Dutch Famine of 1944–1945." *Paediatric and Perinatal Epidemiology* 6 (2): 240–253.

———. 1998. "Compensatory Placental Growth after Restricted Maternal Nutrition in Early Pregnancy." *Placenta* 19 (1): 105–111.

Lumey, L.H., and Aryeh D Stein. 1997. "Offspring Birth Weights after Maternal Intrauterine Undernutrition: A Comparison within Sibships." *American Journal of Epidemiology* 146 (10): 810–819.

Maccani, Matthew A., and Carmen J. Marsit. 2009. "Epigenetics in the Placenta." *American Journal of Reproductive Immunology* 62 (2): 78–89.

Margulis, Lynn, and Dorion Sagan. 2002. *Acquiring Genomes: A Theory of the Origins of Species*. New York: Perseus Books.

Martyn, C.N., D.J.P. Barker, and C. Osmond. 1996. "Mothers' Pelvic Size, Fetal Growth, and Death from Stroke and Coronary Heart Disease in Men in the UK." *The Lancet* 348 (9037): 1264–1268.

McDade, Thomas W., Molly W. Metzger, Laura Chyu, Greg J. Duncan, Craig Garfield, and Emma K. Adam. 2014. "Long-Term Effects of Birth Weight and Breastfeeding Duration on Inflammation in Early Adulthood." *Proceedings of the Royal Society B: Biological Sciences* 281 (1784): 20133116. doi:10.1098/rspb.2013.3116.

Meschia, G., F.C. Battaglia, W.W. Hay, and J.W. Sparks. 1980. "Utilization of Substrates by the Ovine Placenta In Vivo." *Federal Proceedings* 39: 245–249.

Murphy, Susan K., and John W. Hollingsworth. 2013. "Stress: A Possible Link between Genetics, Epigenetics, and Childhood Asthma." *American Journal of Respiratory Critical Care Medicine* 187 (6): 563–564.

Murphy, Vanessa E., Roger Smith, Warwick G. Giles, and Vicki L. Clifton. 2006. "Endocrine Regulation of Human Fetal Growth: The Role of the Mother, Placenta, and Fetus." *Endocrine Reviews* 27 (2): 141–169.

Myatt, Leslie. 2006. "Placental Adaptive Responses and Fetal Programming." *Journal of Physiology* 572 (1): 25–30.

Nelson, D.M., S.D. Smith, T.C. Furesz, Y. Sadovsky, V. Ganapathy, C.A. Parvis, and C.H. Smith. 2003. "Hypoxia Reduces Expression and Function of System A Amino Acid Transporters in Cultured Human Trophoblasts." *American Journal of Physiology—Cell Physiology* 284 (2): C310–C315.

Pepper, Jonathan W., and Matthew D. Heron. 2008. "Does Biology Need an Organism Concept?" *Biology Reviews* 83 (4): 621–626.

Phipps, K., D.J.P. Barker, C.N. Hales, C.H.D. Fall, C. Osmond, and P.M.S. Clark. 1993. "Fetal Growth and Impaired Glucose Tolerance in Men and Women." *Diabetologia* 36 (3): 225–228.

Price, Kimberly C., and Christopher L. Coe. 2000. "Maternal Constraint on Fetal Growth Patterns in the Rhesus Monkey (*Macaca mulatta*): The Intergenerational Link between Mothers and Daughters." *Human Reproduction* 15 (2): 452–457.

Price, Kimberly C., Janet Shibley Hyde, and Christopher L. Coe. 1999. "Matrilineal Transmission of Birthweight in the Rhesus Monkey (*Macaca mulatta*) across Several Generations." *Obstetrics and Gynecology* 94 (1): 128–134.

Redmer, D.A., J.M. Wallace, and L.P. Reynolds. 2004. "Effect of Nutrient Intake during Pregnancy on Fetal and Placental Growth and Vascular Development." *Domestic Animal Endocrinology* 27 (3): 199–217.

Regnault, Timothy H., Barbra de Vrijer, and Frederick Battaglia. 2002. "Transport and Metabolism of Amino Acids in Placenta." *Endocrine* 19 (1): 23–41.

Rodney, Nicole C., and Connie J. Mulligan. 2014. "A Biocultural Study of the Effects of Maternal Stress on Mother and Newborn Health in the Democratic Republic of Congo." *American Journal of Physical Anthropology* 155 (2): 200–209.

Rutherford, Julienne N. 2009. "Fetal Signaling through Placental Structure and Endocrine Function: Illustrations and Implications from a Nonhuman Primate Model." *American Journal of Human Biology* 21 (6): 745–753.

———. 2012. "Toward a Nonhuman Primate Model of Fetal Programming: Phenotypic Plasticity of the Common Marmoset Fetoplacental Complex." *Placenta* 33 (S2): e35–e39.

Rutherford, Julienne N., Victoria de Martelly, Donna G. Layne Colon, Corinna N. Ross, Suzette D. Tardif. 2014. "Developmental Origins of Pregnancy Loss in The Adult Female Common Marmoset Monkey (*Callithrix jacchus*)." *PLOS ONE* 9 (5): e96845. doi:10.1371/journal.pone.0096845.

Sastry, B.V. 1991. "Placental Toxicology: Tobacco Smoke, Abused Drugs, Multiple Chemical Interactions, and Placental Function." *Reproduction, Fertility, and Development* 3 (4): 355–372.

Savage, A., Soto, L., Medina, F., Emeris, G., Soltis, J. 2009. "Litter Size and Infant Survivorship in Wild Groups of Cotton-Top Tamarins (*Saguinus oedipus*) in Colombia." *American Journal of Primatology* 71 (8): 707–711.

Savani, Krishna, Nicole Stephens, and Hazel Rose Markus. 2011. "The Unanticipated Interpersonal and Societal Consequences of Choice: Victim Blaming and Reduced Support for the Public Good." *Psychological Science* 22 (6): 795–802.

Schneiderman, Janet U. 1998. "Rituals of Placenta Disposal." *American Journal of Maternal/Child Nursing* 23 (3): 142–143.

Shank, Gary, and Eric J. Simon. 2000. "The Grammar of the Grateful Dead." In *Deadhead Social Science: "You Ain't Gonna Learn What You Don't Want to Know,"* ed. Rebecca G. Adams and Robert Sardiello, 50–73. Walnut Creek, CA: AltaMira Press.

Sibley, Colin P., Mark A. Turner, Irene Cetin, Paul Ayuk, C.A. Richard Boyd, Stephen W. D'Souza, Jocelyn D. Glazier, Susan L. Greenwood, Thomas Jansson, and Theresa Powell. 2005. "Placental Phenotypes of Intrauterine Growth." *Pediatric Research* 58 (5): 827–832.

Smith, Clement A. 1947. "Effects of Maternal Undernutrition upon the Newborn Infant in Holland (1944–1945)." *Journal of Pediatrics* 30 (3): 229–243.

Søegaard, Fredrik, and Claus Gahrn. 2009. "Soundmapping the Genes." *Journal of Music and Meaning* 8.

Stein, Aryeh D., Anita C.J. Ravelli, and L.H. Lumey. 1995. "Famine, Third-Trimester Pregnancy Weight Gain, and Intrauterine Growth: The Dutch Famine Birth Cohort Study." *Human Biology* 67 (1): 135–150.

Stein, Zena A., and Mervyn Susser. 1975. "The Dutch Famine, 1944–1945, and the Reproductive Process: I. Effects on Six Indices at Birth." *Pediatric Research* 9 (2): 70–76.

Stephens, Nicole M., Hazel Rose Markus, and Stephanie A. Fryberg. 2012. "Social Class Disparities in Health and Education: Reducing Inequality by Applying a Sociocultural Self Model of Behavior." *Psychological Review* 119 (4): 723–744.

Stewart, R.J., Hilda Sheppard, R. Preece, and J.C. Waterlow. 1980. "The Effect of Rehabilitation at Different Stages of Development of Rats Marginally Malnourished for Ten to Twelve Generations." *British Journal of Nutrition* 43: 403–412.

Sullivan, Shannon. 2013. "Inheriting Racist Disparities in Health: Epigenetics and the Transgenerational Effects of White Racism." *Critical Philosophy of Race* 1 (2): 190–218.

Tardif, Suzette D., and Karen Bales. 2004. "Relations among Birth Condition, Maternal Condition, and Postnatal Growth in Captive Common

Marmoset Monkeys (*Callithrix jacchus*)." *American Journal of Primatology* 62 (2): 83–94.

Thayer, Zenata M., and Christopher W. Kuzawa. 2011. "Biological Memories of Past Environments: Epigenetic Pathways to Health Disparities." *Epigenetics* 6 (7): 798–803.

Waddington, C.H. 1942. "The Epigenotype." *Endeavour* 1: 18–20.

Weaver, Ian C.G., Nadia Cervoni, Frances A. Champagne, Ana C. D'Alessio, Shakti Sharma, Jonathan R. Seckl, Sergiy Dymov, Moshe Szyf, and Michal J. Meany. 2004. "Epigenetic Programming by Maternal Behavior." *Nature Neuroscience* 7: 847–854.

Wilson, Robert A., and Matthew Barker. 2014. "The Biological Notion of Individual." In *The Stanford Encyclopedia of Philosophy* (Spring 2014 Edition), ed. Edward N. Zalta. http://plato.stanford.edu/archives/spr2014/entries/biology-individual.

Chapter 2

THE BIOLOGY OF THE FETAL PERIOD
INTERPRETING LIFE FROM FETAL SKELETAL REMAINS

Kathleen Ann Satterlee Blake

While small and often incomplete, fetal remains are a vital component of biological anthropologists' skeletal studies, providing insight into conditions useful in forensic and bioarchaeological investigations as well as osteobiographies (the skeletal changes that reflect biological and cultural influences), which may mirror a group or population as a whole (Saul and Saul 1989). The fetus, within biological anthropology, is assessed primarily by size and development of the skeleton and dentition (Fazekas and Kósa 1978). While there are methodological challenges in fetal remains assessments, omitting fetal remains from analyses limits the interpretations that can be made about an individual's or community's health, social structure, or culture. These assessments provide more than a number within a demographic category; to bioarchaeologists, forensic anthropologists, and osteologists, fetal skeletal remains indicate not only the health of the fetus but, potentially, the health of the mother and population as well (Angel 1966; Baxter 2005a; Brickley 2000; Fazekas and Kósa 1965; Goodman and Armelagos 1989; Kamp 2001; Lewis 2000; Mays 2002; Redfern 2008; Sofaer 2006; Sørensen 2000).

What Is a Fetus and How Is It Defined within Biological Anthropology

Defining the fetal period varies between the medical arena and within biological anthropology. In clinical terms, the embryonic period, a time of development of the body's systems, is followed by the fetal period, a time of marked growth, primarily of nearly complete functional systems. This period concludes at birth, an event that ends life in the womb (Arey 1966). While each system of the body develops independently, several internal factors, such as hormones and genetic controls, and external factors, such as the mother's health and external environment, influence the overall growth and development of the fetus. Hormones, particularly testosterone, influence the development of the embryo as it progresses from an unsexed to a sexed individual at approximately the eighth week (Knickmeyer and Baron-Cohen 2006; Moore and Persaud 1998). While genetic sex is established at conception, primary sex characteristics do not begin to differentiate until testosterone is secreted at high levels. This wash of hormones continues until roughly week 20, differentiating males from the previously undifferentiated individual (Riesenfeld 1972). Without this increase of testosterone, female primary sex characteristics will develop (Knickmeyer and Baron-Cohen 2006).

For the biological anthropologist, difficulties arise in discriminating embryo from fetus or fetus from newborn, and the definitions employed for these periods reflect these issues. As seen in table 2.1, slight variances in terms differ between researchers Louise Scheuer and Sue Black (2000), and Mary Lewis (2007). In addition to term differences, the event of birth varies widely, occurring within a window of many weeks. For the majority of pregnancies, forty weeks is considered full-term when calculating due dates; however, a birth deemed "full-term" may occur at any time between twenty-nine and forty weeks, with most births occurring at roughly thirty-eight weeks (Jukic et al. 2013). Premature or overdue births are not uncommon, as currently 12 percent of babies are born prematurely in poor countries and 9 percent in wealthy countries (Blencowe et al. 2012). Overdue, or postterm, births occur after 42 weeks (WHO 1977). These births have a much lower occurrence, as many women are induced before this time. The rate of postterm birth for modern Europeans range between 0.4 percent and 8 percent (Shea et al. 1998; Zeitlin et al. 2007) and Americans average 5 percent. Premature birth can be spontaneous but may relate to a mother's health, specifically chronic

TABLE 2.1. Time period definitions for *in utero* and birth time frames as defined by anthropological researchers (Scheuer and Black 2000; Lewis 2000).

Time Period	Scheuer and Black	Lewis
Embryo	First two months in utero	First eight weeks in utero
Fetus	Third month to birth	Eight weeks to birth
Still birth	Not listed	Infant born dead after twenty-eight weeks
Perinate	Around time of birth	Around birth to twenty-four weeks gestation to seven postnatal days
Neonate	Birth to end of first month	Birth to twenty-seven postnatal days

Source: Lewis 2000; Scheuer and Black 2000.

health conditions, such as diabetes, or infections (March of Dimes et al. 2012). Postterm gestation can have adverse outcomes, including fetal distress and increased stillbirth and fetal death rates (Shea et al. 1998; Zeitlin et al. 2007), and can lead to increased complications for the mother as well (Caughey and Bishop 2006).

Although the fetal period ends at birth, the assessment of birth is difficult, as this event is not recorded on skeletal remains. Birth can be estimated by examining teeth on a histological level, with birth denoted by the presence of the neonatal line. This line develops as the process of amelogenesis, or enamel formation, is interrupted at the time of delivery (Schour 1936). This indicator can vary due to delivery type, such as caesarean section, or gestational length, such as premature birth (Eli et al. 1989); however, no such correlate is found in skeletal material. Therefore, without dental analysis, the event of birth is estimated based on size (Schaefer et al. 2009). As evidenced in table 2.1, individuals from the time near birth, perinates, newborns, or neonates, and even those who survived the first few weeks after birth, can all appear to be within the fetal time period when evaluated skeletally.

Methodological Approaches

A variety of information can be ascertained from analysis of fetal remains. Age-at-death and health status may be established, but as nearly all remains are skeletal in nature, with the exception of mummified remains, analysis is limited to what information the bones can contribute. Age-at-death is determined principally by bone size and skeletal and dental development, while overall health or pathology indicators are assessed most commonly via macroscopic analysis.

Age-at-Death

Age-at-death is assessed from fetal remains via bone length and dental development. Bone length reference data has been compiled from osteological, radiographic (X-ray), and clinical ultrasound analyses (Jeanty et al. 1982). Research collections of skeletal fetal remains housed in universities or museums often contain age, sex, ancestry, and even medical data about not only the fetus but also the mother (Hunt 1990). Other reference samples are based on coffin plate collections or cemetery excavations with historical records from the community (Saunders 1992). Lastly, some standards are based on ultrasounds or radiographs, where long bone measurements are taken on ultrasound images or x-rays of living individuals with soft tissue present (Scheuer et al. 1980).

While reference and research collections list known ages for fetal material, these ages are estimated based on either the mother's last period or body length measurements by a researcher at time of death or first evaluation. These techniques are not free of bias or discrepancies (Fazekas and Kósa 1965; Huxley 2005). Additionally, some collections provide distinct age-at-death estimates in weeks in utero, while others recorded vague terminology, such as fetus, infant, or stillborn, that may or may not correlate to modern clinical or biological anthropology terms. Oftentimes, these terms are not distinct designators of age but reflect differences between those individuals who experienced a live birth, regardless of the gestational age of the fetus. Therefore, caution should be used when estimating age based on populations not consistent with known samples or without clearly recorded age data.

While bone length is used as a means to calculate fetal age, size is influenced by hereditary factors, as variation in growth is expected between individuals (Johnston and Zimmer 1989). Those individuals who are undersized and less mature due to a developmental disorder, nutritional deficiencies, or maternal illness can be incorrectly assessed for age (Singer et al. 1991). Therefore, chronological age may not be accurate in smaller than normal skeletal remains. Decreases in birth weight are common during periods of stress, such as changes in subsistence or environment (Johnston and Zimmer

1989; Owsley and Jantz 1985). These times of stress may manifest in community health or cultural changes that may lead to premature birth, infanticide, or other cultural practices (Kinaston et al. 2009; Halcrow et al. 2008; Pfeiffer et al. 1989; Stoeckel and Alauddin Chowdhury 1972; Tocheri and Molto 2002). The ability to distinguish between a fetus and a newborn not only provides more accurate demographic data for the population but also can explain cause of death.

When possible, dental development should be assessed, as the formation of teeth begins in utero. Teeth can provide an accurate measure of age because they are less affected by environmental insults, which may alter skeletal growth (Moorrees et al. 1963; Sherwood et al. 2000). When recovered, teeth can provide fetal age estimations as early as twelve weeks through birth (Kraus 1959). While loose teeth may be overlooked in archaeological recovery, fetal teeth are often preserved within the bony crypts in the upper or lower jaws (Moorrees et al. 1963).

One of the challenges with the assessment of fetal remains is the small size of the bones and the incomplete state of their development. Most bones develop from a number of ossification centers within a cartilaginous anlage or template that ossifies at set times beginning by the start of the fetal period (Hill 1939). The pelvis is an excellent example of how this process works. At the end of the embryonic stage at eight weeks, ossification of the three elements of the pelvis—the ischium, ilium, and pubis—begins with the primary ossification center for the three bones at the region of the acetabulum or hip socket (Arey 1966; McAuley and Uhthoff 1990; Moore and Persaud 1998; Noback 1944). Each bone has its own individual primary and secondary centers of ossification and individual timing of growth and development. The ilium appears first and ossifies at eight to nine weeks, while the ischium begins this process slightly later, in the third or fourth months (Delaere and Dhem 1999). The pubis ossifies last, at roughly four to five months in utero, often after the ischium has completely ossified. These bones eventually unite and become one bone mid-childhood (Delaere and Dhem 1999; Hill 1939; McAuley and Uhthoff 1990; Noback 1944; O'Rahilly and Gardner 1975) (see fig. 2.1).

Bone development provides a good indicator of fetal age, as preliminary research has shown that when age on record is compared to bone length age estimates, most individuals fall within the anticipated age categories. Even when affected by an abnormality, growth in the length of long bones is often not as adversely retarded as

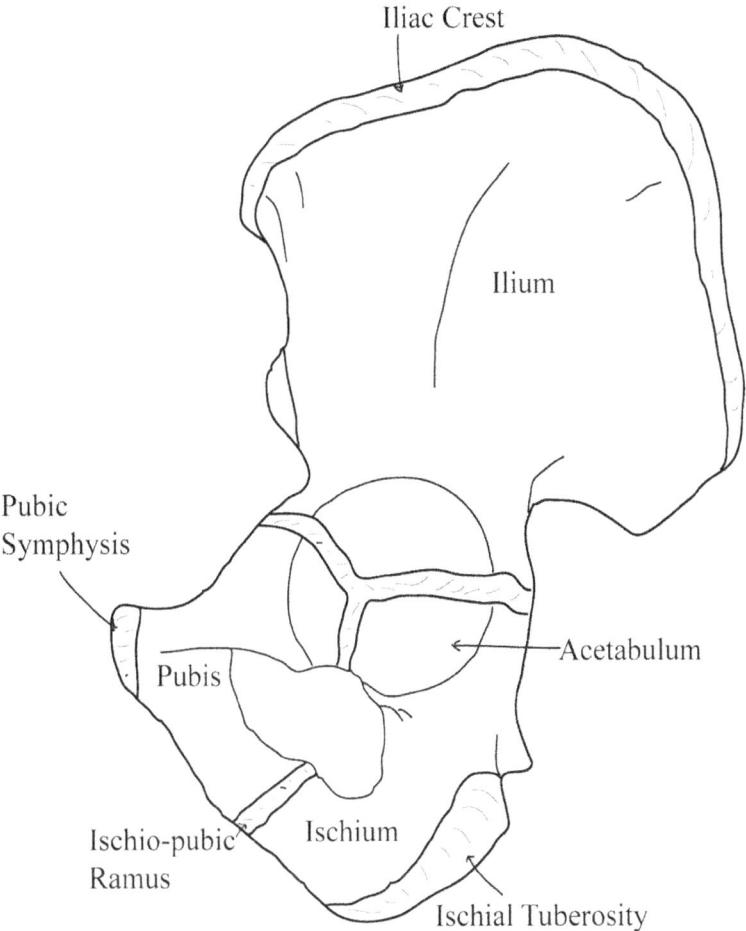

FIGURE 2.1. Features of pelvis, including the three separate components and areas of growth (Blake 2011).

other systems within the body (Mahon et al. 2010; McLean and Usher 1970; Singer et al. 1991). While birth weight is retarded due to maternal disease, fetal infection, or other conditions, birth length is usually within the normal range (Singer et al. 1991). The length of the femur provided the most accurate age estimates, even in individuals experiencing pathology (Sherwood et al. 2000). Studies of fetal remains experiencing growth retardation often found that the epiphyses develop later and are smaller than normal (Philip 1974; Scott and Usher 1964), but infants who experienced growth retardation in utero grew normally after birth (M. Wilson et al. 1967).

When a skeletal population includes fetal remains, comparing those remains to all other subadults can reveal how that specific community grew and developed (Saunders 2000). This examination may illuminate long bone size differences or consistencies within the population that may be masked by those who survive and grow normally to adulthood. When compared to other groups, the examination of subadults can indicate changes in growth velocity or health status (Lovejoy et al. 1990; Mensforth 1985). It should be noted that subadult skeletal samples reflect non-survivors, which may bias analyses; however, the effects of this bias are thought to be negligiable (Saunders and Hoppa 1993). While fetal skeletal remains represent those individuals who did not survive outside the womb, assessments of the population or environment are not complete without them.

Overall size distributions that vary from the norm help researchers establish if fetal remains are small for their age, based on the overall trends in that particular assemblage. This type of comparison can suggest causes of potential deviations, such as maternal disease, malnutrition, environmental influences, or cultural restrictions. While skeletal sample comparisons are not always feasible, it is recommended that they be undertaken when available to explain more completely the physical state of members of a population (Humphrey 2000).

Biological Sex

Analysis of the differences between males and females is a significant component to any bioarchaeological demographic profile. As cultures often treat males and females differently starting at birth, sex may affect overall health and growth or burial treatments. Sex distribution information can reveal circumstances within a community such as infanticide, inheritance of goods, disease rates, and activities (Sofaer 2006). Sex differences are evident in fetal remains as early as eight weeks (Arey 1966; Boucher 1957; Delaere et al. 1992; Nakao 1998; Noback 1944; Reynolds 1945; Rösing 1983; Siiteri and Wilson 1974; L. Wilson et al. 2008). Differences between males and females have been noted since the 1870s—specifically in the pelvic bones, the most dimorphic area of human skeletal anatomy—since the end of the nineteenth century (Fehling 1876; Thomson 1899). Other differences exist in the mandible, phalanges, cranium, and long bone lengths of fetal skeletal remains (Cardoso and Saunders 2008; Galis et al. 2010; Gapert et al. 2009; Holcomb and Konigsberg 1995; Loth and Henneberg 2001; Mittler and Sheridan 1992; Schutkowski 1993; Sutter 2003; Vlak et al. 2008). However, no reliable method to determine biological sex in fetal remains exists without genetic analysis. Although dimorphism is present, overlap in the range of trait differences between the sexes does not allow for sharp distinctions. While several researchers have attempted to determine sex based on elements of the pelvis and mandible (Schutkowski 1993; Weaver 1980), additional research has shown that these traits vary from one population to another (Blake 2011, 2014) or the results could not be duplicated (Cardoso and Saunders 2008; Hunt 1990).

To demonstrate this point, a study of two fetal and newborn skeletal samples (Forensic Fetal Collection, Smithsonian Museum of Natural History; Trotter Fetal Collection, Washington University) found sexual dimorphism in the pelvis was statistically significant for at least one trait in each of the two collections. However, as the dimorphic traits varied in each population, the differences noted were particular to each sample (fig. 2.2). While differentiation between the sexes was evident, methods to quantify and interpret these variations in some meaningful way could not be created. Overlap between males and females as well as variation in trait expression were two major issues. Subsequently, when these two samples were compared to a European collection of subadults, dimorphism differed yet again

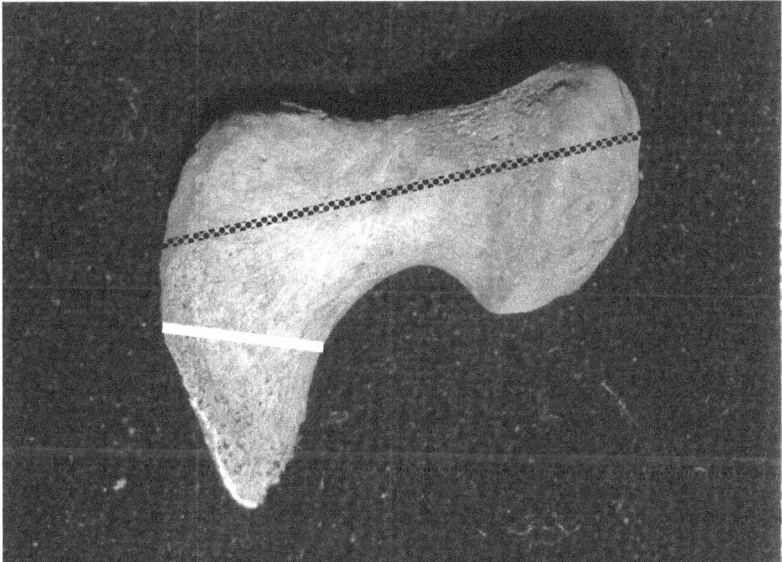

FIGURE 2.2. Right pubic bone. White line demonstrates pubic body width, which is dimorphic between males and females. Dotted line represents pubic width.

(Blake 2011, 2014). When compared with other research studies, the outcomes for levels of dimorphism or sexually dimorphic traits could not be duplicated (Hunt 1990). As dimorphism between sexes varied between groups, indicating a population relationship to the dissimilarity between the sexes in fetal remains, more sample comparisons are needed to better understand how and why these differences occur (Blake 2014; Goodman and Armelagos 1989).

Since dimorphism in these studies was found to vary by population, it is through the examination of skeletal collections as a unit or whole that sex or gender analyses may be considered. Simon Mays (2002) offered one approach to population sex analysis; measurements such as the sciatic notch width, a wide opening on the pelvis, could be plotted and observed for clusters that suggest male or female. Mays found that age obscured dimorphism, as it was less evident in older children's remains than in fetal remains. Research of older children (i.e., over one year) showed that most traits are not significantly dimorphic until nearly puberty (Blake 2011). Mays determined that fetal remains elicited evidence about a population not available from examination of older children.

Health and Pathology

Consideration of health and pathology in fetal remains is primarily macroscopic in nature, although microscopic and radiographic techniques can be utilized. Examination of the bone's surface or its overall shape and morphology can determine if signs of poor health are present. Alterations to the bone's surface, such as porosity, or a widening of the ends of the long bones indicate reaction to disease. These analyses illuminate specific or general disease conditions, such as infections or inflammation. Visual inspection and descriptions of affected bones, with extent and location of the lesions clearly identified, may suggest a potential diagnosis. Descriptions of alterations found on skeletal material should include not only the lesion's condition on a bone but any pattern observed throughout the skeleton (Ortner 2003). For example, does the defect occur on other skeletal elements, and if so, which ones? The patterns observed often suggest a diagnosis for a particular condition or type of ailment (Lovell 2000; Stark 2014). A more complete discussion of pathology can be found in chapter 5.

Whether one examines disease, sexual differences, or size and age, each population should be evaluated as its own unit and compared with other populations to find differences or trends. Comparisons of samples with known temporal makeup to either contemporaneous samples or those from a different time period ensure all potential reasons for the differences in these populations can be explored (Lewis 2002). The technique used by Mays (2002) is currently the only noninvasive means of sorting subadult males from females. Therefore, analysis of differential treatment of boys and girls may be inferred from items within an archaeological analysis, such as grave goods, but must be omitted from osteological fetal analyses (Baxter 2005b). The primary assessments of these individuals from skeletal remains will be age-at-death and pathology.

Study of Fetuses Contributing to Biological Anthropology

From initial skeletal analyses of the fetus, more in-depth population evaluations may be developed to provide insight into the health and well-being of the mother or potentially even the grandmother, or expand our knowledge of the fetal group as a whole. Because disease more adversely affects the developing fetus, these remains can illuminate population disparities or cultural practices not always evident in other skeletal remains. As the mother often survives a condition deadly to the fetus, her skeleton may not demonstrate the problems she experienced during a pregnancy. However, the fetal remains provide a window into a brief time as the insults are captured on the individual who did not survive (Goodman and Armelagos 1989). Therefore, through these small and incompletely formed remains, researchers gain more in-depth knowledge about aspects of health and well-being within a community to eventually incorporate into bioarchaeological, osteobiographical, or forensic analyses.

Lesions on fetal, perinate, and neonate remains can reveal a mother's overall health, genetic issues within the population, and cultural practices and dietary restrictions. Poor nutrition or disease in the mother affects the bones of the fetus, confirming conditions such as rickets, scurvy, and anemia, as well as general inflammation and infection. In modern populations, these deficiencies are observed based on cultural influences; for example, rickets is observed in modern populations where women have limited exposure to sunlight, due to clothing or activity restrictions. Therefore, these lesions in past populations may indicate cultural restrictions on women, such as modes of dress or diet. For example, heavy and layered clothes as well as sunlight avoidance were the cultural norms of Renaissance Europe and contributed to potential cases of rickets in Italian nobility (Giuffra et al. 2015). While not all diseases are evidenced on skeletal material, some conditions in particular are associated with fetal remains. In fact, according to Lewis (2000), as juvenile bones are still growing and remodeling, they are more affected by infections or inflammation to the periosteum (the fibrous covering of bone) than what is evident in adult bones. In addition, clinical evidence shows that bone mineral density is lower in infants born with a lighter weight than anticipated for gestational age, suggesting that bone density is affected by conditions such as fetal malnutrition or intrauterine growth retardation (Petersen et al. 1989).

Low levels of calcium and vitamin D in the fetal period decrease bone density, although overall size, including length and weight of the fetus or infant, is typically less altered with this nutritional deficiency (Hatun et al. 2005; Koo et al. 1982; Prentice 2003). Changes to bony structures, such as the ends of the long bones like the femur, are associated with rickets in newborns (Schultz 1993) and in vitamin D deficiencies (Hewison and Adams 2010). Mineral deficiencies in pregnancy more strongly affect a fetus than similar deficiencies affect infants that breastfeed. In areas with low vitamin D uptake due to a mother's limited exposure to sunlight, preterm babies often suffer from both anemia and rickets (Giuffra et al. 2015, Winston et al. 1989). In addition, neonatal scurvy is diagnosed in newborns and is directly related to the mother's malnutrition during pregnancy (Hirsch et al. 1976).

Vitamin D deficiencies are not solely associated with skeletal changes and lack of bone mineralization, but can reflect nonskeletal health conditions as well. Pregnancy not only demands additional calcium but also alters a mother's immune functions (Liu et al. 2009). Vitamin D deficiencies are associated with many health complications: immunological disorders; conditions such as diabetes, tuberculosis, hypertension; and cognitive and mental changes, including depression (Holick and Chen 2008). In addition, problems with pregnancy, such as preeclampsia, are related to low levels of vitamin D, which in turn leads to a deficient fetus and preterm birth (Bodnar et al. 2007).

New studies of fetal growth suggest that the overall health of a fetus or infant may be an indicator of not only the mother's health but her mother as well (see Rutherford, chapter 1). While developing within the womb, the fetus adjusts to the influences of that environment. Researchers have found that these changes permanently affect the maturing child and are then carried by females into the subsequent generation. For example, individuals who were conceived and born during the famine in the Netherlands in 1944 grew to be normal-sized adults; however, their own pregnancies resulted in low birth weight offspring (Barker 1995). Therefore, small fetal and infant remains substantiate changes in food access in a prior generation; this information could explain population changes noted many years after historically recorded famines or plagues.

Osteologists often consult the clinical literature to gain insight into the mechanisms behind certain conditions and how they manifest in modern living populations and on skeletal elements. For example, a clinical study by Christopher Kovacs determined it was the lack of vitamin D, not calcium deficiencies, that affect the neonate. In addition, in areas where "vitamin D deficiency is endemic and clinical awareness high, clinicians often identify the characteristic changes of rickets soon after birth" (2008: 525S). According to M. Teotia and S.P.S. Teotia (1997), vitamin D deficiency may have both environmental and genetic causation. In addition, twice as many newborns of mothers experiencing preeclampsia were vitamin D deficient compared to those from mothers without this condition (Bodnar et al. 2007). Knowing that vitamin D is a significant issue in modern populations supports the identification of these conditions in past populations as well.

Observations of congenital syphilis, which currently affect 700,000 to 1.5 million pregnancies worldwide, are primarily found in areas such as Sub-Saharan Africa and account for 21 percent of the perinatal deaths in that region (Hossain et al. 2007; Woods 2009). With this condition, skeletal involvement manifests as periostitis (inflammation of the fibrous cover of bone) and osteochondritis (inflammation of bony cartilage) and is found in up to 78 percent of neonates that suffer from congenital syphilis (Woods 2009). These individuals also tend to be small for their age, and 40 percent of the perinatal deaths/stillborn babies were due to an untreated mother (Berry and Dajani 1992; Ortner 2003; Parish 2000; Toohey 1985). Therefore, understanding that overall size changes and conditions, such as periostitis and osteochondritis, in this modern African population are related to syphilis, which may then be identifiable in both modern forensic cases and in past populations, where incidences of fetal mortality due to syphilis are suspected.

Additional issues can occur during pregnancy that may delay fetal maturation. One example is long-bone sclerosis (or increased density or hardening) that can result from lead poisoning during pregnancy (Shannon 2003) and is associated with conditions such as pica (Pearl and Boxt 1980). In addition, neonatal scurvy, usually diagnosed within a few days of birth, reflects maternal scurvy (Hirsch et al. 1976). In a skeletal population from medieval Romania, Anna Osterholtz and colleagues (2014) found evidence of scurvy in both prenatal and perinatal remains, providing insight into the health of individuals from that area.

While the clinical literature aids our evaluation of the causes of these lesions, it is within the archaeological context that these diseases are interpreted. Donald Ortner and colleagues (2001) examined subadult skeletal remains from late prehistoric or early historic Native American sites throughout North America and found variations in food abundance by region, and related the prevalence of scurvy to the use of corn within that population. While these authors examined all juvenile ages, the highest rates of scurvy were found in the category of newborn to three years.[1] This supports the idea that bones of younger individuals are more affected than older ones or more likely to demonstrate the effects of bone mineral insufficiencies (Lewis 2000; Petersen et al. 1989). However, it should be noted that additional factors such as age of weaning and cultural dietary restrictions for subadults might contribute to these levels of vitamin C deficiency in this particular age cohort (Mays 2014). While all causes of this pattern of scurvy must be evaluated, it is clear that without the inclusion of the youngest members of a population, an essential component necessary to completely understand the group would be missing.

Ortner and Mays (1998) examined a medieval English population to assess rickets in infants and determined that signs of the disease were found on the skull, ribs, and long bones. In this population, only these subadults demonstrated signs of rickets; if fetal remains had been omitted from the analysis, the evidence of rickets would have absent as well, altering the interpretation of health and disease at this site. Megan Brickley and Rachel Ives (2006) investigated fetal skeletal remains from an eighteenth- and nineteenth-century cemetery in Birmingham, England, and found scurvy on cranial bones and the scapula. These authors used historical data of the socioeconomic changes in this population to understand how scurvy could be present within the community. Specifically, these researchers determined that this population was affected by a potato famine.

There are limitations to the disease interpretations from skeletal remains. An awareness of the disease process, as well as knowledge of taphonomic conditions that mimic disease, may aid researchers in correctly interpreting the state of remains (Ortner 2003). Only chronic, less acute diseases endure long enough to leave an impression on bone, limiting assessment of ailments affecting populations to specific conditions. Furthermore, health can vary over a person's lifetime, so an individual's assessment of health differs from a community's assessment (Bethard et al. 2014; Steckel 2004).

Forgetting the Fetus in Bioarchaeological Research

Fetal remains have been omitted from studies in the past for several reasons (Saunders 2000). One reason is the assumption that these remains do not survive well after burial. However, lack of bone preservation in fetal remains may not relate directly to the fetal state but reflect a community's burial practices or limitations of archaeological excavations. While subadult remains deteriorate quickly in acidic soils, primarily because of higher cartilage levels in these developing individuals, many researchers have noted well-preserved fetal remains (Buckberry 2000; Gordon and Buikstra 1981; Mays 2002; Walker et al. 1988). A review of Anglo-Saxon burial sites by Jo Buckberry (2000) found preservation varied by site, with poor preservation of those remains buried in sandy soil. However, Amy Scott and Tracy Betsinger (chapter 6) examined fetal remains from seventeenth- and eighteenth-century Poland, which were recovered from sandy soil and exhibited good preservation. Siân Halcrow and colleagues (2008) found variable, although primarily good, preservation; however, as screening was not possible because of the soil type, many small bones were not recovered. Research studies that note well-preserved remains include a study by Shelley Saunders (1992) at the Belleville, Ontario site; an analysis by Theya Molleson (1991) at Poundbury, Dorset; and research by Ortner and Mays (1998) in a South Yorkshire population. In Romania, Osterholtz and colleagues (2014) remarked on the excellent state of bone preservation in the fetal remains they examined (Bethard 2014). Additionally, collections such as the Trotter Fetal Collection of Washington University, created in hospital or research facilities, typically display good to excellent condition. Because taphonomic influences, such as soil color, can cause bone color changes and many infant bones exhibit different morphology than adult forms do, these remains may not be recovered if the archaeologist is unfamiliar with them (Lewis 2007). Often, these bones are misidentified as small stones, dirt, or animal bones (Buckberry 2000).

Moreover, a community often handles fetal burials differently, and the lack of material in an excavation may point to differential burial, not loss of the remains (as is further discussed in chapters 4 and 5). Jane Buikstra and Anna Lagia (2009: 12) suggest using paleodemographics, or "models of population distributions and density," to better interpret the numbers of fetal remains found. According to these authors, the inclusion or exclusion of fetal material from a cemetery population is not the end of the analysis but the beginning. For example, an abundance of fetal/infant graves are interpreted as high numbers of deaths within that category. As previously discussed, preservation of fetal material plays a role, but burial customs may also explain these patterns. Other culturally relevant explanations for a group include high levels of infanticide, or placement of fetal individuals within a separate area of the cemetery (Mays 2002).

Conclusion

Inclusion of fetal remains, such a small component of a group or population, may at first glance appear to provide little understanding, or fail to alter or impact studies of current and/or past populations. However, these tiny individuals offer a more complete analysis of a group's health when perhaps no other method can. As shown, the fetal remains may be the only group either affected by or exhibiting lesions of conditions or diseases absent from older children or adult skeletons. And while adults eventually attain stature within a normal range, the undersized fetal remains can demonstrate population insults, such as famines, both recent and, potentially, in the previous generation. Diseases known to alter fetal bone indicate more than a mild deficiency within a population; they also could potentially show other health ailments, from depression to high blood pressure to immune system failures.

Through an archaeological analysis, researchers can assess lesions and interpret the health of the fetus, the mother, and the community. Practices such as modes of dress or customs to avoid sunlight exposure can be elicited to show how cultural norms affect the health and well-being of a group. Researchers must combine what can be determined from the skeleton, the clinical literature, and historical data. Population trends in size or sex differences over time can also be examined. Evaluation of size and age may demonstrate if individuals are smaller than average for a group, and generalized determination of sex may explain differential treatment of males and females.

Within the modern realm, cultural practices relating to women and pregnancy can be understood in relation to their impact on the developing child. This can be applied within the public health arena, as understanding how conditions such as low vitamin D relate to the mother's health and potential risk factors for conditions, such as preeclampsia, that impact the fetus's health, growth and development. When fetal remains are recovered in the forensic sphere, these conditions may clarify factors contributing to death, such as cultural practices or malnourishment, and ultimately may provide manner and/or cause of death.

Recovery and examination of fetal remains is challenging because of their small size and incomplete state. However, the time and effort invested in determining potential anomalies or pathological conditions could provide insights unavailable without their inclusion. Although it takes more effort, the gains when evaluating this small portion of society are irreplaceable.

Acknowledgments

Sincerest thanks to the organizers of the symposium at the 112th Annual Meeting of the American Anthropological Association that developed into this book, Tracy Betsinger, Amy Scott, and Sallie Han, for the invitation to join them in this topic discussion and their efforts to bring this volume to fruition. I would like to thank the Smithsonian Institute National Museum of Natural History, Cleveland Museum of Natural History, Washington University in St. Louis, and the Centre for Anatomy and Human Identification, University of Dundee, for access to the collection of fetal and subadult material. Photograph credits are to the Hamann Todd Collection at the Cleveland Museum of Natural History and the Forensic Fetal Collection at the National Museum of Natural History. Finally, thank you to our anonymous reviewers for their comments and suggestions.

Kathleen Ann Satterlee Blake, a forensic anthropologist and bioarchaeologist, is currently an assistant professor at the State University of New York at Oswego. Research areas include skeletal growth and development, specifically in the pelvis, and sexual dimorphism of the pelvis, particularly in subadults.

Note

1. Unfortunately, the ages were not more discretely presented, nor were fetal remains mentioned.

References

Angel, J. Lawrence. 1966. "Porotic Hyperostosis, Anemias, Malarias, and Marshes in the Prehistoric Eastern Mediterranean." *Science* 153 (3737): 760–763.
Arey, Leslie B. 1966. *Developmental Anatomy: A Textbook and Laboratory Manual of Embryology*. Philadelphia: W.B. Saunders Co.
Barker, D.J.P. 1995. "The Wellcome Foundation Lecture, 1994. The Fetal Origins of Adult Disease." *Proceedings of the Royal Society B: Biological Sciences* 262 (1363): 37–43.
Baxter, Jane Eva. 2005a. *The Archaeology of Childhood: Children, Gender, and Material Culture*. Walnut Creek, CA: Altamira Press.
———. 2005b. "Introduction: The Archaeology of Childhood in Context." *Archeological Papers of the American Anthropological Association* 15 (1): 1–9.
Berry, M.C., and A.S. Dajani. 1992. "Resurgence of Congenital Syphilis." *Infectious Disease Clinics of North America* 6 (1): 19–29.
Bethard, J., interview by K. Blake. 2014. *In Discussion with Author* (2014 October 1).
Bethard, Jonathan D., Anna J. Osterholtz, Andre Gonciar, and Zsolt Nyaradi. 2014. "A Bioarchaeological Study of Childhood Mortality in 17th Century Transylvania." *American Journal of Physical Anthropology* 153 (S58): 78.
Blake, Kathleen Ann Satterlee. 2011. "An Investigation of Sex Determination from the Subadult Pelvis: A Morphometric Analysis." PhD dissertation. Pittsburgh, PA: University of Pittsburgh.
———. 2014. "Analysis of Non-metric Subadult Sex Determination Traits in Four Samples of Known Age and Sex: Sex Determinants or Population Variants?" In *Proceedings of the American Academy of Forensic Sciences*, vol. 20, 408. Colorado Springs, CO: American Academy of Forensic Sciences.
Blencowe, Hannah, Simon Cousens, Mikkel Z. Oestergaard, Doris Chou, Ann-Beth Moller, Rajesh Narwal, Alma Adler, et al. 2012. "National, Regional, and Worldwide Estimates of Preterm Birth Rates in the Year 2010 with Time Trends since 1990 for Selected Countries: A Systematic Analysis and Implications." *The Lancet* 379 (9832): 2162–2172.
Bodnar, Lisa M., Janet M. Catov, Hyagriv N. Simhan, Michael F. Holick, Robert W. Powers, and James M. Roberts. 2007. "Maternal Vitamin D Deficiency Increases the Risk of Preeclampsia." *The Journal of Clinical Endocrinology and Metabolism* 92 (9): 3517–3522.
Boucher, Barbara J. 1957. "Sex Differences in the Foetal Pelvis." *American Journal of Physical Anthropology* 15 (4): 581–600.

Brickley, Megan. 2000. "The Diagnosis of Metabolic Disease in Archaeological Bone." In *Human Osteology in Archaeology and Forensic Science,* ed. Margaret Cox and Simon Mays, 183–198. Cambridge: Cambridge University Press.

Brickley, Megan, and Rachel Ives. 2006. "Skeletal Manifestations of Infantile Scurvy." *American Journal of Physical Anthropology* 129 (2): 163–172.

Buckberry, Jo. 2000. "Missing, Presumed Buried? Bone Diagenesis and the Under-Representation of Anglo-Saxon Children." *Assemblage: The Sheffield Graduate Journal of Archaeology* 5: 1–17.

Buikstra, Jane, and Anna Lagia. 2009. "Bioarchaeological Approaches to Aegean Archaeology." *Hesperia Supplement* 43: 7–29.

Cardoso, Hugo F.V., and Shelley R. Saunders. 2008. "Two Arch Criteria of the Ilium for Sex Determination of Immature Skeletal Remains: A Test of Their Accuracy and An Assessment of Intra- and Inter-observer Error." *Forensic Science International* 178 (10): 24–29.

Caughey, A.B., and J.T. Bishop. 2006. "Maternal Complications of Pregnancy Increase beyond 40 Weeks of Gestation in Low-Risk Women." *Journal of Perinatology* 26 (9): 540–545.

Delaere, O., and A. Dhem. 1999. "Prenatal Development of the Human Pelvis and Acetabulum." *Acta Orthopaedica Belgica* 65 (3): 255–260.

Delaere, O., V. Kok, C. Nyssen-Behets, and A. Dhem. 1992. "Ossification of the Human Fetal Ilium." *Cells Tissues Organs* 143 (4): 330–334.

Eli, Ilana, Haim Sarnat, and Eliezer Talmi. 1989. "Effect of the Birth Process on the Neonatal Line in Primary Tooth Enamel." *Pediatric Dentistry* 11 (3): 220–223.

Fazekas, István Gyula, and Ferenc Kósa. 1965. "Recent Data and Comparative Studies about the Body Length and Age of the Fetus on the Basis of the Measurements of the Clavicle and Shoulderblade." *Acta Medicinae Legalis et Socialis* 18 (1): 307–325.

———. 1978. *Forensic Fetal Osteology.* Budapest: Akadémiai Kiadó.

Fehling, Herman. 1876. "Die Form des Beckens beim Fötus und Neugeborenen [The Shape of the Pelvis in the Fetus and Newborn]." *Archives of Gynecology and Obstetrics* 10 (1): 1–80.

Galis, Frietson, M. A. Clara, Ten Broek, Stefan Van Dongen, and Liliane C.D. Wijnaendts. 2010. "Sexual Dimorphism in the Prenatal Digit Ratio (2D: 4D)." *Archives of Sexual Behavior* 39 (1): 54–62.

Gapert, Rene, Sue Black, and Jason Last. 2009. "Sex Determination from the Foramen Magnum: Discriminant Function Analysis in an Eighteenth and Nineteenth Century British Sample." *International Journal of Legal Medicine* 123 (1): 25–33.

Giuffra, V., A. Vitiello, D. Caramella, A. Fornaciari, D. Giustini, and G. Fornaciari. 2015. "Rickets in a High Social Class of Renaissance Italy: The Medici Children." *International Journal of Osteoarchaeology* 25 (5): 608–624.

Goodman, Alan H., and George J. Armelagos. 1989. "Infant and Childhood Morbidity and Mortality Risks in Archaeological Populations." *World Archaeology* 21 (2): 225–243.

Gordon, Claire C., and Jane E. Buikstra. 1981. "Soil pH, Bone Preservation, and Sampling Bias at Mortuary Sites." *American Antiquity* 46 (3): 566–571.

Gowland, Rebecca. 2006. "Ageing the Past: Examining Age Identity from Funerary Evidence." In *Social Archaeology of Funerary Remains*, ed. Rebecca Gowland and Christopher Knüsel, 143–150. Oxford: Oxbow Books.

Halcrow, Siân E., Nancy Tayles, and Vicki Livingstone. 2008. "Infant Death in Late Prehistoric Southeast Asia." *Asian Perspectives* 47 (2): 371–404.

Hatun, Sukru, Behzat Ozkan, Zerrin Orbak, Hakan Doneray, Filiz Cizmecioglu, Demet Toprak, and Ali Süha Calikoglu. 2005. "Vitamin D Deficiency in Early Infancy." *The Journal of Nutrition* 135 (2): 279–282.

Hewison, Martin, and John S. Adams. 2010. "Vitamin D Insufficiency and Skeletal Development *In Utero*." *Journal of Bone and Mineral Research* 25 (1): 11–13.

Hill, Alfred H. 1939. "Fetal Age Assessment by Centers of Ossification." *American Journal of Physical Anthropology* 24 (3): 251–272.

Hillson, Simon. 2009. "The World's Largest Infant Cemetery and Its Potential for Studying Growth and Development." *Hesperia Supplement* 43: 137–154.

Hirsch, Menachem, Paul Mogle, and Yehiel Barkli. 1976. "Neonatal Scurvy." *Pediatric Radiology* 4 (4): 251–253.

Holcomb, Susan, and Lyle W. Konigsberg. 1995. "Statistical Study of Sexual Dimorphism in the Human Fetal Sciatic Notch." *American Journal of Physical Anthropology* 97 (2): 113–125.

Holick, Michael F., and Tai C. Chen. 2008. "Vitamin D Deficiency: A Worldwide Problem with Health Consequences." *American Journal of Clinical Nutrition* 87 (4): 1080S–1086s.

Hossain, Mazeda, Nathalie Broutet, and Sarah Hawkes. 2007. "The Elimination of Congenital Syphilis: A Comparison of the Proposed World Health Organization Action Plan for the Elimination of Congenital Syphilis with Existing National Maternal and Congenital Syphilis Policies." *Sexually Transmitted Diseases* 34 (7S): S22–S30.

Humphrey, Louise. 2000. "Growth Studies of Past Populations: An Overview and An Example." In *Human Osteology: In Archaeology and Forensic Science*, ed. Margaret Cox and Simon Mays, 23–38. Cambridge: Cambridge University Press.

Hunt, David R. 1990. "Sex Determination in the Subadult Ilia: An Indirect Test of Weaver's Nonmetric Sexing Method." *Journal of Forensic Sciences* 35 (4): 881–885.

Huxley, Angie Kay. 2005. "Gestational Age Discrepancies due to Acquisition Artifact in the Forensic Fetal Osteology Collection at the National Museum of Natural History, Smithsonian Institution, USA." *American Journal of Forensic Medicine and Pathology* 26 (3): 216–220.

Jeanty, Philippe, Michele Dramaix-Wilmet, N. Elkhazen, Corinne Hubinont, and Nicole Van Regemorter. 1982. "Measurements of Fetal Kidney Growth on Ultrasound." *Radiology* 144 (1): 159–162.

Johnston, Francis E., and Louise O. Zimmer. 1989. "Assessment of Growth and Age in the Immature Skeleton." In *Reconstruction of Life from the Skeleton,* ed. Mehmet Yaşar İşcan and Kenneth A.R. Kennedy, 11–21. New York: Wiley-Liss.

Jukic, A.M., D.D. Baird, C.R. Weinberg, D.R. McConnaughey, and A.J. Wilcox. 2013. "Length of Human Pregnancy and Contributors to Its Natural Variation." *Reproductive Epidemiology* 28 (10): 2848–2855.

Kamp, Kathryn A. 2001. "Where Have All the Children Gone?: The Archaeology of Childhood." *Journal of Archaeological Method and Theory* 8 (1): 1–34.

Kinaston, Rebecca L., Hallie R. Buckley, Siân E. Halcrow, Matthew J.T. Spriggs, Stuart Bedford, Ken Neal, and A. Gray. 2009. "Investigating Foetal and Perinatal Mortality in Prehistoric Skeletal Samples: A Case Study from a 3000-Year-Old Pacific Island Cemetery Site." *Journal of Archaeological Science* 36 (12): 2780–2787.

Knickmeyer, Rebecca Christine, and Simon Baron-Cohen. 2006. "Fetal Testosterone and Sex Differences." *Early Human Development* 82 (12): 755–760.

Koo, W.W., J.M. Gupta, V.V. Nayanar, M. Wilkinson, and S. Posen. 1982. "Skeletal Changes in Preterm Infants." *Archives of Disease in Childhood* 57 (6): 447–452.

Kovacs, Christopher S. 2008. "Vitamin D in Pregnancy and Lactation: Maternal, Fetal, and Neonatal Outcomes from Human and Animal Studies." *American Journal of Clinical Nutrition* 88 (2): 520S–528S.

Kraus, Bertram S. 1959. "Calcification of the Human Decdiuous Teeth." *Journal of the American Dental Association* 59 (6): 1128–1136.

Lewis, Mary E. 2000. "Non-adult Palaeopathology: Current Status and Future Potential." In *Human Osteology in Archaeology and Forensic Science,* ed. Margaret Cox and Simon Mays, 39–57. Cambridge: Cambridge University Press.

———. 2002. "Impact of Industrialization: Comparative Study of Child Health in Four Sites from Medieval and Postmedieval England (AD 850–1859)." *American Journal of Physical Anthropology* 119 (3): 211–223.

———. 2007. *The Bioarchaeology of Children: Perspectives from Biological and Forensic Anthropology.* Cambridge: Cambridge University Press.

Liu, N., A.T. Kaplan, J. Low, L. Nguyen, G.Y. Liu, O. Equils, and M. Hewison. 2009. "Vitamin D Induces Innate Antibacterial Responses in Human Trophoblasts via an Intracrine Pathway." *Biology of Reproduction* 80 (3): 398–106.

Loth, Susan R., and Maciej Henneberg. 2001. "Sexually Dimorphic Mandibular Morphology in the First Few Years of Life." *American Journal of Physical Anthropology* 115 (2): 179–186.

Lovejoy, C. Owen, Katherine F. Russell, and Mary L Harrison. 1990. "Long Bone Growth Velocity in the Libben Population." *American Journal of Human Biology* 2 (5): 533–541.

Lovell, Nancy C. 2000. "Paleopathological Description and Diagnosis." In *Biological Anthropology of the Human Skeleton*, ed. M. Anne Katzenberg and Shelley R. Saunders, 217–248. New York: Wiley-Liss.

Mahon, Pamela, Nicholas Harvey, Sarah Crozier, Hazel Inskip, Sian Robinson, Nigel Arden, Rama Swaminathan, Cyrus Cooper, and Keith Godfrey. 2010. "Low Maternal Vitamin D Status and Fetal Bone Development: Cohort Study." *Journal of Bone and Mineral Research* 25 (1): 14–19.

March of Dimes; Partnership for Maternal, Newborn and Child Health (PMNCH); Save the Children; and World Health Organization (WHO). 2012. *Born Too Soon: The Global Report on Preterm Birth*. Ed. Christopher P. Howson, Mary V. Kinney, and Joy E. Lawn. Geneva: WHO.

Mays, Simon. 2002. *The Archaeology of Human Bones*. London: Routledge.

———. 2014. "The Palaeopathology of Scurvy in Europe." *International Journal of Paleopathology* 5: 55–62.

McAuley, J. P., and Hans K. Uhthoff. 1990. "The Development of the Pelvis." In *The Embryology of the Human Locomotor System*, by and ed. Hans K. Uhthoff, 107–116. Berlin: Springer-Verlag.

McLean, Frances, and Robert Usher. 1970. "Measurements of Liveborn Fetal Malnutrition Infants Compared with Similar Gestation and with Similar Birth Weight Normal Controls." *Biology of the Neonate* 16 (4): 215–221.

Mensforth, Robert P. 1985. "Relative Tibia Long Bone Growth in the Libben and Bt-5 Prehistoric Skeletal Populations." *American Journal of Physical Anthropology* 68 (2): 247–262.

Mittler, Diane M., and Susan G. Sheridan. 1992. "Sex Determination in Subadults Using Auricular Surface Morphology: A Forensic Science Perspective." *Journal of Forensic Sciences* 37 (4): 1068–1075.

Molleson, Theya. 1991. "Demographic Implications of The Age Structure of Early English Cemetery Samples." *Actes des Journees Anthropologiques* 5: 113–121.

Moore, Keith L., and T.V.N. Persaud. 1998. "The Fetal Period." In *The Developing Human: Clinically Oriented Embryology*, 6th ed., ed. Keith L. Moore and T.V.N. Persaud, 107–128. Philadephia: W.B. Saunders Co.

Moorrees, Coenraad F.A., Elizabeth A. Fanning, and Edward E. Hunt. 1963. "Formation and Resorption of Three Deciduous Teeth in Children." *American Journal of Physical Anthropology* 21 (2): 205–213.

Nakao, T. 1998. "A Morphological Study of the Fetal Ilium: Focusing on the Sexual Differences of the Greater Sciatic Notch." *Fukuoka igaku zasshi = Hukuoka acta medica* 89 (2): 56–63.

Noback, Charles R. 1944. "The Developmental Anatomy of the Human Osseous Skeleton during the Embryonic, Fetal and Circumnatal Periods." *The Anatomical Record* 88 (1): 91–125.

O'Rahilly, Ronan, and Ernest Gardner. 1975. "The Timing and Sequence of Events in the Development of the Limbs in the Human Embryo." *Anatomy and Embryology* 148 (1): 1–23.

Ortner, Donald J. 2003. *Identification of Pathological Conditions in Human Skeletal Remains*. San Diego: Academic Press.

Ortner, Donald J., and Simon Mays. 1998. "Dry-Bone Manifestations of Rickets in Infancy and Early Childhood." *International Journal of Osteoarchaeology* 8 (1): 45–55.

Ortner, Donald J., Whitney Butler, Jessica Cafarella, and Lauren Milligan. 2001. "Evidence of Probable Scurvy in Subadults from Archaeological Sites in North America." *American Journal of Physical Anthropology* 114 (4): 343–351.

Osterholtz, Anna J., Jonathan Bethard, Andre Gonciar, and Zsolt Nyaradi. 2014. "Possible Prenatal and Perinatal Scurvy at Telekfalva, Romania." *American Journal of Physical Anthropology* 153 (S58): 201.

Owsley, Douglas W., and Richard L. Jantz. 1985. "Long Bone Lengths and Gestational Age Distributions of Post-contact Period Arikara Indiana Perinatal Infant Skeletons." *American Journal of Physical Anthropology* 68 (3): 321–328.

Parish, Jennifer L. 2000. "Treponemal Infections in the Pediatric Population." *Clinics in Dermatology* 18 (6): 687–700.

Pearl, Marilyn, and Lawrence M. Boxt. 1980. "Radiographic Findings in Congenital Lead Poisoning." *Radiology* 136 (1): 83–84.

Petersen, Sten, Anders Gotfredsen, and Finn Ursin Knudsen. 1989. "Total Body Bone Mineral in Light-for-Gestational-Age Infants and Appropriate-for-Gestational-Age Infants." *Acta Paediatrica Scandinavica* 78 (3): 347–350.

Pfeiffer, Susan, J.Christopher Dudar, and Shawn Austin. 1989. "Prospect Hill: Skeletal Remains from a 19th-Century Methodist Cemetery, Newmarket, Ontario." *Northeast Historical Archaeology* 18 (1): 29–48.

Philip, Alistair G.S. 1974. "Fontanel Size and Epiphyseal Ossification in Neonates with Intrauterine Growth Retardation." *Journal of Pediatrics* 84 (2): 204–207.

Prentice, Ann. 2003. "Micronutrients and the Bone Mineral Content of the Mother, Fetus and Newborn." *The Journal of Nutrition* 133 (5): 1693S–1699S.

Rösing, F.W. 1983 "Sexing Immature Human Skeletons." *Journal of Human Evolution* 12 (2): 149–155.

Redfern, Rebecca. 2008. "A Bioarchaeological Investigation of Cultural Change in Dorset, England (Mid-to-Late Fourth Century B.C. to the End of the Fourth Century A.D.)." *Britannia* 39: 161–191.

Reynolds, Earle L. 1945. "The Bony Pelvic Girdle in Early Infancy: A Roentgenometric Study." *American Journal of Physical Anthropology* 3 (4): 321–354.

Riesenfeld, A. 1972. "Functional and Hormonal Control of Pelvic Morphology in the Rat." *Cells Tissues Organs* 82 (2): 231–253.

Sørensen, Marie Louise Stig. 2000. *Gender Archaeology*. Cambridge: Polity Press.

Saul, Frank P., and Julie Mather Saul. 1989. "Osteobiography: A Maya Example." In *Reconstruction of Life from the Skeleton,* ed. Mehmet Yaşar İşcan and Kenneth A.R. Kennedy, 287–302. New York: Wiley-Liss.

Saunders, Shelley R. 1992. "Subadult Skeletons and Growth Related Studies." In *Skeletal Biology of Past People,* ed. Shelley R. Saunders and M. Anne Katzenberg, 1–20. New York: Wiley-Liss.

Saunders, Shelley R. 2000. "Subadult Skeletons and Growth-Related Studies." In *Biological Anthropology of the Human Skeleton,* ed. M. Anne Katzenberg and Shelley R. Saunders, 135–161. New York: Wiley-Liss.

Saunders, Shelley R., and Robert D Hoppa. 1993. "Growth Deficit in Survivors and Non-survivors: Biological Mortality Bias in Subadult Skeletal Samples." *Yearbook of Physical Anthropology* 36: 127–151.

Schaefer, Maureen, Louise Scheuer, and Sue M. Black. 2009. *Juvenile Osteology: A Laboratory and Field Manual.* London: Academic Press.

Scheuer, J.Louise, Jonathan H. Musgrave, and Suzanne P. Evans. 1980. "The Estimation of Late Fetal and Perinatal Age from Limb Bone Length by Linear and Logarithmic Regression." *Annals of Human Biology* 7 (3): 257–265.

Scheuer, Louise, and Sue M. Black. 2000. "Development and Ageing of the Juvenile Skeleton." In *Human Osteology in Archaeology and Forensic Science,* ed. Margaret Cox and Simon Mays, 9–22. Cambridge: Cambridge University Press.

Schour, Isaac. 1936. "The Neonatal Line in the Enamel and Dentin of the Human Deciduous Teeth and First Permanent Molar." *Journal of the American Dental Association* 23 (10): 1946–1955.

Schultz, Michael. 1993 "Initial Stages of Systemic Bone Disease." In *Histology of Ancient Human Bone: Methods and Diagnosis,* ed. Gisela Grupe and A. Neil Garland, 185–203. Berlin: Springer-Verlag.

Schutkowski, Holger. 1993. "Sex Determinaiton of Infant and Juvenile Skeletons: I. Morphognostic Features." *American Journal of Physical Anthropology* 90 (2): 199–205.

Scott, Kenneth E., and Robert Usher. 1964. "Epiphyseal Development in Fetal Malnutrition Syndrome." *New England Journal of Medicine* 270 (16): 822–824.

Shannon, Michael. 2003. "Severe Lead Poisoning in Pregnancy." *Ambulatory Pediatrics* 3 (1): 37–39.

Shea, Katherine M., Allen J. Wilcox, and Ruth E. Little. 1998. "Postterm Delivery: A Challenge for Epidemiologic Research." *Epidemiology* 9 (2): 199–204.

Sherwood, Richard J., R.S. Meindl, H.B. Robinson, and R.L. May. 2000. "Fetal Age: Methods of Estimation and Effects of Pathology." *American Journal of Physical Anthropology* 113 (3): 305–315.

Siiteri, Pentti K., and Jean D. Wilson. 1974. "Testosterone Formation and Metabolism during Male Sexual Differentiation in the Human Embryo." *The Journal of Clinical Endocrinology and Metabolism* 38 (1): 113–125.

Singer, D.B., C.J. Sung, and J.S. Wigglesworth. 1991. "Fetal Growth and Maturation: With Standards for Body and Organ Development." In *Textbook of Fetal and Perinatal Pathology,* vol. 1, ed. J.S. Wigglesworth and D.B. Singer, 11–47. Oxford: Blackwell.

Sofaer, Joanna. 2006. "Gender, Bioarchaeology and Human Ontogeny." In *Social Archaeology of Funerary Remains,* ed. Rebecca Gowland and Christopher Knüsel, 155–167. Oxford: Oxbow Books.

Stark, Robert J. 2014. "A Proposed Framework for the Study of Paleopathological Cases of Subadult Scurvy." *International Journal of Paleopathology* 5: 18–26.

Steckel, Richard H. 2004. *The Best of Times, the Worst of Times: Health and Nutrition in Pre-Columbian America.* Working Paper 10299. Cambridge, MA: National Bureau of Economic Research.

Stoeckel, John, and A.K.M. Alauddin Chowdhury. 1972. "Neo-Natal and Post-Neo-Natal Mortality in a Rural Area of Bangladesh." *Population Studies* 26 (1): 113–120.

Sutter, Richard C. 2003. "Nonmetric Subadult Skeletal Sexing Traits: I. A Blind Test of the Accuracy of Eight Previously Proposed Methods Using Prehistoric Known-Sex Mummies from Northern Chile." *Journal of Forensic Sciences* 48 (5): 927–935.

Teotia, M., and S.P.S. Teotia. 1997. "Nutritional and Metabolic Rickets." *Indian Journal of Pediatrics* 64 (2): 153–157.

Thomson, Arthur. 1899. "The Sexual Differences of the Foetal Pelvis." *Journal of Anatomy and Physiology* 33 (3): 359.

Tocheri, Matthew W., and J. Eldon Molto. 2002. "Aging Fetal and Juvenile Skeletons from Roman Period Egypt Using Basiocciput Osteometrics." *International Journal of Osteoarchaeology* 12 (6): 356–363.

Toohey, John S. 1985. "Skeletal Presentation of Congenital Syphilis: Case Report and Review of the Literature." *Journal of Pediatric Orthopaedics* 5 (1): 104–106.

Vlak, Dejana, Mirjana Roksandic, and Michael A. Schillaci. 2008. "Greater Sciatic Notch as a Sex Indicator in Juveniles." *American Journal of Physical Anthropology* 137 (3): 309–315.

Walker, Phillip L., John R. Johnson, and Patricia M. Lambert. 1988. "Age and Sex Biases in the Preservation of Human Skeletal Remains." *American Journal of Physical Anthropology* 76 (2): 183–188.

Weaver, David S. 1980. "Sex Differences in the Ilia of a Known Sex and Age Sample of Fetal and Infant Skeletons." *American Journal of Physical Anthropology* 52 (2): 191–195.

Wilson, Laura A., Norman MacLeod, and Louise T. Humphrey. 2008. "Morphometric Criteria for Sexing Juvenile Human Skeletons Using the Ilium." *Journal of Forensic Sciences* 53 (2): 269–278.

Wilson, Miriam G., Harvey I. Meyers, and Allen H. Peters. 1967. "Postnatal Bone Growth of Infants with Fetal Growth Retardation." *Pediatrics* 40 (2): 213–223.

Winston, W.K., Roberta Sherman, Paul Succop, Susan Krug-Wispe, Reginald C. Tsang, Jean J. Steichen, Alvin H. Crawford, and Alan E. Oestreich. 1989. "Fractures and Rickets in Very Low Birth Weight Infants: Conservative Management and Outcome." *Journal of Pediatric Orthopaedics* 9 (3): 326–330.

Woods, Charles R. 2009. "Congenital Syphilis-Persisting Pestilence." *Pediatric Infectious Disease Journal* 28 (6): 536–537.

Word Health Organization (WHO). 1977. "WHO: Recommended Definitions, Terminology and Format for Statistical Tables Related to the Perinatal Period and Use of a New Certificate for Cause of Perinatal Deaths: Modifications Recommended by FIGO as Amended October 14, 1976." *Acta Obstetricia Gynecologica Scandinavica* 56 (3): 247–253.

Zeitlin, J., B. Blondel, S. Alexander, G. Breart, and the PERISTAT Group. 2007. "Variation in Rates of Postterm Birth in Europe: Reality or Artefact?" *BJOG: An International Journal of Obstetrics and Gynaecology* 114 (9): 1097–1103.

Chapter 3

PREGNANT WITH IDEAS
CONCEPTS OF THE FETUS IN THE TWENTY-FIRST-CENTURY UNITED STATES

Sallie Han

What to call "it" was one of the first quandaries I encountered when I began my research with pregnant women more than a decade ago. Was it a child? A baby? A fetus? My aim had been to undertake an anthropological study of pregnancy as a cultural and social experience in the United States. I had posted flyers in doctors' and midwives' offices and on community bulletin boards in bookstores and coffee shops, inviting expectant mothers to participate in my project.[1] Everyone who responded knew what I meant, but as women expecting for the first time, they told me they did not necessarily identify themselves as "mothers"—at least not quite yet—and especially in the early weeks and months of pregnancy, the status of "it" still seemed unreal and uncertain. Thus, they were unclear on how to refer to "it." For example, Dana[2], then nineteen weeks pregnant, had received reassuring results from her amniocentesis but told me: "It's still not like a baby to me yet. It's a little thing inside, that's all. I don't think about it as human." Other women, recalling the ultrasound images they viewed during the early weeks of their pregnancies, referred to it as a "bean" or a "peanut."

A concept of it—whether child, baby, fetus, or embryo—is always a concept of social relations that themselves constitute the subjects enmeshed in those relations. With a child or baby, a woman is a mother, and without one, she is not; there also is no child or baby without a mother. From personal experience, I still recall the dramatic signaling of a ruptured relationship and interrupted identities when a nurse-midwife began to talk carefully to me about "the pregnancy." Moments before, she had been talking with me about "the baby," but, unable to locate its heartbeat with a Doppler device, she performed an ultrasound scan that showed an empty sac. The same nurse who earlier had greeted me as "mom" now returned to counsel me on what to expect during a miscarriage and to schedule a follow-up appointment to check that the "uterine contents" had been expelled. This was at almost twelve weeks, on the cusp of what medical encyclopedias define as a fetus rather than an embryo.

What to call "it" is an important and necessary question to ask. Rather than taking for granted the fetus as a biological fact of life, the anthropology of fetuses seeks to lay them bare as social bodies. In undertaking this project, the chapters in this book perform a kind of archaeology of fetuses, excavating their histories and examining them as cultural artifacts. To this end, the present chapter offers a discussion of the construction of the fetus in both private experience and public culture. It provides an overview of the work of anthropologists, historians, and other feminist scholars and my own ethnographic research on and continuing interest in everyday practices of pregnancy. This chapter builds on cross-cultural and historical perspectives on the unborn, which is not necessarily imaged or imagined as an embryo or fetus, and sifts through the layers of recent history in the United States and recovers the modern fetus that is familiar to us as a concept of social relations in three overlapping contexts.

First, I consider the characterization of fetuses as vulnerable and pregnancies conflicted and tentative during the 1980s. Not only were fetuses politicized in anti-abortion activism, but even more significantly, they were medicalized as patients who should and could be treated (not quite) independently of the women in whose bodies they were gestating. Pregnancy became described in terms of maternal-fetal conflict in which the health and interests of pregnant women were at odds with those of fetuses.

Second, I discuss the personal fetus as made lively through what I call "belly talk" or protoconversations that involve a pregnant woman and an imagined or expected child. The practice of belly talk suggests that during the 1990s and 2000s, pregnancy became experienced less as a period of watchful waiting and more as a time of active preparation and even nascent parenting.

Third, I consider the fetus that has rematerialized during the past decade. Fetuses today are represented in a broad range of media and materials that includes digitally circulated photographs of cookies, cakes, soap, and crocheted dolls shaped as fetuses. The crafting of fetuses coincides with the emergence of an ever more complicated understanding of pregnancy as a maternal-fetal relationship in both the social sciences and biomedical sciences. Yet, at the same time, there have been legislative and legal moves to restrict abortion services and prosecute women whose activities are interpreted as risks to their pregnancies.

Ultimately, the question of what to call "it" is not merely academic but a salient and relevant one that raises concerns also about pregnant women, their communities, and the environments that we inhabit and share.

Fetuses as Artifacts of Life and Deaths

Presented as timeless knowledge—irrefutably and unchangingly true even before modern science enlightened our understanding—the fetus familiar to us today is both culturally particular and historically specific. This modern fetus represents, on the one hand, the origins and beginnings of life, as it has for a number of peoples. It also is understood, on the other hand, to have an undeniably corporeal reality as both a stage of human development (which follows those of the zygote and the embryo) and a biological individual (see Rutherford, chapter 1). The modern fetus is perceived not only as a symbol of potential—it *is* life at its barest, a claim that relies indexically on its tangibility; it does not exist as only a product of the mind but has a body. Thus, an aim of the anthropology of fetuses must be to consider the symbolic and material lives of fetuses and the interactions between them.

In cultural anthropology, the anthropology of the fetus emerges from long-standing concerns with kinship, reproduction, and women and gender. (Other chapters in this book trace the paths of archaeologists and bioarchaeologists.) Early and mid-twentieth-century anthropologists, including Bronislaw Malinowski (1929), undertook the documentation of customs, taboos, and rituals of childbearing and childbirth and their comparison across cultures. (Their interests also included the ritual involvement of men in pregnancy and birth, or couvade.) Anthropologists interpreted these customs as the protection and promotion of the health, whether spiritual or physical, of the woman and the child she was gestating; however, there is not the same interest in pregnancy per se, much less a concept of the child (or embryo or fetus) as evidenced today. The anthropology of the fetus can be traced more recently to the work of feminist scholars, including cultural anthropologists, which I discuss in this chapter. The anthropology of reproduction began in the 1970s by closely examining the significance of pregnancy, birth, and motherhood in women's lives; it gained momentum during the 1980s and 1990s by critically engaging with politics, law, and medicine and science as they affected everyday experiences of reproduction. Notably, anthropologists of reproduction have called attention to the cultural and social constructions of the fetus in and through such routine and ritual practices as fetal ultrasound imaging, which are directed toward the health of a pregnancy, like the rituals of the Trobrianders that Malinowski documented.

Ethnographers have documented the uncertainty with which people cross-culturally and historically have regarded pregnancy. Who or what inhabits the womb is not necessarily acknowledged as a person, a human, or even a fetus. Lynn Morgan (1997) spoke with indigenous women in the Ecuadorean Andes who described *criaturas* (creatures), not babies or fetuses, in their bellies. Historian Barbara Duden notes that medical texts of seventeenth- and eighteenth-century Europe depicted children with their heads and bodies in proportion and adultlike facial expressions and posed standing or even dancing: "Graphics did not represent the tissue inspected by the anatomist as a 'fetus,' but rather as the symbol and emblem of the child-to-be" (1999: 19). In contrast, when a pregnant woman today views a medical illustration of embryonic and fetal growth, she perceives that she is pregnant "with" an embryo or fetus that is also her child or baby.

Where and when fetuses do appear, they are not necessarily ascribed with the same moral, political, or scientific significances currently taken for granted in the United States. Surveying religious cultures around the world, Vanessa Sasson and Jane Marie Law take issue with what they characterize as a much-constricted conception of the fetus in contemporary Western (North American and European) contexts. In contrast, they observe, "Throughout much of human history and across most of the world's cultures, when the fetus was imagined, it enjoyed a much wider range of symbolic and cultural subjectivities, often contributing possibilities of inclusivity, emergence, liminality, and transformation" (2009: 3). In Mexico, at a site called La Venta dating back to three thousand years ago, the

Olmec created colossal stone sculptures of human fetuses standing more than six feet tall and visual images of maize conflated with human embryos. Art historian Carolyn Tate (2009) interprets the Olmec fetuses as cultural symbols of change and growth. Robert Kritzer finds that Buddhist writings produced in India before the third century, along with the texts of classical Indian medicine at the time, offered what appear now as surprisingly accurate and detailed month-by-month descriptions of fetal development. The Buddhist writings were intended as accounts of the cycle of rebirth that "begins at the moment of death in one life, continues through the intermediate existence or *antarabhava*, the moment of conception, and the period of gestation, and culminates in the moment of birth in the next life" (Kritzer 2009: 73).

The accuracy and detail with which fetuses were depicted in Olmec art and in Buddhist writings suggests that the images were based on close observations of fetal bodies. What we know about pregnancy loss among ancient peoples is limited, but Tate (2009) contends it is reasonable to think they were familiar with fetuses due to the incidence of pregnancy loss, which would be comparable with the rates today.[3] It is estimated that between 15 and 20 percent of known or recognized pregnancies spontaneously abort (miscarry) and that the overall rate of pregnancy loss might be as high as 75 percent (Petrozza 2016).

Given that knowledge of fetal life is inferred from fetal deaths, it is not surprising that in both Olmec art and Buddhist writings, human fetuses are discussed in terms of other cultural practices and ideas surrounding life, death, and transformation. Interestingly, cultural anthropologists engaged in scholarship on the fetus have been interested in the work of historians, archaeologists, and bioarchaeologists who have documented and uncovered the mortuary ritual and funerary treatment of fetal remains (such as other chapters in this book describe). In contrast to these ancient fetuses, our modern fetuses of medical science and public culture are read as icons of life. Yet, the establishment of fetuses as biological facts of life in the United States today is based, too, on the artifacts of fetal deaths. Tracing the history of embryology, Morgan (2009) describes the efforts of late nineteenth- and early twentieth-century scientists on behalf of their nascent field of study to collect embryos and fetuses intended to represent every stage of human growth and development. Obstetricians and general practitioners donated the specimens, but it is unclear whether their patients, the women who had lost their pregnancies, were aware or were asked for their permission. In 1965, *Life* magazine published Lennart Nilsson's now iconic photographs of life in the womb, heralding the cover photograph of a fifteen-week-old embryo, taken with a surgical scope, as "the first portrait ever made of a living embryo inside its mother's womb" (*Life* 1965). Sandra Matthews and Laura Wexler note, however, that all of the other photographs had been taken of fetal bodies that had been "surgically removed for a variety of reasons" (2000: 195). They were neither living nor inside the uterus.

What we gain from cross-cultural and historical perspectives on the human fetus is the recognition that a concept of the fetus (or embryo) is not and has not always been relevant to an understanding of pregnancy. The unborn were unseen but not unimagined. Even having accurate and detailed knowledge of fetal anatomy, growth, and development does not necessarily produce a concept of the fetus as the biological fact of life that we take for granted today. What fetuses are, then, is not self-evident, their bodies themselves are socially constructed, and the knowledge and meaning attached to them are culturally ascribed.

The Vulnerable Fetus

Even when separated by centuries, distinct cultural traditions, and social organizations, the various public displays of the fetus share the aim of inspiring and instilling prescribed sets of thoughts and sentiments. La Venta has been identified as an ancient ritual site and civic center or capital of the Olmec, with the sculptures of the fetuses and other monuments arranged on a path leading visitors through a visual narrative of creation and the intertwined destinies of humans and maize (Tate 2009). In the United States during the 1980s, fetal imagery became the medium for a message about the purported "personhood" of fetuses and the wrongfulness of abortion (cf. Luehrmann's discussion of Russian Orthodox anti-abortion activists, chapter 10). The fetus came to be seen as vulnerable, and pregnancy as conflicted and tentative. These perceptions were shaped by both the politicization and intensified medicalization of the fetus and of pregnancy.

Published in the same month as *Life*'s celebratory coverage of the launch of the Gemini rocket, Nilsson's 1965 photographs were accompanied by captions that reminded readers of the "unprecedented photographic feat in color" represented in the images, and trained their attention to "the drama of life before birth" unfolding before their eyes (quoted in Duden 1993: 11). The images are described as portraits of "the fetus as spaceman, floating in a starry sky, connected if at all to a mother only through an umbilical life-support line" (Matthews and Wexler 2000: 195). The fetus was pictured living and growing in metaphorical isolation, but Duden tells us that in fact, the photographs were taken of different bodies, each "removed from a dead woman or a tubal pregnancy," which is not discussed in the magazine. The fetus appears before the eye not as the result of a woman's pregnancy, but an achievement of science and technology. Duden also reminds us that in 1965, "live observation was still a largely unrealized prospect" (1993: 14).

Twenty years later, the sonogram routinely shifted focus from the pregnant woman to the fetus, and placed science and technology in full view. Previously, X-rays had been used to produce in utero images, but they were not used extensively in prenatal medical care, especially as the hazards were understood. Sonar technology, which enables the detection of solid masses in a liquid medium, had been developed for submarine warfare during World War I. Its medical application developed in the decades afterward, and by the 1980s, ultrasound imaging provided glimpses of a living, growing fetus (Oakley 1984). Although fetal ultrasound images themselves appear imprecise and almost unreadable, the awareness of onlookers that the "pictures" are produced during a real-time live observation informs their powerful reception.

In terms of their visual composition, sonograms, like Nilsson's photographs, also depict fetuses in apparent isolation. Notably, the mother's or pregnant woman's uterus appears in ultrasound images in black as a space or void, with the bones and tissue of the developing body appearing in white and gray. This focus on the fetus is described as a result of the technical restrictions and requirements of the sonogram. It is not only the visual qualities of the sonogram but the theater of fetal ultrasound imaging, performed with expensive and sophisticated machinery dominating the exam room, and the expert translation of the sonographers who perform the scans that contribute significantly to the interpretation of the images as "baby pictures" (Han 2009a, 2009b; Mitchell 2001; Taylor 2008).

The politicization and medicalization of fetuses in the United States as vulnerable—and the characterization of pregnancies as contested and tentative—have gone hand in hand, as Rosalind Pollack Petchesky (1987) makes especially clear in her analysis of *The Silent Scream* (Jack Duane Dabner, 1984). Produced by the National Right to Life Committee and seen on network television by about ten million viewers, the film featured footage of what the narrator— an obstetrician and a self-identified anti-abortion activist—describes as a twelve-week-old fetus being aborted. Petchesky noted a disjunction between the ultrasound images, in which she saw "a shadowy, black-and-white, pulsating blob" and "a filmic image of vibrating light and shaded areas," and the voiced narration that refers to "the living unborn child" as "another human being indistinguishable for any of us" (1987: 266).

Notably, the effects that a scan can have on a pregnant woman's perceptions and emotions have been cited almost from the start for both political and medical purposes. Petchesky notes that *The Silent Scream* took its inspiration from an article published in *The New England Journal of Medicine* that claimed early sonograms promoted "maternal bonding" (1987: 265). Thirty years later, in 2012, in a move widely recognized as an effort to deter women from obtaining abortion care, legislators in Virginia passed a law mandating that women receive sonograms—accompanied by narration from a state-sanctioned script—before having their procedures.[4]

In prenatal medical care, too, fetal ultrasound imaging is assumed to have important and meaningful effects on affect. When I talked with doctors and midwives in my own research on pregnancy and pregnancy care, I was told that a benefit of the sonogram was its potential to motivate pregnant women to feel responsibility for an expected child and thus take better care of themselves. Yet, there is no evidence the scans actually produce better outcomes. In 1984, a report issued jointly by the National Institutes of Health and the Food and Drug Administration found "no clear benefit from routine use" for either the expected child or the pregnant woman (Petchesky 1987: 273). In 2015, *The Wall Street Journal* reported that US women received an average of 5.2 scans per delivery,[5] even though a 2010 study had concluded, "routine scans do not seem to be associated with reductions in adverse outcomes for babies" (Helliker 2015). In my own research, I found the question of medical necessity not salient for expectant parents, who viewed the sonogram as a ritual occasion when they and other family members might share an opportunity to "see the baby" (Han 2009a).

It has been observed that the rendering of the fetus as visible—and independently viable[6]—has had the opposite effect on pregnant women and mothers (Stabile 1992). This disappearing act has been explained as resulting from the conditions in which the images of fetuses were made. Yet, an anthropology of fetuses also ought to maintain a healthy skepticism of the claims made about technological innovation and medical necessity. While invention and need are frequently cited as drivers of change, which in turn are characterized as "advances," scholars in the social science of medicine remind us these changes or "advances" must be understood as shifts in cultural ideas and social practices. Tracing the late twentieth-century history of fetology as a field of medicine distinct from obstetrics, gynecology, pediatrics, or even neonatology, Monica Casper contends, "The specialty and its often elusive client [the fetus] have developed in tandem, each serving as a justification for the existence of the other" (1998: 4). She notes that the possibility of "open" fetal surgery—first successfully performed in the United States in 1981 to repair an organ defect—is promoted as "saving" babies' lives and improving their health outcomes. This not only ignores the efforts that women themselves make in order to have healthy pregnancies and healthy babies, but it also seeks a work-around or solution that need not involve mothers at all (and also overlooks the fact that performing surgery on a fetus necessarily entails operating on a woman).

Public health, dating to the late nineteenth and early twentieth centuries, looked to communities as the sites of health and sickness. The rationale had been to improve the conditions of communities in order to improve the conditions of the individuals living in them. While scholars now are rightly critical of the maternalist (and eugenicist) ideologies that guided maternal-child health campaigns (Klaus 1993; Stern 2002), at least it might be said that women (or in any case, mothers) were recognized as central actors in the drama of pregnancy. In contrast, Laury Oaks (2001) notes the recasting of pregnancy as a maternal-fetal conflict during the twentieth century. The placenta, once described in medical texts as a "barrier" that protected the developing fetus, became a "sieve" that permitted not only nutrients but also toxins to pass from mother to child; as metaphors, these descriptions of the placenta themselves can be read not as value-neutral facts derived from scientific progress but rather artifacts of cultural and linguistic practice. Whether it was the medication a doctor prescribed for the relief of nausea, a glass of wine served with dinner (acceptable during pregnancy in Europe but not in North America), or the lilt of another drug on which a woman might have become dependent, a woman's behaviors made her a possible threat to her own pregnancy.

Not least of all was the risk her "choice" represented. Women in the United States today have a protected right to abortion care, which they seek for a range of reasons particular to the circumstances of their lives. Today, about half of all pregnancies in the

United States are unintended (Finer and Zolna 2011). Not all unintended pregnancies are necessarily unwanted, however, and even intended and otherwise much-wanted pregnancies might be terminated. Between 2008 and 2011, the years for which the most recent data are available, the US abortion rate was 16.9 per 1,000 women aged 15 to 44, which is the lowest rate since 1973 (Jones and Jerman 2014). Among the reasons women might seek abortions are the results of prenatal diagnostic testing and the repercussions they can be anticipated to have. Discussing the impact of amniocentesis and prenatal diagnostic testing, Barbara Katz Rothman (1987) found that women did not feel reassured by their bodily experiences of pregnancy—particularly the movements of the fetus that signified as the "quickening" of an expected child—but instead felt "tentative" about their pregnancies until they had received confirmation from the results of the amniocentesis. When the results showed chromosomal anomalies associated with lethal conditions, the "choice" to end a pregnancy is hardly experienced as one, as Rayna Rapp (1999) has documented. Rather, it is seen as preventing pain, hurt, and harm for what would have been an expected child and the expectant family. Rapp movingly describes the decisions that women make as mothers of other children on whom they do not wish to impose the responsibilities of caring for a sibling who will remain always dependent.

In the sense that uncertainty has ever surrounded the status of the unborn, historically and cross-culturally, fetuses could be said to have been always vulnerable and pregnancies conflicted and tentative. An anthropology of fetuses makes clear they are made to be vulnerable, conflicted, and tentative in historically specific and culturally particular ways (see also Cromer's insights into the construction of frozen embryos "leftover" from in vitro fertilization procedures as both valueless and valuable, chapter 8). Especially notable here are the purposefully political and purportedly value-free medical uses of fetal images since the 1980s to inspire and instill a prescriptive set of thoughts and sentiments.

The Personal Fetus

By the 1990s and early 2000s, fetuses had already been well established as key political symbols in no small part because of their depiction as emotionally evocative objects. Public images emphasized the resemblance of a fetus even relatively early in pregnancy to a baby by depicting it, for example, with a thumb in its mouth. Yet, the fetus is a lively central figure because of not only what Petchesky (1987) calls its "visual power"—with seeing itself also recognized as a cultural practice—but also the talk that accompanies our encounters with it. Ultrasound scans had become both a routine practice of prenatal care and a ritual occasion for expectant parents and other family members, who received what amounted to "show-and-tell" tours of fetal anatomy (Han 2009b). The talk of sonographers, doctors, and other care providers were critical to interpreting the black-and-white images flashing across the screen (Nishizaka 2014).

The women who participated in my study were interested in seeing and bonding with not a fetus but their babies. To see a baby was not necessarily to bond with it; that sort of deep attachment came with *talking* to it. Here, I discuss the significance of what I call "belly talk," or the engagement of pregnant women and others (including expectant fathers, siblings, and other family members) in proto-conversations with an expected and imagined child in utero (Han 2013). Actual conversations entail exchanges between two or more participants, but protoconversations involve a single participant assuming additional roles, as when a pet owner engages in talk with a dog or cat. Moreover, pet talk emerges from and expresses an affective relationship or bond between owner and pet; the talk makes the pet a person. In "motherese" or baby talk, a mother voices for not only herself but also her child, whose vocalizations, facial expressions, and bodily movements she interprets.

The protoconversational qualities of belly talk make it a practice unique to the contemporary United States. The grandmothers-to-be I met during my fieldwork frequently expressed their surprise to me about belly talk, which suggests it was historically not a common practice. Cross-culturally, conversing with infants and young children is also not necessarily regarded as appropriate. For example, linguistic anthropologists Elinor Ochs and Bambi Schieffelin (1984) noted that Samoan infants, from birth to five or six months old, were talked about and occasionally sung to but not engaged in conversation because they were considered *pepemeamea* (translated as "baby thing thing") and lacking human sense and understanding. In Japan, Tsipy Ivry (2010) has described a long tradition of *taikyo*, or fetal education in which pregnant women have been expected to talk to their expected children in order to direct their growth and development and even train their impressions. During the late nineteenth and early twentieth centuries, for example, Japanese women were encouraged to recite the tales of national heroes to nurture a sense of patriotism in their children. In the belly talk of the United States, a pregnant woman translates the kicks and rolls felt in her belly as expressions of perception and intention, and adopts what she imagines as her expected child's (individual and independent) perspective, embodying and enlivening it.

Dana, whose story I told in the introduction to this chapter, maintained a calm composure until our conversation turned to her habit of what I call "belly talk" (Han 2009c, 2013). Just a moment earlier, she had referred to "it" and "a little thing inside," but now Dana, who had learned from her amniocentesis that she was expecting a girl, mentioned talking to "her" and became tearful. Talk, I suggest, enlivens the fetus because it is seen as distinctive to human persons. Other animals and even plants are understood to engage in communication more generally, and there is the suggestion that other primates and other species like whales have a capacity for language. Speech, however, has been surmised to be unique to modern humans.[7] Chimpanzees, the primates most closely related to us, produce a wide range of communicative vocalizations but not the vowel and consonant sounds we modern humans can in our utterances. It had been hypothesized that Neanderthals signed rather than spoke (Stokoe 1991), as their different anatomical structure of their larynx presumably did not lend itself to speech (Falk 1975). This understanding has been overturned with new biomechanical analysis suggesting Neanderthals could speak (D'Anastasio et al 2013). Nevertheless, the ability to speak has been claimed as distinctive of *Homo sapiens* as a species, and voice is associated with individuality. Describing the importance and meaning of the human voice as revealing the "inner being" of a person, in contrast to the "outward appearance" apprehended from vision, Tim Ingold suggests: "Where vision places us vis-à-vis one another, 'face-to-face,' leaving each of us to construct an inner representation of the other's mental state on the basis of our observation of outward appearance, voice and hearing establish the possibility of genuine intersubjectivity or a participatory communion of the self and other through shared immersion in the stream of sound" (2000: 247). He notes the significance attached to the oral and aural as messengers of what is authentic, real, and true (cf. Howes-Mischel's discussion of fetal heartbeats, chapter 11).

When Bridget, a woman in my study, described belly talk as a way of "relating" to her expected child, I heard a double meaning: relating meant talking to the belly, specifically telling stories about events from her day. It also meant enacting relationships, with pregnant women and their partners engaging in role-play that appeared to anticipate their real roles as parents. Other women in my study told me about reading aloud to the belly, particularly in the evening, a precursor to the bedtime story for a child. Martina, at twenty-two weeks pregnant, had recruited her husband Daniel's help in talking to their expected child. Martina had been advised that lying supine, on her back, could cut off circulation to the child, so she instructed her husband to lean close to her belly, speak slowly, and tell the baby to "kick Mommy really hard so that she can roll over." Although she and I laughed at her recounting of this moment, I thought it was nevertheless a rather significant one, involving Martina and Daniel's coordinated (and gendered) effort as mother and father; the assignment of authority to Daniel, as the father, to direct the child's behavior; and the identification of Martina as "Mommy" who will accommodate the child's needs. The treatment of their expected child as a conversational partner who can comprehend and respond to their instruction—to communicate with them by kicking—enlivens it.

Bridget and Martina were familiar with the idea that talking, reading aloud, and singing to their bellies were important for bonding with their babies. The practices were promoted in a number of the pregnancy books and other written sources of advice, like magazines and websites, from which they frequently sought information. The authors of *What to Expect When You're Expecting* told their readers, "Any kind of prenatal communication may give you a head start on the long process of parent-baby bonding" (Eisenberg et al. 1996: 187). Other claims were also being made about the benefits of belly talk. A product called BabyPlus offered expectant parents a "prenatal curriculum" of "audio lessons" that its makers promised "improved school readiness and intellectual abilities" as well as "longer attention spans." The advertisement appeared in a pregnancy magazine distributed for free in doctors' and midwives' offices during my fieldwork. In another advertisement—for a sound player and recorder I also saw being sold at a maternity clothing store in an area shopping mall—expectant parents were told, "You can nourish your baby's brain with sounds while listening to and recording your baby's response to the stimuli." The women in my study generally scoffed at such claims. Greta, a high school special education teacher, joked: "I don't sit with headphones on my belly or do anything in particular. Nothing like book reading or going over times tables. No flash cards yet—that'll come later." Although they rejected these portrayals of prenatal learning, which were regarded with skepticism especially because they appeared in advertisements for commercial products, they did not necessarily question the underlying assumptions about the fetus as able to receive and produce messages or, equally significant, the responsibility of parents to care and provide for and teach their children, even during pregnancy. As I have argued elsewhere, pregnancy in the contemporary United States has become conceived less as a period of watchful waiting and more as a time of active preparation and nascent parenting (Han 2013).

The differences we apprehend between seeing and hearing have shaped the particular forms that fetuses—and our conceptions of pregnancy—have taken. We saw earlier a fetus that was vulnerable in a conflicted and tentative pregnancy. Here, the personal fetus is the one with which pregnant women and others engage in the protoconversations and bonding of belly talk. In the next section, I consider some other ways in which fetuses have come to matter recently.

The Rematerialized Fetus

When Meredith Michaels and Lynn Morgan commented on the "proliferation of fetuses in various written and visual forms," which included "obstetric and pediatric journals, ultrasonic imaging, advertising, Hollywood movies, and so on" (1999: 1), they could not have anticipated the representation of fetuses in an even broader range of media and materials during the past ten years. In October 2013, UK artist Damien Hirst unveiled "The Miraculous Journey," described on his website as a series of "fourteen large-scale bronze sculptures that chart the gestation of a foetus from conception to birth," which stand on the grounds of a medical center in Qatar. *The New York Times* described the dramatic unveiling of the monumental sculptures "to the amplified sound of a beating heart" as a balloon of fabric shrouding each figure "opened like a giant flower" (Vogel 2013). In 2013, a pair of entrepreneurs sought support for their company, called 3D Babies, which would produce keepsake figurines based on fetal ultrasound images. According to the company's Indiegogo page: "We use your 3D/4D ultrasound images or newborn baby pictures to create a unique artistic representation of your baby using the latest computer graphics and 3D printing technology."[8] The company promised a figurine with a face individualized from sonograms submitted by the customer, and one of three skin tones (light, medium, and dark) selected presumably to match the child.

The Hirst sculptures were designed to provoke, 3D Babies to make memories tangible—two ways in which representations of the human fetus have figured so significantly in the contemporary United States. Browsing the Web, however, I also found a proliferation of fetus cookies, fetus cakes, fetus soap, and crochet fetuses. To borrow some of the words used in the descriptions and comments about these objects—which were products being sold and bought for occasions like the announcement of a pregnancy, the recently invented tradition of the gender reveal party, and the baby shower—fetuses can be hilarious, creative, cute, wicked cool, delicious, sweet, and appropriate (as in "the perfect gift for my sister"). The fun and funny fetus of the 2000s stands as an index of the personal qualities of the people who baked, crafted, and posted photographs of their creations on their blogs or shared them on Facebook, Pinterest, or other social media. The images invite the viewer to admire the handiwork and laugh at the joke. As one of my undergraduate students said, when she alerted me to the existence of fetus cookies, "I think they're supposed to be cute, not political." In other words, this is a new, improved, hip, and ironic fetus that apparently threatens nobody and nothing (with the possible exception of a feminist anthropologist who came of age during the 1980s, and her sense of humor).

Yet, there is much more at stake than this when we consider that fetuses have proliferated not only in public life and private experience but also as the subjects and objects of study. The crafting of fetuses in new and surprising forms happens to coincide with the rematerializing of fetal bodies in the social sciences and biomedical sciences. While the histories of fetuses are being metaphorically unearthed (Dubow 2011; Morgan 2009), archaeologists and bioarchaeologists are discussing how to literally recover fetal remains and interpret them (see Lewis, chapter 5). Concepts of the unborn and particularly of the fetus do not remain ideas and thoughts; they also come to materialize in our experiences of everyday life, from mortuary rituals (see Scott and Betsinger, chapter 7) to art and craft that range from Olmec statuary to soap sold on Etsy to the practices of prenatal medical care. Fetuses not only "have" materiality—meaning that they matter in human experience in ways other than idea and thought—but they also are material. Ingold, in a recent essay titled "Materials against Materiality," presents us with the puzzle that "the ever-growing literature in anthropology and archaeology that deals explicitly with the subjects of materiality and material culture seems to have hardly anything to say about materials" (2011: 20). In parallel, there is a burgeoning body of work on what we might call "fetality"—the practices and ideas surrounding fetuses and such notions as personhood and kinship that we might infer from them—but we also must be able to talk about fetuses as matter, material, and bodies. If fetuses are real, then the question is real *what*?

In the biomedical sciences, there is the fleshing out of an ever more complicated understanding of what is being called the fetal origins of health and of pregnancy as a relationship between individuals. It is well understood that interactions with and within environments—whether cultural, social, ecological, or material—shape human development and growth in some part because of the influence that those interactions have on gene expression. Given that the fetus represents the most critical stage of development and growth, researchers today have become especially concerned with the links between environmental stimuli in utero and the effects observed in the later life of that offspring (Lane et al. 2014). Notably, how well—or rather, how poorly—nourished a woman is at the time she becomes pregnant has been associated with a range of problems her child later experiences as adult, including elevated risks for cardiovascular disease and type 2 diabetes (Kuzawa and Quinn 2009). In addition, the effects of environmental stresses can be observed not only in the pregnant women directly exposed and their children most immediately affected but also in subsequent generations to follow (see Rutherford, chapter 1). Thus, the framework of the developmental origins of health and development (DOHaD) provides an avenue for investigating issues like racial and ethnic health disparities (Kuzawa and Sweet 2008; Thayer and Kuzawa 2015).

Research of this kind, however, is layered on other recent histories, so the vulnerabilities of the fetus still become discussed in terms of the culpabilities of a pregnant woman, especially when translated from population-level research into personal advice. In 2010, a headline on WebMD cautioned pregnant women, "Too Much Pregnancy Weight Gain Raises Child's Obesity Risk" (Woznicki 2010), while another on Live Science in 2014 warned, "Too Little Weight Gain during Pregnancy Linked to Child's Obesity" (Gholipour 2014). In a 2014 article published in *Nature*, an interdisciplinary team that included a biological anthropologist, a philosopher, and historians commented on the misrepresentations of epigenetic research in public discourse: "A mother's individual influence over a vulnerable fetus is emphasized ... the role of the societal factors is not" (Richardson et al. 2014: 131–132). Addressing their peers in *Trends in Genetics*, a group of geneticists also commented on the responsibilities that both science journalists and scientists themselves have when reporting on research: "Health risk messages like these carry significant ethical and social implications not only because they extrapolate prematurely ... but also because they are targeted to populations that are sensitive, even vulnerable, to health advice that carries the authority of science" (Juengst et al 2014: 427–429). In their words, not only the fetus but entire groups, notably women, are vulnerable. Indeed, the past decade has born witness to legislation restricting women's access to abortion care and the prosecution of women who become deemed risks and threats (still) to their pregnancies. In 2014, Tennessee amended its already standing fetal homicide law to include provisions to prosecute women for the use of narcotic drugs during pregnancy, charging them with aggravated assault (and sentencing them to a maximum of fifteen years in prison) "if they have a pregnancy complication after using illegal drugs" (Rewire 2017). While this law is no longer in effect, similar legislation in other states continues to be challenged.

The mis-messaging of epigenetic research is a problem not only of translation and communication (e.g., journalists need to report more accurately, scientists to write more clearly), but, I suggest, also one of narrative, in which biology, once understood to be deterministic, is increasingly understood to be plastic while culture and society, previously seen as variable, become intractable problems. The fetus is conceived as complex, malleable, and even perfectable, but not in the cultural and social conditions of inequality, economic stress, and so on, which apparently are unchanging and unchangeable. So, it seems as though the story being told now trades biological and genetic determinism for culturally and socially determined defeat.

Conclusion

Of course, the story does not end there. Rather, this latest chapter points to the importance and necessity of a holistic approach that takes into account history, experience, imagination, image, representation, and material presence—not to mention both a pregnant woman and whatever "it" is. Like a child, baby, or mother, an embryo or a fetus both forms and is formed by the relationships of others to it, most significantly a pregnant woman. Attending to the fetus as a concept of social relations is especially urgent in the current context where it is taken for granted as a biological fact of life, with implications in public policy, scientific research, and our everyday experiences of reproduction.

Sallie Han is Professor of Anthropology at State University of New York at Oneonta, and past chair of the Council on Anthropology and Reproduction. She is the author of *Pregnancy in Practice: Expectation and Experience in the Contemporary US* (Berghahn Books, 2013).

Notes

1. I undertook ethnographic research in and around Ann Arbor, Michigan, between October 2002 and January 2004. Central to my study were the repeated, in-depth interviews I recorded with a core group of sixteen pregnant women, all expecting a first child. Most of the women were married to men; one woman had recently ended a relationship with her expected child's father, and another woman, whose previous partners had been women, became a parent on her own (with the aid of donor insemination). Except for one woman who identified as African American, the interviewees were white. All had attended college for at least one year, and many had received postgraduate degrees.
2. Pseudonyms are used for all of the women I quote from my research in order to protect their confidentiality and to respect their privacy.
3. In the case of the Olmec, Tate (2009) notes that the ritual sacrifice of children also has been considered to account for their knowledge of fetuses. The skeletons of fetuses, neonates, and children have been found cached in altars, thrones, and pyramids at sites other than La Venta.
4. Initially, the law mandated a transvaginal ultrasound but was modified to allow women the option of an abdominal scan.
5. The American College of Obstetricians and Gynecologists advises only one scan at twelve weeks and a second one at twenty weeks for low-risk pregnancies.
6. Viability refers to the ability of an expected child to survive outside the uterus, which is determined by the maturation of the lungs. The US Supreme Court included the concept of viability in its decision to protect access to abortion up to the point of viability, at around twenty-four weeks (i.e., first and second trimester). Today, children born at twenty-two weeks gestational age or earlier are considered nonviable. There are, however, questions about children delivered between twenty-two and twenty-four weeks due to the challenges of estimating gestational age and the development of technologies to resuscitate and treat even these extremely premature babies (*Economist* 2015).
7. Linguistic anthropologist Laura Polich (2000) has drawn attention to what she calls orality as a language ideology—that is, the suppositions that languages are "naturally" oral and, as a result, that sign languages are less "real."

8. Although the company is not in operation, the Indiegogo page is available at https://www.indiegogo.com/projects/3d-babies#/.

References

Casper, Monica. 1998. *The Making of the Unborn Patient: A Social Anatomy of Fetal Surgery.* New Brunswick, NJ: Rutgers University Press.

D'Anastasio, Ruggero, Stephen Wroe, Claudio Tuniz, Lucia Mancini, Deneb T. Cesana, Diego Dreossi, Mayoorenda Ravichandiran, Marie Attard, William C.H. Parr, Anne Agur, and Luigi Capasso. 2013. "Micro-Biomechanics of the Kebara 2 Hyoid and Its Implications for Speech in Neanderthals." *PLOS ONE* 8 (12): e82261. doi:10.1371/journal.pone.0082261.

Dubow, Sara. 2011. *Ourselves Unborn: A History of the Fetus in Modern America.* Oxford: Oxford University Press

Duden, Barbara. 1993. *Disembodying Women: Perspectives on Pregnancy and the Unborn.* Cambridge, MA: Harvard University Press.

———. 1999. "The Fetus on the 'Farther Shore': Toward a History of the Unborn." In Morgan and Michaels 1999, 13–25.

Eisenberg, Arlene, Heidi Murkoff, and Sandee Hathaway. 1996. *What to Expect When You're Expecting.* New York: Workman.

Falk, Dean. 1975. "Comparative Anatomy of the Larynx in Man and the Chimpanzee: Implications for Language in Neanderthal." *American Journal of Physical Anthropology* 43 (1): 123–132.

Finer, Lawrence B., and Mia R. Zolna. 2011. "Unintended Pregnancy in the United States: Incidence and Disparities, 2006." *Contraception* 84 (5): 478–485. doi:10.1016/j.contraception.2011.07.013.

Gholipour, Bahar. 2014. "Too Little Weight Gain during Pregnancy Linked to Child's Obesity." *Live Science,* 14 April. http://www.livescience.com/44805-weight-gain-pregnancy-child-obesity.html.

Han, Sallie. 2008. "Seeing the Baby in the Belly: Family and Kinship at the Ultrasound Scan." In *The Changing Landscape of Work and Family in the American Middle Class: Reports from the Field,* ed. Elizabeth Rudd and Lara Descartes, 243–264. New York: Lexington Books.

———. 2009a. "Seeing Like a Family: Fetal Ultrasound Images and Imaginings of Kin." In *Imagining the Fetus: The Unborn in Myth, Religion, and Culture,* ed. Vanessa Sasson and Jane Marie Law, 270–290. Oxford: Oxford University Press.

———. 2009b. "Making Room for Daddy: Men's 'Belly Talk' in the Contemporary United States." In *Reconceiving the Second Sex: Men, Masculinity, and Reproduction,* ed. Marcia Inhorn, Tine Tjornhoj-Thomsen, Helene Goldberg, and Maruska la Cour Mosegaard, 305–326. New York: Berghahn Books.

———. 2013. *Pregnancy in Practice: Expectation and Experience in the Contemporary U.S.* New York: Berghahn Books.

Helliker, Kevin. 2015, "Pregnant Women Get More Ultrasounds, Without Clear Medical Need." *Wall Street Journal,* 17 July. https://www.wsj.com/articles/pregnant-women-get-more-ultrasounds-without-clear-medical-need-1437141219.
Hirst, Damien. 2013. "Monumental New Damien Hirst Sculptures Unveiled. *Damien Hirst,* 8 October. http://www.damienhirst.com/news/2013/miraculous-journey.
Ingold, Tim. 2000. *The Perception of the Environment: Essays in Livelihood, Dwelling, and Skill.* New York: Routledge.
———. 2011. *Being Alive: Essays on Movement, Knowledge, and Description.* London: Taylor & Francis.
Ivry, Tsipy. 2010. *Embodying Culture: Pregnancy in Japan and Israel.* New Brunswick, NJ: Rutgers University Press.
Jones, Rachel K., and Jenna Jerman. 2014. "Abortion Incidence and Service Availability in the United States, 2011." *Perspectives on Sexual and Reproductive Health* 46 (1): 3–14. doi:10.1363/46e0414.
Juengst, Eric T., Jennifer R. Fishman, Michelle L. McGowan, and Richard A. Settersten Jr. 2014. "Serving Epigenetics before Its Time." *Trends in Genetics* 30 (10): 427–429.
Klaus, Alisa. 1993. *Every Child a Lion: The Origins of Maternal and Infant Health Policy in the United States and France, 1890–1920.* Ithaca, NY: Cornell University Press.
Kritzer, Robert. 2009. "Life in the Womb: Conception and Gestation in Buddhist Scripture and Classical Indian Medical Literature." In *Imagining the Fetus: The Unborn in Myth, Religion, and Culture,* ed. Vanessa Sasson and Jane Marie Law, 73–89. Oxford: Oxford University Press.
Kuzawa, Christopher W., and Elizabeth A. Quinn. 2009. "Developmental Origins of Adult Function and Health: Evolutionary Hypotheses." *Annual Review of Anthropology* 38: 131–147.
Kuzawa, Christopher W., and Elizabeth Sweet. 2009. "Epigenetics and the Embodiment of Race: Developmental Origins of US Racial Disparities in Cardiovascular Health." *American Journal of Human Biology* 21 (1): 2–15.
Lane, Michelle, Rebecca L. Robker, and Sarah A. Robertson. 2014. "Parenting from before Conception." *Science* 345 (6198): 756–759.
Life. 1965. "Drama of Life Before Birth." *Life,* 30 April. http://time.com/3876085/drama-of-life-before-birth-landmark-work-five-decades-later/.
"The Limit of Viability." 2015. *Economist,* 20 May. http://www.economist.com/blogs/economist-explains/2015/05/economist-explains-23.
Malinowski, Bronislaw. 1929. *The Sexual Life of Savages in North-West Melanesia.* New York: Eugenics Publishing Company.
Matthews, Sandra, and Laura Wexler. 2000. *Pregnant Pictures.* New York: Routledge.
Michaels, Meredith, and Lynn Morgan. 1999. "Introduction: The Fetal Imperative." In Morgan and Michaels 1999, 1–9.
Mitchell, Lisa. 2001. *Baby's First Picture: Ultrasound and the Politics of Fetal Subjects.* Toronto: University of Toronto Press.

Morgan, Lynn. 1997. "Imagining the Unborn in the Ecuadorian Andes." *Feminist Studies* 23 (2): 323–350.
———. 2009. *Icons of Life: A Cultural History of Human Embryos.* Berkeley: University of California Press.
Morgan, Lynn, and Meredith Michaels. 1999. *Fetal Subjects, Feminist Positions.* Philadelphia: University of Pennsylvania Press.
Nishizaka, Aug. 2014. "Instructed Perception in Prenatal Ultrasound Examinations." *Discourse Studies* 16 (2): 217–246.
Oakley, Ann. 1984. *The Captured Womb: A History of the Medical Care of Pregnant Women.* Oxford: Blackwell.
Oaks, Laury. 2001. *Smoking and Pregnancy: The Politics of Fetal Protection.* New Brunswick, NJ: Rutgers University Press.
Ochs, Elinor, and Bambi Schieffelin.1984. "Language Acquisition and Socialization: Three Developmental Stories and Their Implications." In *Culture Theory: Essays on Mind, Self, and Emotion,* ed. Richard Shweder and Robert Levine, 263–301. Cambridge: Cambridge University Press.
Petchesky, Rosalind Pollack. 1987. "Fetal Images: The Power of Visual Culture in the Politics of Reproduction." *Feminist Studies* 13 (2): 263–292.
Petrozza, John C. 2016. "Early Recurrent Pregnancy Loss." Medscape, 7 October. http://emedicine.medscape.com/article/260495-overview.
Polich, Laura. 2000. "Orality: Another Language Ideology." *Texas Linguistic Forum* 43: 189–199.
Rapp, Rayna. 1999. *Testing Women, Testing the Fetus: The Social Impact of Amniocentesis in America.* New York: Routledge.
Rewire. 2017. "Tennessee Fetal Assault Law (SB1391)." *Rewire,* 10 January. https://rewire.news/legislative-tracker/law/tennessee-pregnancy-criminalization-law-sb-1391.
Richardson, Sarah S., Cynthia R. Daniels, Matthew W. Gillman, Janet Golden, Rebecca Kukla, Christopher Kuzawa, and Janet Rich-Edwards. 2014. "Don't Blame the Mothers." *Nature* 512 (7513): 131–132.
Rothman, Barbara Katz. 1987. *The Tentative Pregnancy.* New York: Viking.
Sasson, Vanessa, and Jane Marie Law, eds. 2009. *Imagining the Fetus: The Unborn in Myth, Religion, and Culture.* Oxford: Oxford University Press.
Stabile, Carol. 1992. "Shooting the Mother: Fetal Photographer and the Politics of Disappearance." *Camera Obscura* 10 (28): 178–205.
Stern, Alexandra Minna. 2002. "Making Better Babies: Public Health and Race Betterment in Indiana, 1920–1935." *American Journal of Public Health* 92 (5): 742–752.
Stokoe, William. 1991. "Signing and Speaking: Competitors, Alternatives, or Incompatibles?" In *Studies in Language Origins,* vol. 2, ed. Walburg von Raffler-Engel, Jan Wind, and Abraham Jonker, 115–122. Amsterdam: John Benjamins Publishing.
Tate, Carolyn E. 2009. "The Colossal Fetuses of La Venta and Mesoamerica's Earliest Creation Story." In *Imagining the Fetus: The Unborn in Myth, Religion, and Culture,* ed. Vanessa Sasson and Jane Marie Law, 223–258. Oxford: Oxford University Press.

Taylor, Janelle. 2008. *The Public Life of the Fetal Sonogram: Technology, Consumption, and the Politics of Reproduction.* New Brunswick, NJ: Rutgers University Press.

Thayer, Zaneta M., and Christopher W. Kuzawa. 2015. "Ethnic Discrimination Predicts Poor Self-Rated Health and Cortisol in Pregnancy: Insights from New Zealand." *Social Science and Medicine* 128: 36–42.

Vogel, Carol. 2013. "Art, From Conception to Birth in Qatar." *New York Times,* 7 October. http://www.nytimes.com/2013/10/08/arts/design/damien-hirsts-anatomical-sculptures-have-their-debut.html.

Woznicki, Katrina. 2010. "Too Much Pregnancy Weight Gain Raises Child's Obesity Risk. *WebMD,* 4 August. http://www.webmd.com/baby/news/20100804/too-much-pregnancy-weight-gain-raises-childs-obesity-risk#1.

Part II

FINDING FETUSES IN THE PAST
ARCHAEOLOGY AND BIOARCHAEOLOGY

Chapter 4

THE BIOARCHAEOLOGY OF FETUSES

Siân E. Halcrow, Nancy Tayles, and Gail E. Elliott

Until relatively recently, fetuses, along with infants and children, were largely overlooked in bioarchaeological research. Over the past twenty years, there has been increasing recognition of the importance of research on subadults in the archaeological context. However, although fetuses are now sometimes included in analyses of population health and isotopic studies of infant weaning and diet in the past, most research focuses on postnatal subadults. The neglect not only of fetuses but also of other immature individuals is a problem because we are missing a crucial part of the skeletal sample well known to be informative for understanding past population demography, stress and adaptation, and social organization factors.

This chapter reviews some of the bioarchaeological research that has been undertaken in this area and starts to build a theoretical framework to conceptualize fetuses from an archaeological context and to identify areas for future research potential. We explore terminological issues and how the fetus is defined in the field, including discerning whether the fetus is in utero. In fact, since finding babies in utero (the medical definition of a fetus) is very rare in an archaeological context, we are effectively using preterm and full-term babies with low birth weight or small for gestational age (SGA) babies, from bioarchaeological samples as proxies for fetuses. Methodological issues with the investigation of fetuses in bioarchaeology include the exclusion of newborns from community cemeteries, and problems with age estimation, including the differentiation of premature from full-term, low birth weight babies. We outline the contribution that the bioarchaeology of fetuses can make to understanding fertility and other demographic information of a population, epidemiology of disease, maternal and infant stress experience, and the consequences of early stress on later life experience, and we outline clues as to cultural or social aspects of personhood and infant loss.

This chapter first places published research on archaeological fetuses in context by reviewing the history of bioarchaeological research not only on fetuses but also on infants and children. This is followed by a review of evidence for fetuses in archaeological contexts, how they might be identified, and problems of differentiating fetuses from premature births and SGA neonates. The contributions that fetal remains can make to research on population demography, population morbidity and mortality, maternal stresses and society, and culture are then discussed.

The Fetus in Bioarchaeology

The concept of fetuses in bioarchaeology probably brings to mind poignant images of tiny bones of a baby in the pelvic cavity of a female adult skeleton, although finds such as these are actually rather rare. "Bioarchaeology" is used in the United States and other parts of the world to refer exclusively to the study of human remains from archaeological contexts. In the United Kingdom, "human bioarchaeology" is becoming more common, although many people in the field continue to identify themselves as "biological anthropologists," while on-site contract biological anthropologists are often referred to as "human osteoarchaeologists."

In practice, many bioarchaeologists apply the description of "fetus" to babies from bioarchaeological samples identified as younger than thirty-seven gestational weeks (e.g., Halcrow et al. 2008; Lewis and Gowland 2007; Mays 2003; Owsley and Jantz 1985). However, as discussed by Kathleen Blake (chapter 2), there are problems associated with estimations of age-at-death of these babies, who may indeed be fetuses, or may also be premature births or SGA, full-term births. If the medical definition of a fetus as an unborn baby is applied (Halcrow and Tayles 2008; Lewis and Gowland 2007; McIntosh et al. 2003; Scheuer and Black 2000), the in utero skeletons would seem to represent the only finds in archaeology that can be confidently identified as fetuses, but see the discussion below about mother-baby mortality and the burial of neonates with their mothers. Even an apparent in utero fetus may in fact have been a neonate death, illustrating the care with which research in this field needs to be completed.

The History of Research on Archaeological Fetuses, Infants, and Children

In reviewing fetuses in the bioarchaeological record, it is important to consider the place of all immature individuals in bioarchaeological research (Buikstra 1977; Larsen 1997: 3). At the beginning of the twentieth century, infants and children from archaeological contexts were often overlooked (Halcrow and Tayles 2008, 2011; Lewis 2007). This can be understood in the context of the research interest at that time of human taxonomy with a focus on description and metrics (Washburn 1951). Physical anthropologists were mainly interested in comparative craniometry, which required the analysis of adult crania (Gould 1996; Hooton 1930; Hrdlicka 1924). Comparatively, infant and child crania were deemed useless because they were often found disarticulated in archaeological contexts, as the bones of skull are not yet fused together.

Earnest Hooton (1930: 15) typifies the disinterest in the analysis of infants and children at the time: "In the case of infants and immature individuals, the cartilaginous state of epiphyses and the incomplete ossification of sutures, as well as the fragility of the bones themselves usually results in crushing and disarticulation. In any event, the skeletons of young subjects are of comparatively little anthropological value." This lack of interest of infants and children has also been documented in anthropology and archaeology (Bluebond-Langner and Korbin 2007; Gottlieb 2000; Halcrow and Tayles 2008). When Grete Lillehammer published her landmark article "A Child Is Born" in 1989, researchers appeared to have answered her pleas to include infants and children in anthropological work. Since then, there has been a substantial increase in archaeological and anthropological research on children and childhood (Baxter 2005, 2008; Crawford and Lewis 2008; Halcrow and Tayles 2008, 2011; Kamp 2001; Lancy 2008; Lewis 2007; Lillehammer 1989, 2015; Schwartzman 2005; Sofaer Derevenski 2000; Wileman 2005).

Francis Johnston (1961, 1962) was a pioneer in the study of investigations of growth, development, and mortality from the Indian Knoll skeletal sample, who reports nine fetuses, along with a high number of infant burials from the site. Since this time, numerous bioarchaeological studies have investigated infant and child mortality, growth, and growth disruption from infant and child skeletal collections (Halcrow et al. 2008; Humphrey 2000, 2003; Kamp 2001; Lewis 2000, 2002, 2004; Lewis and Gowland 2007; Lewis and

Roberts 1997; Lovejoy et al. 1990; Mays 1995, 1999; Robbins Schug 2011; Saunders 2000). It is now acknowledged that infant and child human remains are particularly useful for the study of patterns of health and disease in prehistory, in that they are the most sensitive indicators of biocultural change (Buikstra and Ubelaker 1994: 39; Halcrow and Tayles 2011; Halcrow et al. 2007; Lewis 2007; Van Gerven and Armelagos 1983: 39).

Generally, little bioarchaeological research considers fetuses. For example, some growth studies and demographic analyses do not include preterm infants because of lack of comparative fetal bone size data (e.g., Johnston 1961). Moreover, the attention afforded to purported evidence for infanticide, based primarily on the reported high number of perinates in some skeletal assemblages, has deflected interest away from the contributions that fetuses can make to understanding bioarchaeological questions, including maternal health and disease, and social organization from mortuary ritual analyses (Bonsall 2013; Faerman et al. 1998; Gilmore and Halcrow 2014; Mays 1993; Mays and Eyers 2011; Mays and Faerman 2001; Smith and Kahila 1992).

Approximately three in ten pregnancies are spontaneously aborted, the majority of which occur in the first trimester as the result of genetic abnormalities (Fisher 1951). First-trimester fetuses are very unlikely to be recovered in the bioarchaeological context. Bone development does not start until approximately six to eight gestational weeks, and any bone formation before the second trimester would unlikely be preserved because of the low level of mineralization and/or would be extremely difficult to identify in an archaeological context. The only first-trimester fetus published from an archaeological context is from the Libben sample in Ohio, a Late Woodland site occupied from the ninth to the twelfth centuries CE (White 2000: 20). There are published instances of preserved fetal individuals from the second trimester, for example, the well-preserved fetus of twenty gestational weeks from the Kellis 2 site, Dakhleh Oasis, Egypt (Wheeler 2012: 223). A female from the postmedieval site of Chelsea Old Church in London had a fetus in utero aged twenty to twenty-two gestational weeks (Rebecca Redfern, personal communication). Douglas Owsley and Richard Jantz (1985) have found three fetuses younger than twenty-eight gestational weeks at Arikara sites in South Dakota. Simon Hillson (2009) has also reported the findings of fetuses as young as twenty-four gestational weeks from a large Classical period infant cemetery at Kylindra on Astypalaia in Greece.

Defining Types of Fetus Burials

This section reviews evidence for fetuses in bioarchaeological research, and methods used to distinguish fetuses found in utero, postbirth, and postmortem ("coffin"). Differentiating these burial contexts can potentially contribute to research on maternal health and the cause of death for the mother and child. For example, a premature birth is more likely to indicate poor health and/or nutritional status of a woman, compared with a baby who died around full-term from obstructed labor. Distinguishing the type of fetal death and burial, as well as whether the baby was full-term or a preterm or SGA baby, in conjunction with evidence for stress and diet of both the mother and baby may give insights into overall health in past populations (fig. 4.1). Age estimation of non-survivors may underestimate their

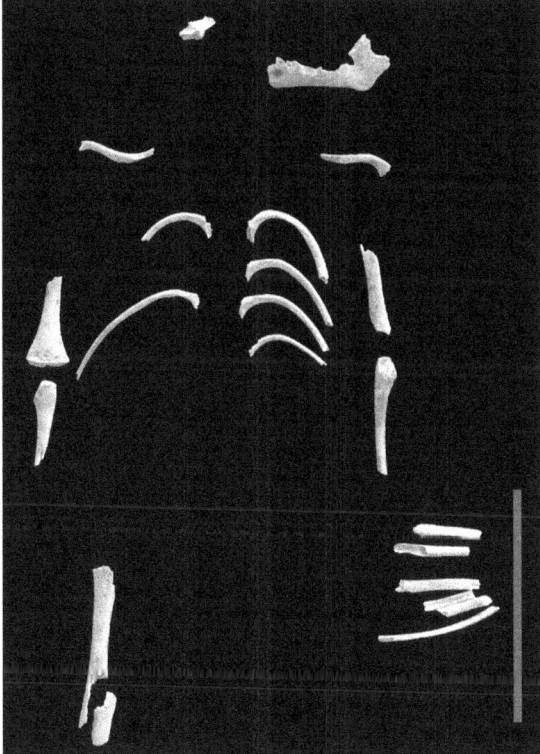

FIGURE 4.1. Second trimester fetal burial from the late Iron Age site of Non Ban Jak, northeast Thailand. This burial has a gestational age of approximately 22–24 weeks and if a live birth, would have died shortly after birth (photograph by author, red scale = 5cm).

age, as there is a bias for SGA babies to die prematurely (Bukowski et al. 2014). Age estimation of fetuses in bioarchaeology and the complexities involved are discussed in Blake (chapter 2).

In Utero Fetuses

If the skeletal remains of a baby are found crouched in a fetal position within the pelvic cavity of an adult female, the mother died while the fetus was in utero, before, or during labor. The pregnant woman may therefore have died because of pregnancy or labor complications (Lewis 2007: 34). As noted, there is very little evidence for in utero fetuses in the bioarchaeological context. More than twenty cases of pregnant or laboring females (i.e., interred with fetal remains in situ) have been published in the archaeological literature, being argued to represent complications from childbirth (e.g., Ashworth et al. 1976; Alduc-Le Bagousse and Blondiaux 2002; Connell et al. 2012; Cruz and Codinha 2010; de Miguel 2008; Hawkes and Wells 1975; Högberg et al. 1987; Smith and Wood-Jones 1910, in Lewis 2007; Lieverse et al. 2015; Malgosa et al. 2004; O'Donovan and Geber 2010; Owsley and Bradtmiller 1983; Persson and Persson 1984; Pounder et al. 1983; Sjovold et al. 1974; Roberts and Cox 2003; Wells 1978).

The dearth of literature on in utero fetuses in bioarchaeology may be due not to absence of evidence but rather to the small bones being missed or misidentified during excavation, or reported only in the gray literature. Although there are very detailed and useful books on identifying fetal and infant remains (e.g., Baker et al. 2005; Cunningham et al. 2016; Fazekas and Kosa 1978; Scheuer and Black 2000, 2004), there are numerous accounts of fetuses being misidentified as animal bones during excavation (e.g., Ingvarsson-Sundström 2003). For example, Charlotte Roberts and Margaret Cox (2003) have reported at least twenty-four unpublished cases of fetuses from British excavations. There are further instances of fetal bones being found comingled with adult burials post-excavation, which may represent a baby in utero, or a possible mother and baby postbirth burial. To this end, Mary Lewis (2007: 34) has stressed the importance of careful excavation to establish the position of fetal burials.

As noted, there has been a focus in fetal bioarchaeology on birth trauma and obstructed labor that did not take into account other factors of population health, maternal diet, fetus age-at-death, and infant burial type. For example, Douglas Owsley and Bruce Bradt-miller (1983) investigated maternal physical stress and birth trauma as an explanation for the high levels of female mortality in Arikara skeletal series by determining the frequency of females who died with fetal remains in utero. Because only two females (0.9 percent of the total females in the sample) were identified with in utero fetal remains, they argue that this does not provide evidence for stress of childbearing was the cause of death of the high number of young females in the samples. One of these was a case of in utero twins from the site of Mobridge, and another was a purported singleton in utero baby from Larson, but no further information could be gained from the fetus because of poor preservation (Owsley and Bradtmiller 1983). This interpretation, however, dismisses the high perinate mortality rate at the sites, and the high female mortality may have arisen from postnatal hemorrhage and other complications (e.g., sepsis), which are the leading causes of birth-related deaths for mothers today (WHO 2017b).

Bioarchaeologists have reported on cases of purported obstructed labor causing maternal and fetal perinatal death based on positioning of the fetus in the pelvic cavity or the finding of preterm mummified remains in utero (Arriaza et al. 1988; Ashworth et al. 1976; Lieverse et al. 2015; Luibel 1981; Malgosa et al. 2004; Wells 1975). There is now some effort to consider paleopathological evidence with cases (or purported cases) of in utero death (e.g., Willis and Oxenham 2013), including the early work by Owsley and Bradtmiller (1983).

Postbirth "Fetuses"

If a perinate is found buried alongside an adult, with the same head orientation, then the infant has been buried postbirth, whether naturally or by caesarian section (Lewis 2007: 34) (fig. 4.2). In some contexts, it is very common for newborns to be placed on the chest of adult women (presumably their mother) (Standen et al. 2014). To identify this archaeologically, if the majority of the infant remains are in the pelvic cavity of the adult, yet the legs are extended and/or the cranium lies among the ribcage, then the baby may have been delivered and then placed on top of the mother's (or other adult's) torso during burial (Lewis 2007: 34). As both mother and baby bodies' skeletonize, the baby's bones can become settled among the mother's ribs and vertebrae. This is important to note as these neo-

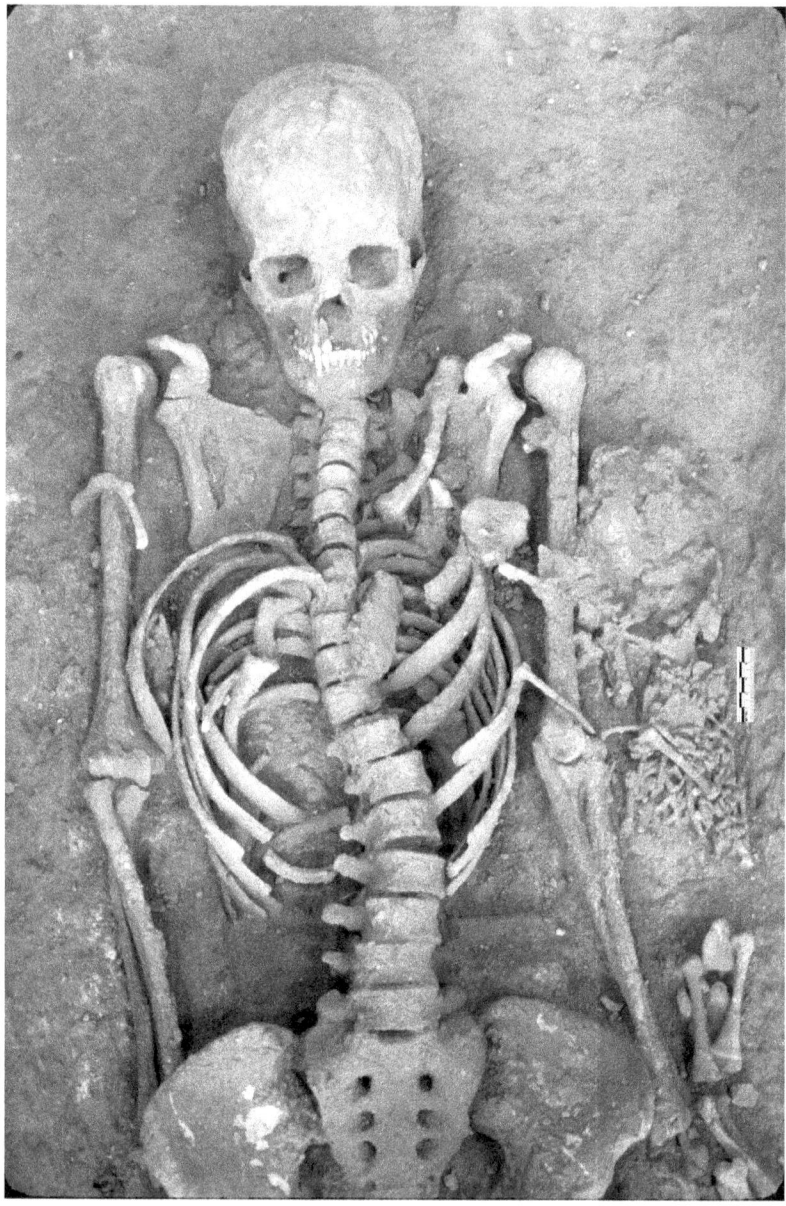

FIGURE 4.2. Full-term neonate (burial 48) buried alongside an adult female (burial 47) from Khok Phanom Di, Thailand. This could possibly represent a perinate and mother who died from complications in childbirth (photographs courtesy of C.F.W. Higham).

nates may be mistaken for breech, obstructed labors in the archaeological context. Anna Willis and Marc Oxenham (2013: 678), for example, describe an "in utero breech" presentation of a fetus aged thirty-eight gestational weeks from Neolithic southern Vietnam. They describe the cranium as "below the mother's right lower ribs" (it is not clear if they mean inside the abdominal/thoracic cavity or inferior to the right lower ribs) and the postcranial skeleton as "extended down toward the mother's pelvis" with the left femur "positioned within the mother's pelvic cavity and a tibia ... positioned beside [lateral] the lesser trochanter of the mother's right femur." They also state the "right pars lateralis [part of the base of the occipital bone of the cranium] was concreted to the anterosuperior portion of the shaft of the 10th right rib of the mother, near the sternal end." (Willis and Oxenham 2013: 678). Given this partially extended (nonfetal) positioning and the part of the cranial base being found anterior to the rib cage), it could be possible that the baby was not in the abdominal cavity but rather placed on top of the mother's torso after birth.

Ancient DNA analyses may be used to assess the relationship of the adult and fetal burials where the fetus has been placed on the purported mother or where the archaeological context is unclear. Lewis (2007: 35) has argued that this is important to distinguish these relationships, as in some contexts—for example, in the Anglican burial tradition, babies were interred with nonmaternal women in instances of coinciding death (Roberts and Cox 2003: 253).

Multiple Fetal Pregnancies and Births

There have been two reported instances of twin fetuses in utero in the bioarchaeological literature (Lieverse et al. 2015; Owsley and Bradtmiller 1983), with others found in a postbirth context. Interest in multiple births in bioarchaeology has recently increased, including an investigation of social identity and concepts of personhood through the investigation of mortuary treatment (e.g., Einwögerer et al. 2006; Halcrow et al. 2012). Human twins are rare, with approximately one occurrence for every hundred births (Ball and Hill 1996). However, they appear in the literature more commonly than expected, compared with singleton fetuses (e.g., Black 1967; Chamberlain 2001; Crespo et al. 2011; Einwögerer et al. 2006; Flohr 2014; Halcrow et al. 2012; Lieverse et al. 2015; Owsley and Bradtmiller

1983). This is probably because the archaeologist sees them as more significant.

An example of a possible twin burial was found in an Upper Paleolithic site of Krems-Wachtberg, Austria (Einwögerer et al. 2006). The infants from this double burial were identified as twins from their identical age (as estimated from their dentition), similar femora size, and simultaneous interment (both estimated at full-term age-at-death). Interestingly, the bodies lay under a mammoth scapula and a part of a tusk and were associated with thirty ivory beads. Thomas Einwögerer and colleagues (2006) suggest, based on this mortuary evidence, that these newborns were an important part of their community. Another case of a twin burial is from the mid-fourth-century site of Olèrdola in Barcelona, Spain (Crespo et al. 2011). The two newborns were found at the same stratigraphic level with their lower limbs entwined, indicating they were buried simultaneously. Siân Halcrow and colleagues (2012) presented an extremely rare finding of at least two and possibly four twin burials from a three-thousand- to four-thousand-year-old BP Southeast Thailand site, offering a methodological approach for the identification of archaeological twin (or other multiple birth) burials and a social theoretical framework to interpret these in the past.

Postmortem Birth ("Coffin Birth")

Postmortem birth or "coffin birth" refers to the occurrence of fetuses that were in utero when the mother died and were expelled after burial (O'Donovan and Geber 2010). Postmortem birth by fetal extrusion has been documented in rare forensic cases from the buildup of gas within the abdominal cavity, resulting in the emission of the fetus (Lasso et al. 2009; Schultz et al. 2005). Lewis (2007: 34–37, 91) and Edmond O'Donovan and Jonny Geber (2009) argue that if fetal remains are complete and in a position inferior to and in line with the pelvis outlet, with the head oriented in the opposite direction to the mother, then there is the possibility of coffin birth. If they lie within the pelvic outlet, this means there was partial extrusion during decomposition (Hawkes and Wells 1975). However, partial extrusion could also be the result of an obstructed labor of a baby in the breech position, but this would likely result in extrusion of the lower limbs. Duncan Sayer and Sam Dickenson (2015) argue that postmortem fetal extrusion is implausible under some burial conditions, and thus, decomposition of the baby in utero would mean it isn't likely to be birthed from an undilated cervical canal. This, however, assumes there was no dilation at the time of death of the mother.

Etiology of Infant Death

Because samples in bioarchaeology are inherently biased in that we are dealing with the non-survivors in a population (DeWitte and Stojanowski 2015; Wood et al. 1992), it is important to consider the causes of death in the past. As the birth process and the first few days after birth constitute the most critical time in a perinate's life after the first trimester (Kelnar et al. 1995), it is little wonder that most enter the archaeological record around the time of birth (Halcrow and Tayles 2008). The age-at-death distributions of modern stillbirths and the age-at-death of live births dying within seven days of birth both peak at thirty-eight to forty weeks (Butler and Alberman 1969; Gibson and McKeown 1951; Hoffman et al. 1974; Mays 1993). The main causes of newborn deaths are prematurity, low birth weight, infections, asphyxia, and birth trauma (WHO 2016a).

Because of the evolutionary development of bipedalism and large brains in humans, the anterior-posterior dimension of the pelvic inlet is actually smaller than the length of the neonatal cranium (Kurki 2011; Rosenberg and Trevathan 2002). As a result, obstruction of the infant during labor can occur and occasionally result in death of the mother and infant. Today, in developing countries, one in a hundred births result in maternal death mostly because of obstructed labor or hemorrhage (Say et al. 2014).

Although thirty-eight weeks has traditionally been considered full-term and is used as such in bioarchaeological estimates, recent clinical literature has shown a significant increase in morbidity and mortality in infants born between thirty-seven and thirty-eight weeks compared with those born at thirty-nine to forty gestational weeks (Clark and Fleischman 2011). Every year, an estimated fifteen million babies are born preterm (before thirty-seven completed weeks of gestation). The leading reasons for preterm death include multiple pregnancies, infections, and chronic conditions (WHO 2016c). A main factor resulting in death from preterm infants is respiratory distress syndrome owning to their immature lungs (Kramer et al. 2000).

Infanticide

Infanticide has been considered in the interpretation of a high number of perinates in the archaeological context. As noted, the focus on infanticide results in a lack of consideration of important factors to bioarchaeological research, including fetal and maternal health. For example, Simon Mays and Marina Faerman (2001) have reported that the Ashkelon perinatal skeletal sample from Roman-period Israel was the result of unwanted babies being dumped in a well associated with a brothel (Gilmore and Halcrow 2014; Smith and Kahila 1992). Besides the problems with the authors' arguments that the archaeological remains and artifacts support that the "perinate" remains were associated with a brothel, the long bone length data show the inclusion of premature fetal individuals who would not have survived postbirth (Gilmore and Halcrow 2014). Similarly, Simon Mays and Jill Eyers (2011) have argued for evidence of infanticide based on the age distribution of ninety-seven "perinates" from the Yewden Roman villa site of Hambledon in Buckinghamshire, England. Again, this sample contains preterm fetal individuals from thirty-two gestational weeks, who were unlikely to survive postbirth (Gilmore and Halcrow 2014).

By focusing on an argument of infanticide, other causes for the entry of fetuses into the mortality profile, which may give information on heritable disease, maternal health and infection, and general socioeconomic and living conditions for women and different social groups, are overlooked. Very little evidence from the bioarchaeological record shows abuse of neonates causing death, though there is evidence for neonatal birth trauma, commonly fractures to the clavicle (Lewis 2013, chapter 5). However, most historical evidence for killing of infants has been argued to result from exposure, drowning, or smothering, which would not leave skeletal evidence. Also, infanticide victims are more likely to be buried covertly rather than in normal cemetery plots (Gilmore and Halcrow 2014).

The Potential of Fetuses and Fetal Life Records in Bioarchaeology

As noted, bioarchaeological research of infants is becoming recognized as a worthwhile endeavor for a number of reasons (Halcrow et al. 2014; Halcrow and Tayles 2008), even though fetuses are often excluded from these analyses. This section reviews the contribution that the bioarchaeology of fetuses can make to understanding fertility, demographic characteristics of a population, epidemiology of disease, maternal stress, and social aspects of personhood and age (Gilmore and Halcrow 2014; Halcrow et al. 2008; Halcrow and Tayles 2008; Lewis 2007; Tocheri et al. 2005). Lewis (chapter 5) reviews the field of fetal paleopathology in deciphering infectious and genetic disease in the past from this subsample, which may be used in conjunction with the approaches and methods reviewed below.

Fetal Bioarchaeology and Demography

Demographic analyses in bioarchaeology are important for understanding population growth, health, and maternal stress (Bocquet-Appel 2008; Larsen 2015). Typically in bioarchaeology, these analyses exclude individuals younger than five years of age, as they are often assumed to be underrepresented in a cemetery sample (Halcrow and Tayles 2011; Lewis 2007). However, this is not the case at many sites with many under five-year-olds represented (Angel 1971; Boric and Stefanovic 2004; Halcrow et al. 2008; Owsley and Jantz 1985; Tayles and Halcrow 2016; Wheeler 2012). Given that populations with high fertility usually have high rates of infant and fetal mortality (Gurven 2012), it could be argued that fetuses should also be included in demographic analyses if they are present in a sample. Gwen Robbins (2011; Robbins Schug 2011) is one of the only bioarchaeologists who has presented methods for the development of fertility estimates using perinatal and infant mortality data.

Maternal and Fetal Mortality and Stress

As noted, the stresses infants and children experience are widely accepted as good measures of overall population health (Halcrow and Tayles 2008; Lewis 2007). Likewise, fetal and maternal health is arguably the most sensitive measure of population health (see Rutherford, chapter 1), given the increased energetic requirements placed on pregnant and lactating mothers, as well as the energy requirements for fast-growing fetuses. The major direct causes of maternal morbidity and mortality today include hemorrhage, infection, high blood pressure, unsafe abortion, and obstructed labor (WHO 2017a).

In bioarchaeology, differences in mortality between young adult males and females are often explained by childbirth hazards (e.g.,

Domett 2004; Owsley and Bradtmiller 1983; Willis and Oxenham 2013). Factors related to obstetric death may include compromised maternal growth resulting in stunted growth in the pelvis (Nwogu-Ikojo 2008. However, Helen Kurki (2011) cautions that modern clinical standards do not take into account the variation in human body size and morphology, as well as the effect this variation may have on obstetric capacity without compromising obstetric function.

Although obstetric issues can affect mortality and morbidity of mothers, the general implications that pregnancy can have for maternal health are often overlooked in bioarchaeology. Pregnancy can have a general negative effect on women already compromised by malnutrition and disease. For example, malnutrition during pregnancy can lead to iron deficiency, which can increase the risk of maternal mortality (Black et al. 2008). Also, the effects of some infections can be exacerbated during pregnancy. For example, infection with *Plasmodium vivax* or *Plasmodium falciparum* during pregnancy leads to chronic anemia and placental malaria infection, reducing the birth weight and increasing the risk of neonatal death (Brabin et al. 2004; WHO 2016b).

The main causes of fetal and newborn death has been reviewed. Some bioarchaeologists make the interpretive distinction between the timing of perinatal death as a reflection of contributing endogenous and exogenous factors toward mortality (e.g., Halcrow 2006; Lewis 2002; Lewis and Gowland 2007). This is based on the interpretive distinction that demographers and clinicians make between neonatal and post-neonatal infant mortality—the former seen as a consequence of endogenous causes, including low birth weight and trauma, and the latter from exogenous, postpartum environmental factors, including infection (Wiley and Pike 1998: 318). Lewis (2002) applies this interpretative tool in assessing infant mortality in the past where neonatal mortality (within the first four weeks after birth) is reflective mostly of the infant's genetic factors and maternal health (such as nutrition and disease burden), whereas death occurring after this period in infancy is generally indicative of the external environment (Pressat 1972). However, insults in utero, if the fetus does not die at birth, can have a lasting effect on the health of an infant (Barker 1994; Binkin et al. 1988; Duray 1996; Furmaga-Jablonska et al. 1999; Gunnell et al. 2001; McCarron et al. 2002; Rewekant 2001). This blurs the apparent distinction demographers make between death in the neonatal period arising from endogenous factors and death in the postneonatal period resulting from exogenous factors.

Fetal and infant bioarchaeology may also inform and be conceptualized in light of the developmental origins of health and disease (DOHaD) theory (Barker and Osmond 1986), which describes how the environmental impact on a mother induces physiological changes in fetal and child growth and development that have long-term consequences on later health and disease risk for the child (Gluckman and Hanson 2006). There has been a recent surge of interest in this work in the biomedical context, with animal work in epigenetics providing a mechanism for explaining changes in prenatal and infant development for future life outcomes (Gluckman and Hanson 2006; Hochberg et al. 2011, Rutherford, chapter 1). This is only just starting to be considered within bioarchaeology research (Gowland 2015; Klaus 2014). Although it has been argued that the bioarchaeological record may be for the most part "mute" on offering insight into epigenetics (Klaus 2014: 300), the consideration of maternal and fetal health through isotopic studies and/or early life stress indicators from teeth developing during the prenatal environment is an obvious place to consider these factors in the past (Gowland 2015). For example, George Armelagos and colleagues (2009), in a study of dental enamel defects from published bioarchaeological research, found that individuals with this evidence for systemic stress during the in utero and early infant and childhood period die earlier in life. These findings support the DOHaD hypothesis. Investigating early and later life events in this way is complementary to the development of osteobiographies that use an individual life course approach (e.g., Sofaer 2006).

Research on developmental dental enamel defects holds much promise for understanding stress during the maternal and fetal period. Certainly, the examination of dental enamel defects is already a major area of bioarchaeological investigation of health and stress (e.g., Goodman et al. 1980; Hillson 1996; Larsen 2015). This is, in part, because of the relative good preservation of teeth due to their robust nature, and because they do not remodel, they can provide an individual record of stress during tooth development (Halcrow and Tayles 2011). We cannot assess the teeth of archaeological fetuses and young infants for developmental defects, as their teeth have not completed maturation, therefore leaving any enamel deficiencies unobservable. However, we can look retrospectively at dental enamel defects formed during the fetal and early infancy period by assessing the dentition of infants and children in the same skeletal sample who have survived past this time. Although there is a dearth of bioarchaeological research assessing stress using deciduous dentition (e.g., Blakey and Armelagos 1985; Halcrow et al. 2008), there have been advances in the field. These developments include the microscopic analyses of developmental enamel defects in deciduous dentition to build up a detailed chronology of stress during in utero and postnatal development. Assessing the width of the neonatal line, a microscopic hypomineralized area of enamel forming because of stress during birth, has been used to assess the severity of this stress (Żądzińska et al. 2015). The presence of the neonatal line can also assist to ascertain if an infant was alive or stillborn at the time of birth and has been used to support arguments for the presence of infanticide (Smith and Kahila 1992).

Developments in bioarchaeological isotope techniques have great potential for understanding stress and diet of the fetus and mother. Traditionally, the focus of stable isotope studies of infants has been around the investigation of breastfeeding and supplementation of food (e.g., Fuller et al. 2003 Richards et al. 2002). However, more isotopic research of diet and stress is beginning to focus on fetuses and perinates (e.g., Beaumont et al. 2015; Kinaston et al. 2009). Recent experimental isotopic work assessing incremental section sampling of deciduous teeth has been able to assess the relationship between nitrogen isotopes and the maternal signature (Beaumont et al. 2015). Although δ^{15} nitrogen values in archaeological infants have traditionally been interpreted as providing a breastfeeding signal, Julia Beaumont and colleagues (2015) argue that high nitrogen values in perinates reflect poor maternal health. This is explained by the fact that if the mother is ill or malnourished, the body will metabolize proteins, which leads to higher nitrogen values. The bone chemical (isotopic) examination of fetal and maternal pairs, potentially in conjunction with evidence of stress or pathology, offers a new method to assess the experience of past life experiences. Further, using new isotopic sampling techniques over the dental development of an individual (e.g., Beaumont et al. 2013, 2015), the early life experiences of the mother, and the potential relationship with fetal outcome can be assessed.

Social Identity

The mortuary treatment of fetuses may provide information on social identity and in turn the organization of a community (Boric and Stefanovic 2004; Parker Pearson 1982; Scott and Betsinger, chapter 7; Tainter 1978). For example, if a young child is ascribed grave wealth, this may indicate inherited social status and therefore some type of hierarchical social organization (Tainter 1978). In modern and past cemetery samples, infants are often buried in special burial plots, which has been argued as a sign of their marginalization in society (Cannon and Cook 2015; Lewis 2007: 92; cf. Murphy 2011). However, numerous cases in different regions of the world indicate infants and fetuses are buried within the community cemetery and make up a large proportion of the skeletal sample (e.g., Angel 1971; Boric and Stefanovic 2004; Gowland and Chamberlain 2002; Lewis and Gowland 2007; Owsley and Jantz 1985).

In some societies, infants needed to live for a certain time post-birth before they were given certain burial rites, as has been documented in historical records from pre-Christian Rome where babies under nine days old were not grieved over nor buried in cemeteries (Soren and Soren 1995: 43–44). Louise Steel (1995: 200) has interpreted the lack of energy in terms of burial ritual and wealth for infant and perinatal remains in Iron Age Cyprus as an indication that these individuals were not viewed as members of the community. Tulsi Patel (1994) noted that in a contemporary Rajasthan village in India, when an infant dies there is a lack of elaborate ritual. This was attributed to the fact that infants had not acquired the social personality or status that adults have in that society (Patel 1994: 142). Contrary to this finding, research at the New Kingdom Egyptian workers' village of Deir el Medina argues that infants (full-term and premature perinates) were perceived as part of the community. Lynn Meskell (2000: 425) states, citing Erika Feucht's (1995: 94) interpretations from ancient text and artistic representations, that "while already in the womb, the unborn child was considered a living being and as such required protection in the social realm." Furthermore, various studies have noted different burial treatments and places of burial. For example, in the Ashanti Hinterland on the Gold Coast of Africa, infants under eight days of age were buried in pots in the town center, while in some tribes in the region there was an absence of burial ritual for these babies (Ucko 1969: 271).

The investigation of mortuary treatment of pregnant women may give us information on social identity related to childbearing. For example, the discovery of a thirty-four- to thirty-six-week-old fetus cremated with the circa 850 BCE "Rich Athenian Lady" led to a recognition that her grave wealth may have been related to her dying while pregnant or during childbirth, rather than primarily her social status (Liston and Papadopoulos 2004).

Research of the archaeology of grief is starting to consider community members' responses to infant and fetal death (e.g., Cannon and Cook 2015; Murphy 2011). The purported marginalization of fetuses along with infants in the archaeological record, including location and simplified mortuary treatment, has led some scholars to interpret that they were of little concern beyond immediate family members (Cannon and Cook 2015). Considering literature on intense grief after miscarriage and infant death starts to challenge the notion that their loss was of little consequence (Murphy 2011).

Conclusion

This chapter has provided the first review of fetuses within bioarchaeology. Although there is an increasing recognition that immature individuals are informative of central bioarchaeological questions of health and development, the potential of research of fetuses has not been fully explored. We have provided a short discussion on the definition of fetuses as commonly applied in bioarchaeology and a methodological approach to identify fetuses in the past. We have highlighted how research of fetuses can answer central bioarchaeological questions of population demography; fetal, maternal, and population health; and social aspects of pregnancy and death.

Acknowledgments

The authors would like to thank the editors for their thoughtful comments on our chapter.

Siân E. Halcrow is a senior lecturer in biological anthropology at the University of Otago. She has a research interest primarily in the bioarchaeology of infants and children in prehistoric Southeast Asia and South America.

Nancy Tayles is an honorary associate professor, retired from biological anthropology at the University of Otago. She continues to research bioarchaeology and anatomy of Southeast Asia.

Gail E. Elliott is a PhD candidate in biological anthropology at the University of Otago, researching human growth as an indicator of physiological stress in past populations. She also has a research interest in gross anatomy and forensic anthropology.

References

Alduc-Le Bagousse, A., and J. Blondiaux. 2002. "Maternal Death and Perinatal Pathology at Lisieux (Calvados, France) during the First Millennium." [In French.] *Bulletins et Mémoires de la Société d'Anthropologie de Paris* 14 (3–4): 295–309.

Angel, J. Lawrence. 1971. *The People of Lerna: Analysis of a Prehistoric Aegean Population*. Washington, DC: Smithsonian Institution Press.

Armelagos, Geroge J., Alan H. Goodman, Kristin N. Harper, and Michael L. Blakey. 2009. "Enamel Hypoplasia and Early Mortality: Bioarchaeological Support for the Barker Hypothesis." *Evolutionary Anthropology: Issues, News, and Reviews* 18 (6): 261–271.

Arriaza, Bernardo, Marvin Allison, and Enrique Gerszten. 1988. "Maternal Mortality in Pre-Columbian Indians of Arica, Chile." *American Journal of Physical Anthropology* 77 (1): 35–41.

Ashworth, Joel Thomas, Marvin Jerome Allison, Enrique Gerszten, and Alejandro Pezzia. 1976. "The Pubic Scars of Gestation and Parturition in a Group of Pre-Columbian and Colonial Peruvian Mummies." *American Journal of Physical Anthropology* 45 (1): 85–89.

Baker, Brenda J., Tosha L. Dupras, and Matthew Tocheri. 2005. *The Osteology of Infants and Children*. College Station: Texas A&M University Press.

Ball, Helen L. and Catherine M. Hill. 1996. "Reevaluating Twin Infanticide." *Current Anthropology* 37 (5): 856–863.

Baxter, Jane Eva 2005. *The Archaeology of Childhood: Children, Gender, and Material Culture*. Gender and Archaeological Series. Walnut Creek, CA: Altamira Press.

Baxter, Jane Eva 2008. "The Archaeology of Childhood." *Annual Review of Anthropology* 37: 159–175.

Barker, David. 1994. *Mothers, Babies, and Disease in Later Life*. London: BMJ Publishing Group.

Barker, D.J.P, and C. Osmond. 1986. "Infant Mortality, Childhood Nutrition, and Ischaemic Heart Disease in England and Wales." *The Lancet* 327 (8489): 1077–1081.

Beaumont, Julia, A. Gledhill, Julia Lee-Thorp, and Janet Montgomery. 2013. "Child Diet: A Closer Examination of the Evidence from Dental Tissues Using Stable Isotope Analysis of Incremental Human Dentine." *Archaeometry* 55 (2): 277–295.

Beaumont, Julia, Janet Montgomery, Jo Buckberry, and Mandy Jay. 2015. "Infant Mortality and Isotopic Complexity: New Approaches to Stress, Maternal Health, and Weaning." *American Journal of Physical Anthropology* 157 (3): 441–457.

Binkin, Nancy J., Ray Yip, Lee Fleshood, and Frederick L. Trowbridge. 1988. "Birth Weight and Childhood Growth." *Pediatrics* 82 (6): 828–834.

Black, Glenn A. 1967. *Angel Site: An Archaeological, Historical, and Ethnological Study.* Indianapolis: Indiana Historical Society.

Black, Robert E., Lindsay H. Allen, Zulfiqar A. Bhutta, Laura E. Calufield, Mercedes de Onis, Majid Ezzati, Colin Mathers, and Juan Rivera, for the Maternal Child Undernutrition Study Group. 2008. "Maternal and Child Undernutrition: Global and Regional Exposures and Health Consequences." *The Lancet* 371 (9608): 243–260.

Blakey, Michael L. and Geroge J. Armelagos. 1985. "Deciduous Enamel Defects in Prehistoric Americans from Dickson Mound: Prenatal and Postnatal Stress." *American Journal of Physical Anthropology* 66 (4): 371–380.

Bluebond-Langner, Myra, and Jill E. Korbin. 2007. "Challenges and Opportunities in the Anthropology of Childhoods: An Introduction to Children, Childhoods, and Childhood Studies." *American Anthropologist* 109 (2): 241–246.

Bocquet-Appel, Jean-Pierre, ed. 2008. *Recent Advances in Palaeodemography.* Dordrecht: Kluwer Academic Publishers Group.

Bonsall, Laura. 2013. "Infanticide in Roman Britain: A Critical Review of the Osteological Evidence." *Childhood in the Past: An International Journal* 6 (2): 73–88.

Boric, Dušan, and Sofija Stefanović. 2004. "Birth and Death: Infant Burials from Vlasac and Lepenski Vir." *Antiquity* 78 (301): 526–546.

Brabin, B.J., C. Romagosa, S. Abdelgalil, C. Mendéndez, F.H. Verhoeff, R. McGready, K. Fletcher, et al. 2004. "The Sick Placenta: The Role of Malaria." *Placenta* 25 (5): 359–378.

Buikstra, Jane E. 1977. "Biocultural Dimensions of Archaeological Study: A Regional Perspective." In *Biocultural Adaptation in Prehistoric America,* ed. Robert L. Blakely, 67–85. Athens: University of Georgia Press.

Buikstra, Jane E., and Douglas H. Ubelaker. 1994. *Standards for Data Collection from Human Skeletal Remains.* Arkansas Archaeological Survey Report No. 44. Fayetteville: Arkansas Archeological Survey.

Bukowski, Radek, Nellie I. Hansen, Marian Willinger, Uma M. Reddy, Corette B. Parker, Halit Pinar, Robert M. Silver, et al., for the Eunice Kennedy Shriver National Institute of Child Health and Human Development Stillbirth Collaborative Research Network. 2014. "Fetal Growth and Risk of Stillbirth: A Population-Based Case–Control Study." *PLOS Medicine* 11: e1001633. http://dx.doi.org/10.1371/journal.pmed.1001633.

Butler, Neville R., and Eva D. Alberman, eds. 1969. *Perinatal Problems: The Second Report of the 1958 British Perinatal Survey Under the Auspices of the National Birthday Trust Fund.* Edinburgh: Livingstone.

Cannon, Aubrey, and Katherine Cook. 2015. "Infant Death and the Archaeology of Grief." *Cambridge Archaeological Journal* 25 (2): 399–416.

Chamberlain, Geoffrey. 2001. "Two Babies That Could Have Changed World History." *Historian* 72: 6–10.

Clark, S.L., and A.R. Fleischman. 2011. "Term Pregnancy: Time for a Redefinition." *Clinical Perinatology* 38 (3): 557–564.

Connell, Brian, Amy Gray Jones, Rebecca Redfern, and Don Walker. 2012. *A Bioarchaeological Study of Medieval Burials on the Site of St Mary Spital: Excavations at Spitalfields Market, London E1, 1991–2007.* Monograph Series 60. London: Museum of London Archaeology.

Crawford, Sally, and Carenza Lewis. 2008. "Childhood Studies and the Society for the Study of Childhood in the Past." *Childhood in the Past: An International Journal* 1 (1): 5–16.

Crespo, L., M.E. Subirà, and J. Ruiz. 2011. "Twins in Prehistory: The Case from Olerdola (Barcelona, Spain; 2. IV II BC)." *International Journal of Osteoarchaeology* 21 (6): 751–756.

Cruz, C.B., and S. Codinha. 2010. "Death of Mother and Child Due to Dystocia in 19th Century Portugal." *International Journal of Osteoarchaeology* 20 (4): 491–496.

Cunningham, Craig, Louise Scheuer, and Sue Black. 2016. *Developmental Juvenile Osteology,* 2nd ed. London: Academic Press.

de Miguel Ibáñez, María Paz. 2008. "Gestantes en contextos funerarios altomedievales navarros." *Lucentum* 27: 233–242.

DeWitte, Sharon N., and Christopher M. Stojanowski. 2015. "The Osteological Paradox 20 Years Later: Past Perspectives, Future Directions." *Journal of Archaeological Research* 23 (4): 397–450.

Domett, K.M. 2004. "The People of Ban Lum Khao." In *The Origins of the Civilization of Angkor, Volume 1: The Excavation of Ban Lum Khao,* ed. C.F.W Higham and R. Thosarat, 113–118. Bangkok: Thai Fine Arts Department.

Duray, Stephen M. 1996. "Dental Indicators of Stress and Reduced Age at Death in Prehistoric Native Americans." *American Journal of Physical Anthropology* 99 (2): 275–286.

Einwögerer, Thomas, Herwig Friesinger, Marc Händel, Christine Neugebauer-Maresh, Ulrich Simon, and Maria Teschler-Nicola. 2006. "Upper Palaeolithic Infant Burials." *Nature* 444 (7117): 285.

Faerman, Marina, Gila Kahila Bar-Gal, Dvora Filon, Charles L. Greenblatt, Lawrence Stager, Ariella Oppenheim, and Patricia Smith. 1998. "Determining the Sex of Infanticide Victims from the Late Roman Era through Ancient DNA Analysis." *Journal of Archaeological Science* 25 (9): 861–865.

Fazekas, István Gyula, and F. Kosa. 1978. *Forensic Fetal Osteology.* Budapest: Akademiai Kiado.

Feucht, Erika. 1995. *Das Kind im Alten Ägypte: Dis Stellung des Kindes in Familie und Desellschaft nach alt-ägyptischen Texten und Darstellungen.* Frankfurt: Campus Verlag.

Fisher, Russell S. 1951. "Criminal Abortion." *Journal of Criminal Law and Criminology* 42: (2) 242.

Flohr, S. 2014. "Twin Burials in Prehistory: A Possible Case From the Iron Age of Germany." *International Journal of Osteoarchaeology* 24 (1): 116–122.

Fuller, B.T., M.P. Richards, and S.A. Mays. 2003. "Stable Carbon and Nitrogen Isotope Variations in Tooth Dentine Serial Sections from Wharram Percy." *Journal of Archaeological Science* 30 (12): 1673–1684.

Furmaga-Jablonska, W., H. Chrzastek-Spruch, and M. Kozlowska. 1999. "Longitudinal Analysis of Head Growth in Children Born With Low Birth Weight." *Perspectives in Human Biology* 4 (2): 23–32.

Gibson, J.R., and Thomas McKeown. 1951. "Observations On All Births (23,970) in Birmingham, 1947: III. Survival." *British Journal of Social Medicine* 5 (3): 177–183.

Gilmore, Helen F., and Siân E. Halcrow. 2014. "Sense or Sensationalism? Approaches to Explaining High Perinatal Mortality in the Past." In *Tracing Childhood: Bioarchaeological Investigations of Early Lives in Antiquity*, ed. Jennifer L. Thompson, Marta P. Alfonso-Durruty, and John J. Crandall, 123–138. Gainesville: University Press of Florida.

Gluckman, Peter D., and Mark A. Hanson, eds. 2006. *The Developmental Origins of Health and Disease*. Cambridge: Cambridge University Press.

Goodman, Alan H., George J. Armelagos, and Jerome C. Rose. 1980. "Enamel Hypoplasias as Indicators of Stress in Three Prehistoric Populations from Illinois." *Human Biology* 52 (3): 515–528.

Gottlieb, Alma. 2000. "Where Have all the Babies Gone? Towards an Anthropology of Infants (and Their Caretakers)." *Anthropological Quarterly* 73: 121–132.

Gould, Stephen J. 1996. *The Mismeasure of Man*. New York: Norton.

Gowland, Rebecca L. 2015. "Entangled Lives: Implications of the Developmental Origins of Health and Disease Hypothesis for Bioarchaeology and the Life Course." *American Journal of Physical Anthropology* 158 (4): 530–540.

Gowland, Rebecca L. and Andrew T. Chamberlain. 2002. "A Bayesian Approach to Ageing Perinatal Skeletal Material From Archaeological Sites: Implications for the Evidence for Infanticide in Roman-Britain." *Journal of Archaeological Science* 29 (6): 677–685.

Gunnell, D., J. Rogers, and P. Dieppe. 2001. "Height and Health: Predicting Longevity From Bone Length in Archaeological Remains." *Journal of Epidemiology and Community Health* 55: 505–507.

Gurven, M. 2012. "Infant and Fetal Mortality Among a High Fertility and Mortality Population in the Bolivian Amazon." *Social Science and Medicine* 75: 2493–2502.

Halcrow, Siân E. 2006. "Subadult Health and Disease in Late Prehistoric Mainland Southeast Asia." PhD diss., University of Otago.

Halcrow, Siân E., and Nancy Tayles 2008. "The Bioarchaeological Investigation of Childhood and Social Age: Problems and Prospects." *Journal of Archaeological Method and Theory* 15: (2) 190–215.

———. 2011. "The Bioarchaeological Investigation of Children and Childhood." In *Social Bioarchaeology*, ed. Sabrina C. Agarwal and Bonnie A. Glencross, 333–360. Chichester: Wiley-Blackwell.

Halcrow, Siân E., Nancy Tayles, and Hallie R. Buckley. 2007. "Age Estimation of Children in Prehistoric Southeast Asia: Are the Standards Used Appropriate?" *Journal of Archaeological Science* 34 (7): 1158–1168.

Halcrow, Siân E., Nancy Tayles, and Vicki Livingstone. 2008. "Infant Death in Late Prehistoric Southeast Asia." *Asian Perspectives* 47 (2): 371–404.

Halcrow, Siân, Nancy Tayles, Raelene Inglis, and Charles Higham. 2012. "Newborn Twins from Prehistoric Mainland Southeast Asia: Birth, Death and Personhood." *Antiquity* 86 (333): 838–852.

Halcrow, S.E., N.J. Harris, N. Beaven, and H.R. Buckley. 2014. "First Bioarchaeological Evidence of Probable Scurvy in Southeast Asia: Multifactorial Etiologies of Vitamin C Deficiency in a Tropical Environment." *International Journal of Paleopathology* 5: 63–71.

Hawkes, Sonia Chadwick, and Calvin Wells. 1975. "Crime and Punishment in an Anglo-Saxon Cemetery?" *Antiquity* 49 (194): 118–122.

Hillson, Simon. 1996. *Dental Anthropology.* Cambridge: Cambridge University Press.

———. 2009. "The World's Largest Infant Cemetery and Its Potential for Studying Growth and Development." *Hesperia Supplement* 43: 137–154.

Hochberg, Z., R. Feil, M. Constancia, M. Fraga, C. Junien, J-C. Carel, P. Boileau, Y Le Bouc, C. Deal, and K. Lillycrop. 2011. "Child Health, Developmental Plasticity, and Epigenetic Programming." *Endocrine Reviews* 32: 159–224.

Hoffman, H.J., C.R. Stark, F.E. Lundin Jr, and J.D. Ashbrook. 1974. "Analysis of Birth Weight, Gestational Age, and Fetal Viability, U.S. Births, 1968." *Obstetrical and Gynecological Survey* 29: 651–681.

Högberg, U., E. Iregren, C-H. Siven, and L. Diener. 1987. "Maternal Deaths in Medieval Sweden: an Osteological and Life Table Analysis." *Journal of Biosocial Science* 19: 495–503.

Hooton, Earnest A. 1930. *Indians of Pecos Pueblo: A Study of Their Skeletal Remains.* New Haven: Yale University Press.

Hrdlicka, Ales, ed. 1924. *Catalogue of Human Crania in the United States National Museum Collections: The Eskimos, Alaska and Related Indians, Northeastern Asiatics.* Washington, DC: Government Printing Office.

Humphrey, Louise T. 2000. "Growth Studies of Past Populations: An Overview and an Example." In *Human Osteology in Archaeology and Forensic Science*, ed. Margaret Cox and Simon Mays, 23–38. London: Greenwich Medical Media.

Humphrey, L.T. 2003. "Linear Growth Variation in the Archaeological Record." In *Patterns of Growth and Development in the Genus Homo*, ed. G.E. Krovitz and A.J. Nelson, 144–169. Cambridge: Cambridge University Press.

Ingvarsson-Sundström, A. "Children Lost and Found: A Bioarchaeological Study of the Middle Helladic Children in Asine With a Comparison to Lerna." PhD diss., Uppsala University.

Johnston, Francis E. 1961. "Sequence of Epiphyseal Union in a Prehistoric Kentucky Population from Indian Knoll." *Human Biology* 33 (1): 66–81.

Johnston, Francis E. 1962. "Growth of the Long Bones of Infants and Young Children at Indian Knoll." *American Journal of Physical Anthropology* 20 (3): 249–254.

Kamp, Kathryn A. 2001. "Where Have All the Children Gone?: The Archaeology of Childhood." *Journal of Archaeological Method and Theory* 8 (1): 1–34.

Kelnar, Christopher J.H., David Harvey, and Carol Simpson. 1995. *The Sick Newborn Baby.* London: Baillière Tindall.

Kinaston, Rebecca L., Hallie R. Buckley, Siân E. Halcrow, Matthew J.T. Spriggs, Stuart Bedford, Ken Neal, and A. Gray. 2009. "Investigating Foetal and Perinatal Mortality in Prehistoric Skeletal Samples: A Case Study from a 3000-Year-Old Pacific Island Cemetery Site." *Journal of Archaeological Science* 36 (12): 2780–2787.

Klaus, Haagen D. 2014. "Frontiers in the Bioarchaeology of Stress and Disease: Cross-disciplinary Perspectives from Pathophysiology, Human Biology, and Epidemiology." *American Journal of Physical Anthropology* 155 (2): 294–308.

Kramer, Michael S., Kitaw Demisse, Hong Yang, Robert W. Platt, Reg Sauvé, and Robert Liston, for the Fetal and Infant Health Study Group for the Canadian Perinatal Surveillance System. 2000. "The Contribution of Mild and Moderate Preterm Birth to Infant Mortality." *JAMA* 284 (7): 834–849.

Kurki, Helen K. 2011. "Compromised Skeletal Growth? Small Body Size and Clinical Contraction Thresholds for the Female Pelvic Canal." *International Journal of Paleopathology* 1 (3–4): 138–149.

Lancy, David F. 2008. *The Anthropology of Childhood: Cherubs, Chattel, Changelings.* Cambridge: Cambridge University Press.

Larsen, Clark Spencer. 1997. *Bioarchaeology: Interpreting Behaviour from the Human Skeleton.* Cambridge: Cambridge University Press.

———. 2015. *Bioarchaeology: Interpreting Behaviour from the Human Skeleton.* 2nd ed. Cambridge: Cambridge University Press.

Lasso, E., M. Santos, A. Rico, J.V. Pachar, and J. Lucena. 2009. "Expulsión fetal postmortem [Postmortem fetal extrusion]." *Cuadernos de Medicina Forense* 15 (155): 77–81.

Lewis, Mary E. 2000. "Non-adult Palaeopathology: Current Status and Future Potential." In *Human Osteology in Archaeology and Forensic Science,* ed. Margaret Cox and Simon Mays, 39–57. London: Greenwich Medical Media.

———. 2002. "Impact of Industrialization: Comparative Study of Child Health in Four Sites from Medieval and Postmedieval England (AD 850–1859)." *American Journal of Physical Anthropology* 119 (3): 211–223.

———. 2004. "Endocranial Lesions in Non-adult Skeletons: Understanding Their Aetiology." *International Journal of Osteoarchaeology* 14 (2): 87–97.

———. 2007. *The Bioarchaeology of Children: Perspectives from Biological and Forensic Anthropology.* Cambridge: Cambridge University Press.

———. 2013. "Sticks and Stones: Exploring the Nature and Significance of

Child Trauma in the Past." In *The Bioarchaeology of Human Conflict: "Traumatized Bodies" from Early Prehistory to the Present,* ed. Chriostpher Knüsel and Martin Smith, 39–63. New York: Routledge.

Lewis, Mary E., and Rebecca Gowland. 2007. "Brief and Precarious Lives: Infant Mortality in Contrasting Sites from Medieval and Post-medieval England (AD 850–1859)." *American Journal of Physical Anthropology* 134 (1): 117–129.

Lewis, Mary E., and Charlotte Roberts. 1997. "Growing Pains: The Interpretation of Stress Indicators." *International Journal of Osteoarchaeology* 7 (6): 581–586.

Lieverse, Angela R., Vladimir Ivanovich Bazaliiskii, and Andrzej W. Weber. 2015. "Death by Twins: A Remarkable Case of Dystocic Childbirth in Early Neolithic Siberia." *Antiquity* 89 (343): 23–38.

Lillehammer, Grete. 1989. "A Child Is Born: The Child's World in an Archaeological Perspective." *Norwegian Archaeological Review* 22 (2): 89–105.

———. 2015. "25 Years with the 'Child' and the Archaeology of Childhood." *Childhood in the Past: An International Journal* 8 (2): 78–86.

Liston, Maria A., and John K. Papadopoulos. 2004. "The 'Rich Athenian Lady' Was Pregnant: The Anthropology of a Geometric Tomb Reconsidered." *Hesperia* 7 (1): 7–38.

Lovejoy, C. Owen, Katherine F. Russell, and Mary L. Harrison. 1990. "Long Bone Growth Velocity in the Libben Population." *American Journal of Human Biology* 2 (5): 533–541.

Luibel, A.M. 1981. *Use of Computer Assisted Tomography in the Study of a Female Mummy from Chihuahua, Mexico.* Annual Meeting of the Southwestern Anthropological Associatation, Santa Barbara, CA.

Malgosa, A., A. Alesan, S. Safont, M. Ballbé, and M.M. Ayala. 2004. "A Dystocic Childbirth in the Spanish Bronze Age." *International Journal of Osteoarchaeology* 14 (2): 98–103.

Mays, Simon. 1993. "Infanticide in Roman Britain." *Antiquity* 67 (257): 883–888.

———. 1995. "The Relationship between Harris Lines and Other Aspects of Skeletal Development in Adults and Juveniles." *Journal of Archaeological Science* 22 (4): 511–520.

———. 1998. *The Archaeology of Human Bones.* London: Routledge.

———. 1999. "Linear and Appositional Long Bone Growth in Earlier Human Populations: A Case Study from Mediaeval England." In *Human Growth in the Past: Studies From Bones and Teeth,* ed. Robert D. Hoppa, and Charles M. Fitzgerald, 290–312. Cambridge: Cambridge University Press.

Mays, Simon A. 2003. "Bone Strontium: Calcium Ratios and Duration of Brestfeeding in a Mediaeval Skeletal Population." *Journal of Archaeological Science* 30 (6): 731– 741.

Mays, Simon, and Jill Eyers. 2011. "Perinatal Infant Death at the Roman Villa Site at Hambleden, Buckinghamshire, England." *Journal of Archaeological Science* 38 (5): 1931–1938.

Mays, Simon, and Marina Faerman. 2001. "Sex Identification in Some Putative Infanticide Victims from Roman Britain Using Ancient DNA." *Journal of Archaeological Science* 28 (5): 555–559.

McCarron, Peter, Mona Okasha, James McEwan, and Geroge Davey Smith. 2002. "Height in Young Adulthood and Risk of Death From Cardiorespiratory Disease: A Prospective Study of Male Former Students of Glasgow University, Scotland." *American Journal of Epidemiology* 155 (8): 683–687.

McIntosh, Neil, Peter J. Helms, and Rosalind L. Smyth, eds. 2003. *Forfar and Arneil's Textbook of Pediatrics.* 6th ed. Edinburgh: Churchill Livingstone.

Meskell, Lynn. 2000. "Cycles of Life and Death: Narrative Homology and Archaeological Realities." *World Archaeology* 31 (4): 423–441.

Murphy, Eileen M. 2011. "Parenting, Child Loss and the Cilliní of Post-Medieval Ireland." In *(Re)thinking the Little Ancestor: New Perspectives on the Archaeology of Infancy and Childhood,* BAR International Series 2271, ed. Mike Lally and Alison Moore, 63–74. Oxford: British Archaeological Reports.

Nwogu-Ikojo, E.E., S.O. Nweze, and H.U. Ezegwui. 2008. "Obstructed Labour in Enugu, Nigeria," *Journal of Obstetrics and Gynaecology* 28 (6): 596–599.

O'Donovan, Edmond, and Jonny Geber. 2009. "Archaeological Excavations on Mount Gamble Hill: Stories from the First Christians in Swords." In *Axes, Warriors and Windmills: Recent Archaeological Discoveries in North Fingal,* ed. Christine Baker, 64–74. Fingal: Fingal County Council.

O'Donovan, Edmond, and Jonny Geber. 2010. "Archaeological Excavations on Mount Gamble Hill, Swords, Co. Dublin." In *Death and Burial in Early Medieval Ireland in Light of Recent Excavations,* ed. Christiaan Corlett and Michael Potterton, 64–74. Dublin: Wordwell.

Owsley, Douglas W., and Bruce Bradtmiller. 1983. "Mortality of Pregnant Females in Arikara Villages: Osteological Evidence." *American Journal of Physical Anthropology* 61 (3): 331–336.

Owsley, Douglas W., and Richard L. Jantz. 1985. "Long Bone Lengths and Gestational Age Distributions of Post-contact Period Arikara Indian Perinatal Infant Skeletons." *American Journal of Physical Anthropology* 68 (3): 321–328.

Parker Pearson, Michael. 1982. "Mortuary Practices, Society and Ideology: An Ethnoarchaeological Study." In *Symbolic and Structural Archaeology,* ed. Ian Hodder, 99–103. Cambridge: Cambridge University Press.

Patel, Tulsi. 1994. *Fertility Behaviour: Population and Society in a Rajasthan Village.* Delhi: Oxford University Press.

Persson, O., and E. Persson. 1984. *Anthropological Report on the Mesolithic Graves from Skateholm, Southern Sweden. I: Excavation Seasons 1980–1982.* Lund: University of Lund.

Pounder, Derrick J., M. Prokopec, and G.L. Pretty. 1983. "A Probable Case of Euthanasia Amongst Prehistoric Aborigines at Roonka, South Australia." *Forensic Science International* 23 (2–3): 99–108.

Pressat, Roland. 1972. *Demographic Analysis: Methods, Results, Applications.* Trans. Judah Matras. Chicago: Aldine-Atherton.

Rascón Pérez, Josefina, Óscar Cambra Moo, and Armando Martín Gonzalez. 2007. "A Multidisciplinary Approach Reveals an Extraordinary Double Inhumation in the Osteoarchaeological Record." *Journal of Taphonomy* 5 (2): 91–101.

Rewekant, Artur. 2001. "Do Environmental Disturbances of an Individual's Growth and Development Influence the Later Bone Involution Processes? A Study of Two Mediaeval Populations." *International Journal of Osteoarchaeology* 11 (6): 433–443.

Richards, M.P., S. Mays, and B.T. Fuller. 2002. "Stable Carbon and Nitrogen Values of Bone and Teeth Reflect Weaning Age at the Medieval Wharram Percy Site, Yorkshire, UK." *American Journal of Physical Anthropology* 119 (3): 205–210.

Robbins, Gwen. 2011. "Don't Throw Out the Baby with the Bathwater: Estimating Fertility From Subadult Skeletons." *International Journal of Osteoarchaeology* 21 (6): 717–722.

Robbins Schug, Gwen. 2011. *Bioarchaeology and Climate Change: A View from South Asian Prehistory.* Gainesville: University Press of Florida.

Roberts, Charlotte, and Margaret Cox. 2003. *Health and Disease in Britain: From Prehistory to the Present Day.* Thrupp: Sutton Publishing.

Rosenberg, Karen, and Wenda Trevathan. 2002. "Birth, Obstetrics and Human Evolution." *BJOG: An International Journal of Obstetrics and Gynaecology* 109 (11): 1199–1206.

Saunders, Shelley R. 2000. "Subadult Skeletons and Growth-Related Studies." In *Biological Anthropology of the Human Skeleton,* ed. M. Anne Katzenberg and Shelley R. Saunders, 135–162. New York: Wiley-Liss.

Say, Lale, Doris Chou, Alison Gemmill, Özge Tunçalp, Ann-Beth Moller, Jane Daniels, A. Metin Gülmezoglu, Marleen Temmerman, and Leontine Alkema. 2014. "Global Causes of Maternal Death: A WHO Systematic Analysis." *The Lancet Global Health* 2 (6): e323–e333.

Sayer, Duncan, and Sam D. Dickson. 2013. "Reconsidering Obstetric Death and Female Fertility in Anglo-Saxon England." *World Archaeology* 45 (2): 285–297.

Scheuer, Louise, and Sue Black. 2000. *Developmental Juvenile Osteology.* San Diego: Academic Press.

Scheuer, Louise, and Sue. Black. 2004. *The Juvenile Skeleton.* San Diego: Academic Press.

Schulz, Friedrich, Klaus Püschel, and Michael Tsokos. 2005. "Postmortem Fetal Extrusion in a Case of Maternal Heroin Intoxication. *Forensic Science, Medicine, and Pathology* 1 (4): 273–276.

Schwartzman, Helen B. 2005. "Materializing Children: Challenges for the Archaeology of Childhood." *Archaeological Papers of the American Anthropological Association* 15 (1): 123–131.

Sjøvold, T., I. Swedborg, and I. Diener. 1974. "A Pregnant Woman from the Middle Ages with Exostosis Multiplex." *Ossa* 1: 3–23.

Smith, Patricia, and Gila Kahila. 1992. "Identification of Infanticide in Archaeological Sites: A Case Study from the Late Roman-Early Byzante Periods at Ashkelon, Israel." *Journal of Archaeological Science* 19 (6): 667–675.

Sofaer Derevenski, Johanna, ed. 2000. *Children and Material Culture.* London: Routledge.

Sofaer, Johanna R. 2006. *The Body as Material Culture: A Theoretical Osteoarchaeology.* Cambridge: Cambridge University Press.

Soren, David, and Noelle Soren. 1995. "Who Killed the Babies of Lugnano?" *Archaeology* 48 (5): 43–48.

Standen, Vivien G., Bernaro T. Arriaza, and Calogero M. Santoro. 2014. "Chinchorro Mortuary Practices on Infants: Northern Chile Archaic Period (BP 7000–3600)." In *Tracing Childhood: Bioarchaeological Investigations of Early Lives in Antiquity,* ed. Jennifer L. Thompson, Marta P. Alfonso-Durruty, and John J. Crandall, 58–74. Gainesville: University of Florida Press.

Steel, Louise. 1995. "Differential Burial Practices in Cyprus at the Beginning of the Iron Age." In *Archaeology of Death in the Ancient Near East,* ed. Stuart Campbell and Anthony Green, 199–204. Exeter: The Short Run Press.

Tainter, Joseph A. 1978. "Mortuary Practices and the Study of Prehistoric Social Systems." In *Advances in Archaeological Method and Theory,* vol. 1, ed. Michael B. Schiffer, 105–141. San Diego: Academic Press.

Tayles, Nancy, and Siân E. Halcrow. 2016. "Age-at-Death Estimation in a Sample of Prehistoric Southeast Asian Adolescents and Adults." In *Bioarchaeology in Southeast Asia and the Pacific,* ed. Marc Oxenham and Hallie Buckley, 239–256. London: Routledge.

Tocheri, M.W., T.L. Dupras, P. Sheldrick, and J.E. Molto. 2005. "Roman Period Fetal Skeletons From the East Cemetery (Kellis 2) from Kellis, Egypt." *International Journal of Osteoarchaeology* 15 (5): 326–341.

Tocheri, Matthew W., and J. Eldon Molto. 2002. "Aging Fetal and Juvenile Skeletons from Roman Period Egypt Using Basiocciput Osteometrics." *International Journal of Osteoarchaeology* 12 (6): 356–363.

Tompkins, Robert L. 1996. "Human Population Variability in Relative Dental Development." *American Journal of Physical Anthropology* 99 (1): 79–102.

Ucko, Peter J. 1969. "Ethnography and Archaeological Interpretation of Funerary Remains." *World Archaeology* 1 (2): 262–280.

Van Gerven, Dennis P., and Geroge J. Armelagos. 1983. ""Farewell to Paleodemography?" Rumours of Its Death have been Greatly Exaggerated." *Journal of Human Evolution* 12 (4): 353–360.

Washburn, S.L. 1951. "The New Physical Anthropology." *Transactions of the New York Academy of Sciences* 13 (7): 298–304.

Wells, C. 1975. "Ancient Obstetric Hazards and Female Mortality." *Bulletin of the New York Academy of Medicine* 51 (11): 1235.

Wells, C. 1978. "A Medieval Burial of a Pregnant Woman." *The Practitioner* 221: 442–444.

Wheeler, S.M. 2012. "Nutritional and Disease Stress of Juveniles from the Dakhleh Oasis, Egypt." *International Journal of Osteoarchaeology* 22 (2): 219–234.

White, Tim D. 2000. *Human Osteology.* San Diego: Academic Press.

Wileman, Julie. 2005. *Hide and Seek: The Archaeology of Childhood.* Stroud: Tempus.

Wiley, Andrea S., and Ivy L. Pike. 1998. "An Alternative Method for Assessing Early Mortality in Contemporary Populations." *American Journal of Physical Anthropolology* 107 (3): 315–330.

Willis, Anna, and Marc F. Oxenham. 2013. "A Case of Maternal and Perinatal Death in Neolithic Southern Vietnam, c. 2100–1050 BCE." *International Journal of Osteoarchaeology* 23 (6): 676–684.

Wood, James W., George R. Milner, Henry C. Harpending, and Kenneth M. Weiss. 1992. "The Osteological Paradox: Problems of Inferring Prehistoric Health From Skeletal Samples." *Current Anthropology* 33 (4): 343–370.

World Health Organization (WHO). 2016a. "Newborns: Reducing Mortality." Reviewed January 2016. http://www.who.int/mediacentre/factsheets/fs333/en.

———. 2016b. "Malaria in Pregnant Women." Last updated 25 April. http://www.who.int/malaria/areas/high_risk_groups/pregnancy/en.

———. 2016c. "Preterm Birth." Reviewed November 2016. http://www.who.int/mediacentre/factsheets/fs363/en.

———. 2017a. "Maternal Health." Accessed 3 May. http://www.who.int/topics/maternal_health/en.

———. 2017b. "Maternal, Newborn, Child and Adolescent Health." Accessed 3 May. http://www.who.int/maternal_child_adolescent/topics/maternal/maternal_perinatal/en.

Żądzińska, Elżieta, Wiesław Lorkiewicz, Marta Kurek, and Beata Borowska-Strugińska. 2015. "Accentuated Lines in the Enamel of Primary Incisors from Skeletal Remains: A Contribution to the Explanation of Early Childhood Mortality in a Medieval Population from Poland." *American Journal of Physical Anthropology* 157 (3): 402–441.

Chapter 5

FETAL PALEOPATHOLOGY
AN IMPOSSIBLE DISCIPLINE?

Mary E. Lewis

This chapter introduces the concept of fetal paleopathology in archaeological material, highlighting the limitations and potential of such research to inform us about the lives of mothers and their babies in the past. Problems with preservation and recognizing lesions in such tiny skeletal remains are discussed, before potential new sources of research are highlighted.

Today, an estimated four million babies die each year worldwide within the first month of life (neonatal period), with the first day being the most critical period, and two-thirds of these babies dying within a week. The most common causes of death are prematurity, sepsis and pneumonia, asphyxia, tetanus, and diarrhea. None of these conditions are traditionally recognized in skeletal remains, and of the 14 percent who died of "other" causes, only 7 percent were from congenital conditions. These figures relate to death rates in industrialized countries that have clear social and economic divides, with children from poor backgrounds most at risk (Lawn et al. 2005). Other factors influencing neonatal mortality include low birth weight, complications during delivery, limited access to specialist care, maternal anemia, syphilis, and fever (Black et al. 2010).

But we should be cautious about directly transferring these data onto societies of different cultures, levels of development, and economies, especially in the past. Clinical studies have consistently identified a higher neonatal mortality rate for males, suggesting that females are more buffered against environmental insults in the womb and cope better with the stress of childbirth (Bekedam et al. 2002). While the precise mechanisms behind this phenomenon are still unknown, differential treatment where boys are favored can reverse this survivability pattern (McMillen 1979; Lawn et al. 2005). Application of neonatal and postneonatal death ratios may be used to explore whether children were dying from endogenous or exogenous causes (Lewis and Gowland 2007), but our imprecise age divisions make it a crude way of assessing perinatal causes of death.

Training in non-adult osteology has improved over the past decade, advancing an area of research that was previously underdeveloped. However, the identification of fetal remains and familiarity with their normal anatomy is still limited and may result in many subtle pathological lesions going unnoticed. Given all of these issues, it is understandable that paleopathologists rarely choose to study perinates, and a review of their potential to provide information on disease and trauma is long overdue.

Prenatal Pathology

The fetus is most vulnerable to environmental disruption during periods of rapid cell differentiation, particularly during the first two weeks of development when disruptive agents (teratogens) can result in spontaneous abortion of the embryo (Moore 1988). Skeletal development begins between eight to twelve weeks with the formation of the skeletal matrix, followed by intramembranous (in membrane) ossification of the clavicle, mandible, and bones of the skull vault. Endochondral (in cartilage) mineralization forms the rest of the skeletal structure with the ossification of the appendicular skeleton, ilium, and scapula by sixteen weeks, metacarpals and tarsals by sixteen to twenty gestational weeks (Krakow et al. 2009). Successful ossification relies on a good maternal oxygen supply through the bloodstream during fetal development (Waldron 2009). The homeobox (Hox) genes regulate the differentiation process of the skeleton into limb or spinal elements, and mutations in these genes are responsible for malformations during embryonic development (Scheuer and Black 2000: 172).

Infections can spread to the fetus via the placenta. The placenta is highly vascular, and any breach of its integrity may lead to large amounts of a pathogen reaching the fetus. Pathogens may also spread through the umbilical vein, through ingestion of infected amniotic fluid, or during birth with exposure to an infected birth canal or maternal fluids (Zeichner and Plotkin 1988). The fetus may suffer damage or disruption to their developing cells, or may mount an autoimmune response, but because the fetus has a reduced inflammatory reaction to infection (Holt and Jones 2000), skeletal signs of disease are limited before birth. Rubella, smallpox, tuberculosis, syphilis, leprosy, chickenpox, mumps, measles, and scarlet fever can all be transmitted transplancentally (Lorin et al. 1983; Naeye and Blanc 1965; Al-Qattan and Thomson 1995) and may be evident on the perinatal skeleton, either through signature (pathognomonic) signs, or generalized new bone formation. While exposure to influenza is responsible for many congenital defects, leprosy and syphilis may instead cause early spontaneous abortion or, if transmitted toward the end of the third trimester, remain latent until later infancy and childhood (Dorfman and Glaser 1990, Melsom et al. 1982). While congenital syphilis may result in characteristic notches on the unerupted deciduous upper incisors on radiograph (Hutchinson 1857), cranial deformities caused by malaria or maternal smoking (Lampl 2003) are all but impossible to identify on thin and unfused perinatal cranial bones.

Enigmatic Skeletal Lesions

New Bone Formation

Traditionally, paleopathologists interpret gray bone deposits (fiber bone) on the outer bone surface as a sign of trauma or infection. For the perinatal paleopathologist, things are less straightforward. P. de Silva and colleagues (2003) warned clinicians against misdiagnosing what they termed "physiological" periostitis as abuse in infants. Periostitis is a term used to denote inflammation (or inflammitis) of the fibrous sheath (periosteum) surrounding the bones in life. They noted symmetrical new bone formation on the long bones, especially the femora, humeri, and tibiae in babies aged one to six months old. While the new bone was usually concentric, the tibia was most commonly affected on the medial aspect, and in most cases, bone formation was confined to the long bone shaft (or diaphysis).

Charles Shopfner (1966) examined the radiographic appearance of the long bones of 335 healthy premature and full-term infants, and he noted periosteal new bone in 35 percent of cases. The bone deposits were thick, but not multilayered, and appeared on radiograph as double contours before they became incorporated into the underlying bone surface. K. Gleser (1949) warned that increased formation and mineralization of the long bones during the normal growth process might mimic pathological features in a two- to five-month-old month infant, when signs of congenital syphilis, scurvy, and rickets may be suspected. For the paleopathologist examining dry bone, perinates with widespread new bone formation on the cranium, long bones, and ilia are common findings, but we are ill equipped to determine if this indicates one of the many infections that may be responsible for neonatal death or if this signals the child was experiencing a growth "spurt" when they died. Common sense may dictate that a child on the brink of death is unlikely to be undergoing rapid growth, but this does not account for accidental deaths or deaths that occur shortly after a growth spurt but before the bone can remodel.

The bone turnover rate in neonates has been estimated to be high before birth and in the first forty-eight hours of life, and greater in the preterm neonate than full-term babies, with bone turnover rates of infants being several times higher than in adults (Mora et al. 1997). While the extent of bone turnover in perinates may be difficult to quantify, remodeling of the femur, for example, significantly changes bone density within the first six months. Hence, any new bone formation in perinates will be rapid, with several days needed for the newly deposited organic matrix to be mineralized (Rauch and Schoenau 2002). It would also be reasonable to expect any trauma or inflammation experienced during the birth process to be remodeled within six months of the child's life, making identification of lesions in the perinates all the more crucial.

In some cases, localized or profuse new bone deposits that are gray may point to specific conditions. Infantile cortical hyperostosis (or Caffey disease) is an inflammatory disorder of unknown etiology causing profuse new bone formation on the long bones and mandible that heal spontaneously (Caffey and Silverman 1945). A. Alduc-Le Bagousse and J. Blondiaux (2001) suggested a possible fetal case, found within the abdomen of a female in Lisieux, France. All the surviving long bones have profuse new bone formation, and the tibiae and cranium are most severely affected. Layers of new bone are often identified on the internal cranial surfaces of perinatal material, confined to the occipital. As this part of the cranium is undergoing rapid growth after birth, many of these lesions are likely part of the normal bone remodeling process (Lewis 2004), but they may also be signs of inflammation or calcified blood pools (hematomas).

J.P. Crozer Griffith (1919) states that perinatal intracranial hemorrhage occurred in 15 percent of premature births and may result from cerebral palsy. Birth trauma may cause localized bleeding and, if a child were laid on their back, then blood or pus would pool to the occipital area (Mitchell 2006). A. González Martin and colleagues (1997) raised the possibility that the irregular and porotic appearance of the pars basilaris (the bone that forms part of the base of the skull) seen in children from birth to six years was potentially pathological, calling for more research into this area. Without more detailed information on the pattern and timing of perinatal growth in individual skeletal elements, we may never be able to untangle these issues.

Lytic Cranial Lesions

Tania Kausmally and Rachel Ives (2007) highlight problems with the interpretation of destructive (or lytic) lesions in 7.4 percent (seven out of ninety-four) of perinates from postmedieval London. The lesions caused holes in the cranium but only occurred on a few of the perinates from the same site and context, suggesting postmortem damage was not to blame. Kausmally and Ives identified several possible causes including cancer—for example, infantile chordoma, infantile myofibromatosis, Langerhans cell histiocytosis (LCH), and tuberculosis (TB). As the lesions were too common to support very rare cancerous conditions, they favor TB or LCH as a cause.

John Caffey (1978: 32) had previously noted these lesions, calling them "lacunar skull" and describing them as most marked on the parietal and frontal bones. He linked them to spina bifida, hydrocephalus (cranial enlargement due to "water on the brain"), meningocele, and meningioencephocele (protrusion of the brain lining through the cranial vault), but he also saw them on the radiographs of apparently health newborns. As the lytic lesions tended to heal spontaneously, Caffey considered them the result of delayed development of the membrane that eventually forms the cranial vault. Sheila Mendonça de Souza and colleagues (2008) noted destructive lesions on the skull of a six-month-old Peruvian mummy that were accompanied by new bone formation on the internal surface of the cranium, active bone formation on the frontal and parietal bones, and a flattened occipital. They suggest cranial modification through head binding, followed by bone death and secondary infection as the cause.

Potential Perinatal Paleopathology: Infections

Rubella

Congenital rubella results from transmission of the rubella virus during the first trimester and would have had serious consequences for the survival of a newborn in the past. Pregnancies ending in a spontaneous abortion or stillbirth would result in perinates entering the skeletal record without pathological changes. Those born alive suffer congestive heart failure, low birth weight, and difficulty in feeding, along with deafness and cerebral palsy in older child (Cooper et al. 1969). Arnold Rudolf and colleagues report that 45.3 percent of perinates with exposure to maternal rubella display bone lesions between the ages of one to eight weeks and that 76 percent of the thirty-four perinates are male. Osseous changes include wide radiolucent bands and "beaklike projections" at the ends of the bone shafts (metaphyses) during healing, coupled with enlarged anterior fontanelles at the areas where the cranial sutures meet (Rudolf et al. 1965: 430).

The rubella virus's ability to inhibit cell multiplication of bone and fiber forming cells (i.e., osteoblasts and fibroblasts) and other tissues in the body is well known (Naeye and Blanc 1965; Reed 1969). These poorly defined zones of calcification are similar to what paleopathologists might see in children with rickets or congenital syphilis. However, once virus excretion has ceased, bone lesions can disappear in several months (Sekeles and Ornoy 1975). While slightly older than perinates, three infants aged between three and six months were identified with unusually large fontanelles in South Africa dating to the twentieth century, leading M. Steyn and colleagues (2002) to suggest rubella as a possible differential diagnosis. Although rubella lesions are rare and transient in the perinatal period, it should be considered a differential diagnosis in early rickets cases.

Neonatal Osteomyelitis

Neonatal osteomyelitis, an abscess-forming infection affecting multiple bones, was a common cause of death in the past (Trueta 1959) but has yet to be identified in perinates from archaeological contexts. Clinically, multiple bone and joint involvement occurs in 41 percent and 70 percent of cases, respectively (Weissberg et al. 1974), assisted by the presence of open transphyseal vessels that allow the spread of infection across the growth plate (Trueta 1959). John Ogden (1979) describes abscess and new sheath (sequestrum) formation in the first few days after birth due to a blood-born spread of infection and from a localized infection such as a burn. The humerus in the upper arm and the knee are recognized as common sites of infection, and radiographs of these areas may reveal abscesses inside the bone, while enlarged nutrient foramina may suggest involvement of transphyseal vessels.

Sheaths of new bone should be readily identified in the perinatal skeleton, but long bone epiphyses (the growing end plates) are not ossified at birth, and smooth-based localized lesions on the metaphyseal surface caused by abscess may be more difficult to distinguish from postmortem damage or a normal undulating surface. Caution is needed when using clinical cases to reference diseases from the pre-antibiotic era. Ogden's (1979) neonates had chemotherapy that may have allowed development of chronic lesions where more rapid death would normally occur. How common neonatal osteomyelitis was in the past is difficult to judge as the emergence of penicillin-resistant *Staphylococcus* meant a resurgence of osteomyelitis in a more virulent form and a higher number of neonatal cases from the 1950s to 1970s (Gilmour 1962).

Early Onset Congenital Syphilis

Congenital syphilis develops in the fetus secondary to venereal syphilis in the mother. The causative organism, *Treponema pallidum*, can be transmitted as early as the ninth week of gestation. The pathogen enters the fetal bloodstream and spreads to almost every bone in the body. Toxins released from dead microorganisms may invoke an allergic response and uterine contractions in the mother, resulting in fetal death and spontaneous abortion (17 percent) in the first half of the pregnancy (Genç and Ledger 2000). These tiny skeletal remains, if they survive into the archaeological record, will show no signs of disease. At term, a child may be stillborn (23 percent), premature, weak, and sickly, or between 39 and 66 percent may appear perfectly healthy (Harman 1917; Hollier and Cox 1998). About 21 percent of the latter will go on to display signs of infection around two years of age ("late congenital syphilis") (Harman 1917).

A trio of skeletal lesions occurring together is indicative of congenital syphilis in the perinate: joint disruption (osteochondritis), osteomyelitis, and profuse new bone formation (Caffey 1939). György Pálfi and colleagues (1992) describe a seven-month-old fetus recovered from the abdomen of a woman from Costebelle, France. The fetus displayed profuse new bone on the long bones, maxilla, ribs, and cranial vault; possible destructive lesions on the parietal bones; and a characteristic Wimberger's sign (thick dark band) on a radiograph. Convinced they had a case of congenital syphilis, the authors argued that the mother was in the early stages of syphilis, where clinical evidence suggests nearly all pregnancies will involve the spread of infection to the developing child. More recently, aDNA analysis (Montiel et al. 2012) confirmed syphilis in two perinates from postmedieval Huelva, Spain (Malgosa et al. 1996). This was the first time DNA successfully identified a subspecies of the treponeme and may have been because of the abundance of spirochetes known to invade the bone cells of neonates (Montiel et al. 2012).

Potential Perinatal Paleopathology: Congenital Defects and Skeletal Dysplasia

The prenatal ossification process begins around eight to twelve weeks in utero and from this point, osteologists have the potential to recognize congenital defects in the skeleton. In particular, irregularities in the formation of bones and replacement of the fetal spinal cord, or notochord, mean we have the opportunity to identify lesions that may signal more serious soft tissue defects resulting in perinatal death (fig. 5.1). Bruce Anderson (1989) describes axial

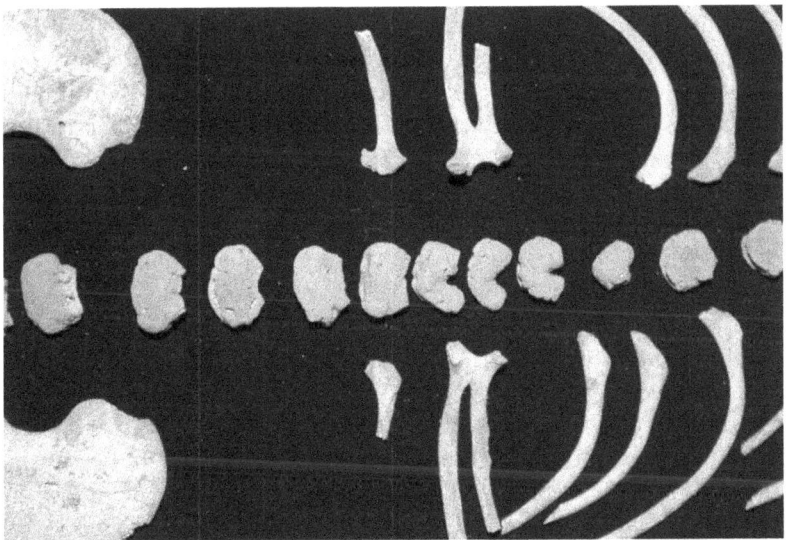

FIGURE 5.1. Congenital defects in the thorax of a thirty-nine-week-old perinate from St. Oswald's Priory in Gloucester, England. The poorly mineralized centra and fused ribs may signal a more serious soft tissue condition that led to the child's death. (Photo courtesy of author.)

congenital anomalies and a possible familial relationship in three neonates from Homol'ovi III, Arizona. The first child presented fused thoracic (chest) vertebrae and asymmetry of the maxilla and mandible with dental overcrowding, suggesting marked facial asymmetry. The second was a perinate with fused second and third cervical (neck) vertebrae and two mid-thoracic vertebrae, possibly indicating type 2 Klippel-Feil syndrome. The child also had a malformed rib. Finally, a younger perinate also demonstrated fusion (or non-separation) of the spinous processes of the second and third cervical vertebrae and the fourth and fifth thoracic vertebrae. M. Hinkes (1983) noted flared sternal ends at the front of ribs 1–8 in a neonate from the Grasshopper Pueblo, and Don Brothwell and R. Powers (2000) recorded merged ribs in a neonate from early medieval Lechlade, England. A tiny example of a fused radius and ulna in the arm (radio-ulnar synostosis) has been reported in a perinate from a double burial at El Molon, Spain (Lorrio et al. 2010).

Cleft, butterfly, and block vertebra, as well as anterior and posterior spina bifida, all have the potential to be identified in the perinate (Müller et al. 1986), and the presence of extra cervical ribs may reveal information about the potential cause of death. In a study of 318 perinates born in Utah between 2006 and 2009, Larissa Furtado and colleagues (2011) reported a significantly higher prevalence of cervical ribs in stillborns compared to live-born children who died within the first year (43 percent compared to 12 percent). They conclude that cervical ribs signal a disadvantageous fetal environment that leads to a greater likelihood of stillbirth with similar results presented elsewhere (Bots et al. 2011). This would require identification and careful examination of the neural arches for the seventh cervical vertebrae for facets. Although, Sue Black and Louise Scheuer (1997) note that cervical ribs themselves will not be found in children under the age of ten years, as they do not fully develop until fusion of the posterior arch to the body of the vertebrae.

More than 350 forms of skeletal dysplasia have the potential to be identified in perinatal remains. Differentiating between them is clinically problematic so is likely to be impossible in archaeological circumstances, although some features such as glabella bossing at the top of the nose, flattened nasal bridge, vertebral (centrum) morphology, and poor mineralization of the cranium and skeleton may signal dysplasia (Krakow et al. 2009). Osteofibrous dysplasia of the neonate, while infrequent, may present as an expansile lytic lesion at the midshaft of a single tibia causing bulging of the bone surface macroscopically and, potentially, a pathological fracture (Hindman et al. 1996). Mark Mooney and colleagues (1992) presented a series of fetal facial measurements that can be used to identify cleft lip and palate, defects that are hard to recognize in the tiny unfused perinatal maxillae. These include the premaxillary length and nasal opening length, which can theoretically be identified on dry bone, but techniques to test and apply this to archaeological material have yet to be developed. Although there is potential for more research to date, congenital skeletal defects are the most common form of pathology identified in archaeological perinates.

T. Sjøvold and colleagues (1974) identified a full-term fetus with multiple bony projections (osteochondromas) from St. Clement, Visby in Gotland, found within the abdominal area of a seventeen- to twenty-year-old female with the same condition. The deformities may have caused an obstruction leading to the death of the mother and child. Kenneth Bennett (1967) describes a case of multiple cranial suture fusion in a perinate from Utah that would have led to a cloverleaf deformity if the child had lived. Darcy Cope (2008) suggests the presence of a meningocele in the fragile cranial bones of a neonate from the Dakhleh Oasis, and a perinate with an even more severe cranial malformation (holoprosencephaly) (Tomasto-Cagigao 2011) buried in an urn in Palpa, Peru. The burial was normal except for the presence of another apparently normal neonate. Although E. Tomasto-Cagigao (2011) does not describe the rest of the skeleton, segmental errors in the spine are often associated with this cranial deformity (O'Rahilly et al. 1980, 1983).

The oldest archaeological case of a perinate with anencephaly (a fatal condition where the brain and the bones of the cranial vault fail to form) comes from an Egyptian catacomb in Hermopolis, built to house the mummies of sacred monkeys and ibises (Saint-Hillaire 1826). J. Christopher Dudar (2010) discusses a possible case of anencephaly in a child from Elmbank Cemetery in nineteenth-century Toronto with associated fused ribs. Detailed studies of the perinates from the Dakhleh Oasis, Egypt, have revealed a variety of remarkable cranial malformations (encephalocele, iniencephaly, and anencephly) (Cope 2008). Stevie Mathews (2008) provides a useful description of known anencephaly in perinates housed at the Smithsonian Institution in Washington, DC. Deformities included malformed sphenoid lesser and greater wings, lack of frontal and parietal bones with isolated orbital rims, deformed squama of the temporal bones, and early fusion of the elements with an elliptical rather than round tympanic ring (a small ring of bone that forms at the external opening of the ear).

A.L. East and J. Buikstra (2001) discuss an achondroplastic female from Elizabeth Mounds, Tennessee, with an in utero fetus they also suspected had achondroplasia based on metaphyseal flaring, disproportionately short limbs, and cranial and long bone measurements that were outside the normal range of the forty-six perinates they compared it to. This case has not been fully published, and the authors caution other dysplasias have yet to be ruled out. Arthur Keith (1913) provides a very useful comparison of the individual bones of an achondroplastic child against an unaffected child of the same age. The affected child displayed a reduced foramen magnum, absent suture mendosa on the developing occipital bone, premature fusion of the basioccipital suture, and short and broad wing of the pars lateralis. In the full-term child, the basiocciput and presphenoid fused limiting further expansion of the brain. Frontal bossing was evident because of the brain seeking compensatory space.

Down Syndrome

Trisomy 21 is the most common chromosomal abnormality among live-born infants and is related to increasing maternal age (e.g., thirty-five years). Today, Down syndrome is estimated to occur in one in seven hundred to one in a thousand births in the United States (Benacerraf 1996). The vast array of skeletal and dental features associated with this syndrome has always attracted the attention of paleopathologists, and while five suspected cases of Down have been identified, none of these are perinates. This may be because children with the syndrome tend to survive into older childhood, but given mothers of increasing age are more likely to suffer from complications at birth, we might expect some cases to enter our perinatal sample. Ernest Hook (1979) estimated 21 percent of trisomy 21 children died between midgestation and full-term. Sonographs of second-trimester fetuses reveal some skeletal features used to predict the presence of trisomy 21 at birth. These include a shortened femur and humerus when measured against the maximum width of the skull and in comparison to the population norm, a shortened iliac length, reduced limb bone length in comparison to the axial skeleton, and shortening of the middle phalanx of the fifth digit of the hand (Keeling et al. 1997; Stempfle et al. 1999). Unfortunately, these features require either an intact cadaver or a large enough series of well-preserved perinates to gauge the normal dimensions. Although hand phalanges ossify around twenty-four gestational weeks, they are so tiny they are rarely excavated, making measurement or an assessment of agenesis impossible. In addition, the short and stumpy fifth-finger phalanx, if recovered, may be misidentified as belonging to the foot. Clinically, the accuracy of these measurements in identifying Down syndrome is between 40 and 50 percent, only rising to 74 percent when the soft tissue nuchal fold of the neck is included (Benacerraf 1996).

Nevertheless, some features hold promise. Absence of the nasal bone is commonly reported in Down syndrome neonates (Dedick and Caffey 1953; Keeling et al. 1997; Otaño et al. 2002; Stempfle et al. 1999) and should ossify between fifteen and forty gestational weeks. L. Otaño and colleagues (2002) found absent nasal bones in three of five (60 percent) of Down syndrome fetuses compared to only 0.6 percent of non–Down syndrome cases. A related condition, trisomy 13, can result in malformations in the lumbosacral and thoracic spine, where small and irregular bones have been noted (Kjær et al. 1997). A fetus with triplody, a rare and lethal condition where there are three sets of chromosomes, can present with cranial base malformations including extra ossification centers, and fusion of two or more vertebral bodies, or a disproportion in the size of the cervical bodies. Another chromosomal disorder (aneuploidy) is a possible explanation for the cranial deformations identified in a thirty-eight- to forty-week perinate from Andover Road in Winchester, United Kingdom. The wing of the pars lateralis is bipartite, and the posterior condyle canal is in two halves (fig. 5.2). The thickened and regular edges suggest this was a congenital anomaly rather than a basilar linear fracture of the occipital (Falys 2010).

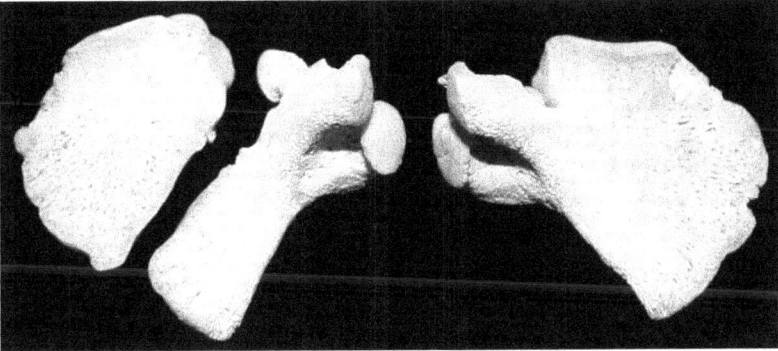

FIGURE 5.2. Possible aneuploidy in a thirty-eight- to forty-week-old perinate from Andover Road in Winchester, United Kingdom (reproduced with kind permission from Ceri Falys, TVAS).

Osteogenesis Imperfecta

Osteogenesis imperfecta (or "brittle bone disease," a fatal congenital condition leading to multiple fractures) is a probable diagnosis for a thirty-eight-week-old perinate from Dakhleh Oasis, Egypt, with severe bowing of all the surviving long bones and pathological fractures of the left ulna, femur, and tibia (Cope and Dupras 2011). The baby was buried on its side in contrast to the normal supine extended burials of other children in the cemetery. Although the basilar fragments of the skull were preserved, they did not show the malformations (for example, a triangular pars basilaris) identified in a forensic case from Guatemala (Lewis 2007: 107).

Potential Perinatal Paleopathology: Trauma

Perimortem cut marks on neonatal remains have been interpreted as indicating surgical removal (or an embryotomy). Only three cases have so far been identified; two are Romano-British examples from Poundbury Camp in Dorset and Yewden villa, Buckinghamshire. The Poundbury perinate was decapitated and has extensive cut marks throughout the long bones (Mays et al. 2012; Molleson and Cox 1988). A possible nineteenth-century embryotomy was identified in L'Aquila, Italy (Capasso et al. 2016), based on severe jumbling of the bones in a wrapped and mummified twenty-nine-week-old fetus.

Birth Injuries

Perinates who die shortly after birth are too young for us to assess any paralysis that may occur as the result of trauma during childbirth, but some fractures may be evident. In 1950s New York, 6 percent of all neonatal hospital admissions of newborns were for injuries sustained at birth (Montagu 1950). Andrew P. Dedick and John Caffey (1953) reported fractured clavicles in 1.2 percent of their 1,030 newborns; these were always unilateral and occurred more commonly on the left side. Caffey (1978) describes hematomas (cephalhematomas) on the ectocranial surface as the result of bruising during breech birth usually positioned away from midline sutures, a detail that may aid in their differentiation from meningocele.

Ossified hematomas may persist for months or years. Skull fractures as a result of birth injury present more of a challenge because of the fragile and fragmentary nature of the perinatal cranium. Linear and depressed fractures may not survive fragmentation in the ground, or if the child dies shortly after birth, perimortem fractures may be indistinguishable from postmortem breaks. Caffey (1978) also suggests that the maxilla is a frequent site for infection in the first few weeks of life because of birth trauma and may be visible as reactive new bone formation around the developing dental germs. Three published cases of possible birth injuries have been identified in paleopathology, involving a skull fracture in a thirty-eight-week perinate (Baxarias et al. 2010) and two cases of unilateral clavicular fractures (Brickley et al. 2006; Soren et al. 1995).

Conclusions

This chapter has reviewed the challenges and potential of examining perinatal remains from archaeological contexts in order to identify skeletal pathology. Although the majority of perinates likely died from infectious or innate conditions, identifying pathology based on new bone on subperiosteal and endocranial surfaces is problematic, as we have yet to develop criteria that allow us to distinguish pathological lesions from the normal growth process. Nevertheless, there has been increasing success in the identification of congenital malformations from archaeological contexts, with fifteen of the twenty-nine (52 percent) reported perinatal cases describing congenital disorders. While fetal remains recovered from the pelvic cavities of female graves hint at obstetric hazards, individual perinatal burials have the potential to tell us much about the health of the fertile maternal population, as well as the environmental factors that affect the survival of newborns. A review of the clinical literature has allowed for the identification of skeletal features that may help us recognize new conditions in the perinate, including rubella, Down syndrome, and other chromosomal disorders, and birth trauma and cleft palate, but further research is needed to understand the range of pathology that can be identified on these tiny remains. Appreciating how far we still have to go in our analyses will help us set the agenda for future studies.

Mary E. Lewis is an associate professor at the University of Reading and specializes in non-adult skeletal pathology. She is author of *The Bioarchaeology of Children* (2007) and *Paleopathology of Children* (2017). She is an associate editor of the *American Journal of Physical*

Anthropology and the *International Journal of Paleopathology*, and she is on the board of the *International Journal of Osteoarchaeology*.

References

Alduc-Le Bagousse, Armelle, and Joel Blondiaux. 2001. "Hyperostoses corticales foetal et infantile à Lisieux (IVe s.): Retour à Costebelle." *Centre archaéologique du Var*: 60–64.

Al-Qattan, M.M., and H.G. Thomson. 1995. "Congenital Varicella of the Upper Limb: A Preventable Disaster." *Journal of Hand Surgery (British & European Volume)* 20 (1): 115–117.

Anderson, Bruce Edward. 1989. "Immature Human Skeletal Remains from Homol'ovi III." *KIVA* 54 (3): 231–244.

Baxarias, J., V. Fontaine, E. Garcia-Guixé, and R. Dinarés. 2010. "Perinatal Cranial Fracture in a Second Century BC Case from Ilturo (Cabrera de Mar, Barcelona, Spain)." Paper presented at the 23rd European Meeting of the Paleopathology Association, Vienna, Austria.

Bekedam, Dick J., Simone Engelsbel, Ben W.J. Mol, Simone E. Buitendijk, and Karin M. van der Pal-de. 2002. "Male Predominance in Fetal Distress during Labor." *American Journal of Obstetrics & Gynecology* 187 (6): 1605–1607.

Benacerraf, Beryl.R. 1996. "The Second-Trimester Fetus with Down Syndrome: Detection Using Sonographic Features." *Ultrasound in Obstetrics & Gynecology* 7 (2): 147–155.

Bennett, Kenneth A. 1967. "Craniostenosis: A Review of the Etiology and A report of New Cases." *American Journal of Physical Anthropology* 27 (1): 1–10.

Black, Robert E., Simon Cousens, Hope L. Johnson, Joy E. Lawn, Igor Rudan, Diego G. Bassani, Prabhat Jha, Harry Campbell, Christa Fischer Walker, and Richard Cibulskis. 2010. "Global, Regional, and National Causes of Child Mortality in 2008: A Systematic Analysis." *The Lancet* 375 (9730): 1969–1987.

Black, Sue, and Louise Scheuer. 1997. "The Ontogenetic Development of the Cervical Rib." *International Journal of Osteoarchaeology* 7 (1): 2–10.

Bots, Jessica, Liliane C.D. Wijnaendts, Sofie Delen, Stefan Van Dongen, Kristiina Heikinheimo, and Frieston Galis. 2011. "Analysis of Cervical Ribs in a Series of Human Fetuses." *Journal of Anatomy* 219 (3): 403–409.

Brickley, Megan, Helena Berry, and Gaynor Western. 2006. "The People: Physical Anthropology." In *St. Martin's Uncovered: Investigations in the Churchyard of St. Martin's-in-the-Bull-Ring, Birmingham, 2001*, ed. Megan Brickely, Simon Buteux, Josephine Adams, and Richard Cherrington, 90–151. Oxford: Oxbow Books.

Brothwell, Don Reginald, and R. Powers. 2000. "The Human Biology." In *Cannington Cemetery*, ed. Philip Rath, Sue Hirst, and Susan M. Wright, 135–161. London: English Heritage.

Caffey, John. 1939. "Syphilis of the Skeleton in Early Infancy." *American Journal of Roentgenology Radium Therapy* 42 (5): 637–655.

Caffey, John. 1978. *Pediatric X-Ray Diagnosis.* 7th ed. Chicago: Year Book Medical Publishers.

Caffey, John, and William A. Silverman. 1945. "Infantile Cortical Hyperostosis. Preliminary Report on a New Syndrome." *American Journal of Roentgenology and Radium Therapy* 54 (1): 1–16.

Capasso, L., M. Sciubba, Q. Hua, V. Levchenko, J. Viciano, R. D'Anastasio, and F. Bretuch, 2016. "Embryotomy in the 19th Century of Central Italy." *International Journal of Osteoarchaeology* 26 (2): 345–347

Cooper, Louis Z., Philip R. Ziring, Albert B. Ockerse, Barbara A. Fedun, Brian Kiely, and Saul Krugma. 1969. "Rubella: Clinical Manifestations and Mangement." *American Journal of Disease in Childhood* 118 (1): 18–29.

Cope, Darcy J. 2008. "Bent Bones: The Pathological Assessment of Two Fetal Skeletons from the Dakhleh Oasis, Egypt." MA thesis. Orlando: University of Central Florida.

Cope, Darcy J., and Tosha L. Dupras. 2011. "Osteogenesis Imperfecta in the Archaeological Record: An Example from the Dakhleh Oasis, Egypt." *International Journal of Paleopathology* 1 (2–3): 188–199.

Dedick, Andrew P., and John Caffey. 1953. "Roentgen Findings in the Skull and Chest in 1,030 Newborn Infants." *Radiology* 61 (1): 13–20.

de Silva, P., G. Evans-Jones, A. Wright, and R. Henderson. 2003. "Physiological Periostitis: A Potential Pitfall." *Archives of Diseases in Childhood* 88 (12): 1124–1125.

Dorfman, David .H., and Joy H. Glaser. 1990. "Congenital Syphilis Presenting in Infants after the Newborn Period." *New England Journal of Medicine* 323 (19): 1299–1302.

Dudar, J. Christopher. 2010. "Qualitative and Quantitative Diagnosis of Lethal Cranial Neural Tube Defects from the Fetal and Neonatal Human Skeleton, with a Case Study Involving Taphonomically Altered Remains." *Journal of Forensic Science* 55 (4): 877–883.

East, A.L., and J.E. Buikstra. 2001. "Is the Fetus from Elizabeth Mound 3, Lower Illinois River Valley an Achrondroplastic Dwarf?" *American Journal of Physical Anthropology* 114 (S32): 61.

Falys, Ceri G. 2010. "Human Bone." In *Land Adjacent to 135-7 Andover Road, Winchester: Post-Excavation Assessment,* ed. S. Hammond and S. Preston. Unpublished report.

Furtado, Larissa, Harshwardhan Thaker, Lance K. Erickson, Brian H. Shirts, and John M. Opitz. 2011. "Cervical Ribs Are More Prevalent in Stillborn Fetuses Than in Live-Born Infants and are Strongly Associated with Fetal Aneuploidy." *Pediatric and Developmental Pathology* 14 (6): 431–437.

Genç, Mehmet, and William J. Ledger. 2000. "Syphilis in Pregnancy." *Sexually Transmitted Infections* 76 (2): 73–79.

Gilmour, William N. 1962. "Acute Haematogenous Osteomyelitis." *Journal of Bone and Joint Surgery* 44-B (4): 841–853.

Gleser, Kurt. 1949. "Double Contour, Cupping and Spurring in Roentgenograms of Long Bones in Infants." *American Journal of Roentgenology and Radium Therapy* 61 (4): 482–492.

González Martin, A., F.J. Robles Rodriguez, M.C. García Martín, and M. Campo Martín. 1997. "Problematica IV: What Is This? A New Trait Found on Infant Skulls?" *Journal of Palaeopathology* 9 (1): 61–62.

Griffith, John Price Crozer 1919. *The Diseases of Infants and Children,* London: W.B. Saunders Co.

Harman, Nathaniel Bishop. 1917. *Staying the Plague.* London, Methuen & Co.

Hindman, B.W., S. Bell, T. Russo, and C.W. Zuppan. 1996. "Neonatal Osteofibrous Dysplasia: Report of Two Cases." *Pediatric Radiology* 26 (4): 303–306.

Hinkes, Madeleine J. 1983. "Skeletal Evidence of Stress in Subadults: Trying to Come of Age at Grasshopper Pueblo." PhD dissertation. Tuscon: University of Arizona.

Hollier, Lisa M., and Susan M. Cox. 1998. "Syphilis." *Seminars in Perinatology* 22 (4): 323–331.

Holt, P.G., and C.A. Jones. 2000. "The Development of the Immune System during Pregnancy and Early Life." *Allergy* 55 (8): 688–697.

Hook Ernest B., and Susan Harlap. 1979. "Differences in Maternal Age-Specific Rates of Down Syndrome between Jews of European Origin and of North African or Asian Origin." *Teratology* 20 (2): 243–248.

Hutchinson, Jonathan. 1857. "On the Influence of Hereditary Syphilis on the Teeth." *Transactions of the Odontological Society (Great Britain)* 2: 95–106.

Kausmally, Tania, and Rachel Ives. 2007. "Lytic Lesions in the Skull: Problems of Diagnosis in Infantile Human Remains." Poster presented at the 7th British Association of Biological Anthropology and Osteoarchaeology, University of Reading, 14 September.

Keeling, Jean W., Birgit Fischer Hansen, and Inger Kjær. 1997. "Pattern of Malformations in the Axial Skeleton in Human Trisomy 21 Fetuses." *American Journal of Medical Genetics* 68 (4): 466–471.

Keith, Arthur. 1913. "Abnormal Crania: Achondroplastic and Acrocephalic." *Journal of Anatomy and Physiology* 47 (2): 189–206.

Kjær, Inger, Jean W. Keeling, and Birgit Hansen. 1997. "Pattern of Malformations in the Axial Skeleton in Human Trisomy 13 Fetuses." *American Journal of Medical Genetics* 70 (4): 421–426.

Krakow, Deborah, Ralph S. Lachman, and David Rimoin. 2009. "Guidelines for the Prenatal Diagnosis of Fetal Dysplasias." *Genetic Medicine* 11 (2): 127–133.

Lampl, Michelle. 2003. "Fetus to Infant in Biomedical Perspective." *American Journal of Physical Anthropology* 120 (S36): 135.

Lawn, Joy E., Simon Cousens, and Jelka Zupan, for the Lancent Neonatal Surival Steering Team. 2005. "4 Million Neonatal Deaths: When? Where? Why?" *The Lancet* 365 (9462): 891–900.

Lewis, Mary E. 2004. "Endocranial Lesions in Non-Adult Skeletons: Understanding Their Aetiology." *International Journal of Osteoarchaeology* 14 (2): 82–97.

Lewis, Mary E. 2007. *The Bioarchaeology of Children: Perspectives from Biological and Forensic Anthropology.* Cambridge: Cambridge University Press.

Lewis, Mary E., and Rebecca Gowland. 2007. "Brief and Precarious Lives: Infant Mortality in Contrasting Sites from Medieval and Post-medieval England (AD 850–1859)." *American Journal of Physical Anthropology* 134 (1): 117–129.

Lorin, Martin I., Katharine H.K. Hsu, and Susan C. Jabob. 1983. "Treatment of Tuberculosis in Children." *Pediatric Clinics of North America* 30 (2): 333–348.

Lorrio, Alberto J. 2010. "Enterramientos infantiles en el oppidum en El Mólón (Camporrobles, Valencia)." *Cuadernos de Arqueología de la Universidad de Navarra* 18: 201–262

Malgosa, A., M. Aluja, and A. Isidro. 1996. "Pathological Evidence in Newborn Children from the Sixteenth Century in Huelva (Spain)." *International Journal of Osteoarchaeology* 6 (4): 388–396.

Mathews, Stevie L. 2008. "Diagnosing Anaencephaly in Archaeology: A Comparative Analysis of Nine Clinical Specimens from the Smithsonian Institution National Museum of Natural History, and One from the Archaeological Site of Kellis 2 Cemetery in Dakhleh Oasis, Egypt." MA thesis. Orlando: University of Central Florida.

Mays, S., K. Robson-Brown, S. Vincent, J. Eyers, H. King, and A. Roberts. 2012. "An Infant Femur Bearing Cut Marks from Roman Hambleden, England." *International Journal of Osteoarchaeology* 24 (1): 111–115.

McMillen, Marilyn M. 1979. "Differential Mortality by Sex in Fetal and Neonatal Deaths." *Science* 204 (4388): 89–91.

Melsom, R., M. Harboe, and M.E. Duncan. 1982. "IgA, IgM and IgG Anti-M. Leprae Antibodies in Babies of Leprosy Mothers during the First 2 Years of Life." *Clinical & Experimental Immunology* 49 (3): 532–542.

Mendonça de Souza, Sheila M.F., Karl J. Reinhard, and Andrea Lessa. 2008. "Cranial Deformation as the Cause of Death for a Child from the Chillion River Valley, Peru." *Chungara, Revista de Antropología Chilena* 40 (1): 41–53.

Mitchell, Piers D. 2006. "Child Health in the Crusader Period Inhabitants of Tel Jezreel, Israel." *Levant* 38 (1): 37–44.

Mollcson, Theya, and Margaret Cox. 1988. "A Neonate with Cut Bones from Poundbury Camp, 4th Century AD, England." *Bulletin de la Société royale Belge d'Anthropologie et de Préhistoire* 99 : 53–59.

Montagu, Ashley. 1950. "Constitutional and Prenatal Factors in Infant and Child Health." In *Symposium on the Healthy Personality: Transactions of Special Meetings of Conference on Infancy and Childhood June 8–9 and July 3–4, 1950,* ed. Milton J.E. Senn, 148–175. New York: Joshua Macy Jr. Foundation.

Montiel, Rafael, Eduvigis Solórzano, Nancy Díaz, Brenda A. Álvarez-Sandoval, Mercedes González-Ruiz, Mari Pau Cañadas, Nelson Simões, et al. 2012. "Neonate Human Remains: A Window of Opportunity to the Molecular Study of Ancient Syphilis." *PLOS ONE* 7 (5): e36371. doi:10.1371/journal.pone.0036371.

Mooney, Mark P., Michael I. Siegel, Kyle R. Kimes, John Todhunter, and Janine Janosky. 1992. "Multivariate Analysis of Second Trimester midfacial Morphology in Normal and Cleft Lip and Palate Human Fetal Specimens," *American Journal of Physical Anthropology* 88 (2): 203–209.
Moore, Keith L. 1988. *Essentials of Human Embryology.* Totonto: Decker.
Mora, S., C. Prinster, A. Bellini, G. Weber, M. Proverbio, M. Puzzovio, C. Bianchi, and G. Chiumello. 1997. "Bone Turnover in Neonates: Changes of Urinary Excretion Rate of Collagen Type I Cross-Linked Peptides during The First Days of Life and Influence of Gestational Age." *Bone* 20 (6): 563–566.
Müller, Fabiola, Ronan O'Rahilly, and D.R. Benson. 1986. "The Early Origins of Vertebral Anomalies, as Illustrated by a 'Butterfly Vertebra.'" *Journal of Anatomy* 149: 157–169.
Naeye, Richard L., and William Blanc. 1965. "Pathogenesis of Congenital Rubella." *JAMA* 194 (12): 1277–1283.
Ogden, John A. 1979. "Pediatric Osteomyelitis and Septic Arthritis: The Pathology of Neonatal Disease." *Yale Journal of Biology and Medicine* 52 (5): 423–448.
O'Rahilly, Ronan, Fabiola Müller, and D.B. Meyer. 1980. "The Human Vertebral Column at the End of the Embryonic Period Proper: 1. The Column as a Whole." *Journal of Anatomy* 131 (3): 565–575.
O'Rahilly, Ronan, Fabiola Müller, and D.B. Meyer. 1983. "The Human Vertebral Column at the End of the Embryonic Period Proper: 2. The Occipitocervical Region," *Journal of Anatomy* 136 (1): 181–195.
Otaño, L., H. Aiello, L. Igarzábal, T. Matayoshi, and E. Gadow. 2002. "Association between First Trimester Absence of Fetal Nasal Bone on Ultrasound and Down Syndrome." *Prenatal Diagnosis* 22 (10): 930–932.
Pálfi, György, Olivier Dutour, Marc Borreani, Jean-Pierre Brun, and Jacques Berato. 1992. "Pre-Columbian Congenital Syphilis from the Late Antiquity in France." *International Journal of Osteoarchaeology* 2 (3): 245–261.
Rauch, F., and E. Schoenau. 2002. "Skeletal Development in Premature Infants: A Review of Bone Physiology beyond Nutritional Aspects." *Archives of Disease in Childhood, Fetal and Neonatal Edition* 86 (2): 82–85.
Reed, George B. Jr. 1969. "Rubella Bone Lesions." *Journal of Pediatrics* 74 (2): 208–213.
Rudolf, Arnold, Edward Singleton, Harvey S. Rosenberg, Don B. Singer, and C. Alan Phillips. 1965. "Osseous Manifestations of the Congenital Rubella Syndrome." *American Journal of Disease in Childhood* 110 (4): 428–433.
Saint-Hillaire, Geoffrey. 1826. "Note about an Egyptian Monster Found in the Ruins of Thebes, in Egypt, by M. Passalascqua." *Bulletin des Sciences Medicales* 7: 105–108.
Scheuer, Louise, and Sue Black. 2000. *Developmental Juvenile Osteology.* London: Academic Press.
Sekeles, E., and A. Ornoy. 1975. "Osseous Manifestations of Gestational Rubella in Young Human Fetuses." *American Journal of Obstetrics & Gynecology* 122 (3): 307–312.

Shopfner, Charles E. 1966. "Periosteal Bone Growth in Normal Infants: A Preliminary Report." *American Journal of Roentgenology* 97 (1): 154–163.

Sjøvold, T., I. Swedborg, and L. Diener. 1974. "A Pregnant Woman from the Middle Ages with Exostosis Multiplex." *Ossa* 1: 3–23.

Soren, David, Todd Fenton, and Walter Birkby. 1995. "The Late Roman Infant Cemetery Near Lugnano in Teverina, Italy: Some Implications." *Journal of Paleopathology* 7 (1): 13–42.

Stempfle, Noelle, Yolene Huten, Catherine Fredouille, Herve Brisse, and C. Nessmann. 1999. "Skeletal Abnormailites in Fetuses with Down's Syndrome: A Radiographic Postmortem Study." *Pediatric Radiology* 29 (9): 682–688.

Steyn, M., J. H. Meiring, W. Nienaber, and M. Loots. 2002. "Large Fontanelles in an Early 20th Century Rural Population from South Africa." *International Journal of Osteoarchaeology* 12 (4): 291–296.

Tomasto-Cagigao, Elsa. 2011. "A Holoprosencephaly (Cyclopia) Case from the Nasca Culture, Peru." Poster presented at the 4th Palaeopathology Association Meeting in South America, Lima, Peru, 2–5 November.

Trueta, J. 1959. "The Three Types of Haematogenous Osteomyelitis." *Journal of Bone and Joint Surgery* 41-B (4): 671–680.

Waldron, Tony. 2009. *Palaeopathology.* Cambridge: Cambridge University Press.

Weissberg, Eleanor D., Arnold L. Smith, and David H. Smith. 1974. "Clinical Features of Neonatal Osteomyelitis." *Pediatrics* 53 (4): 505–510.

Zeichner, Steven L., and Stanley A. Plotkin. 1988. "Mechanisms and Pathways of Congenital Infections." *Clinics in Perinatology* 15 (2): 163–188.

Chapter 6

THE NEOLITHIC INFANT CEMETERY AT GEBEL RAMLAH IN EGYPT'S WESTERN DESERT

Jacek Kabaciński, Agnieszka Czekaj-Zastawny, and Joel D. Irish

In 2001 and 2003, the first Neolithic cemeteries in the Egyptian part of the Western Desert were discovered and excavated by members of the Combined Prehistoric Expedition (CPE). These sites were located near Gebel Ramlah Playa,[1] a paleolake some 150 kilometers west of the Nile Valley in far southern Egypt. In 2009, a new project concentrated exclusively on the recognition and recovery of Neolithic burial grounds. The unique result of this work was the discovery of a cemetery dated 4500 to 4300 BC[2] specifically for the burial of infants—the earliest infant cemetery yet known in this region. This chapter discusses infant burial practices during the Neolithic in Northern Africa and the unique mortuary behaviors devoted strictly to the newborns in this region.

In archaeological interpretations, one of the most important determinations is whether the remains (e.g., fetus) were intentionally put into a grave or were just discarded. In most cases, we can establish this relatively easily through the course of excavation. However, in some cases—for instance, when a skeleton is found in a refuse pit among waste—we may assume it is not a grave but a remnant of a disposed body. When fetal remains are deposited in a grave, a number of basic features are analyzed, including grave construction, body position and orientation of the skeleton, presence of grave goods and their pattern of distribution, specific ceremonial behaviors (e.g., powdering the body with colorant), and spatial relation to other burials and settlements. Comparison of these mortuary traditions across age groups within a population help to identify differences that may be associated with fetal status and shifting ideologies (Scott and Betsinger, chapter 7).

Research Background

Societies that used burial grounds in the Gebel Ramlah region are considered Neolithic. The term "Neolithic" was introduced in the nineteenth century by Sir John Lubbock (1865) and generally refers to societies with a farming economy (land cultivating and husbandry) in contrast to hunters-gatherers with subsistence based on wild resources. Neolithic societies appeared more or less simultaneously in various parts of the world beginning in the Middle East, around 9000 to 8500 BC and spread into Europe and Northern Africa (Chapman 1994; Gronenborn 2003; Zvelebil 2001). In the Western Desert, the Neolithic period is prominent between 9300 and 3600 BC and is also the case of Gebel Ramlah (Czekaj-Zastawny et al. 2017). The basis of their economy was pastoralism with supplemental foraging of plants (including sorghum) and hunting (Wendorf and Schild 2001a). These groups lived in small settlements located on edges of lakes and buried their deceased nearby.

From the early 1970s, this region of the Sahara Desert was a primary research focus of the CPE, originally directed by Fred Wendorf and later by Roman Schild (Schild and Wendorf 2002). For the first three decades, research activity was centered on the paleolake of Gebel Nabta called Nabta Playa and since 2001 has also focused on Gebel Ramlah. Both areas, separated by about twenty kilometers, provide completely different evidence and, based on archaeological data, point to a striking difference in mortuary practices. Around Gebel Nabta, there is evidence of continuous human occupation starting with the early Holocene El Adam phase dated around 9300 to 8050 BC (Jórdeczka et al. 2011). Subsequent settlement stages confirm the presence of pastoral peoples here to the end of the Neolithic (Wendorf and Schild 2001b; Schild and Wendorf 2013). What is striking in comparison with the neighboring Gebel Ramlah area is that only a few human burials were recorded in the Nabta Playa region. Until around mid of the 6th millennium BC, only single burials were used where the deceased were placed within pits usually in a contracted position, and the burials were all located near or within settlements. In the Late Neolithic, after circa 5500 BC, another form of a grave appeared known as the tumuli, in which the body was placed in a pit under an earthen cover surrounded by circles of stone (Bobrowski et al. 2014; Schild and Wendorf 2012).

Similar burial patterns have been identified at Gebel Ramlah during the Early and Middle Neolithic (ca. 9300–5550 BC). However, by the Late Neolithic, people in Gebel Ramlah were buried in single graves (occasionally two or three graves located in close proximity to one another) around associated settlements. The pattern had changed remarkably at the beginning of the Final Neolithic (ca. 4600–3600 BC) when the very first cemeteries appeared. From this period on, the southwestern edges of the Gebel Ramlah Playa became an extensive burial ground where people were interred in specific areas devoted to deceased members of the society (Czekaj-Zastawny and Kabaciński 2015; Kobusiewicz et al. 2010).

Until now, no known infant burials were dated before the Final Neolithic in the Western Desert. Inhumations of very young children appear only in Fnal Neolithic cemeteries at Gebel Ramlah, but there is archaeological evidence from other areas, varying in time and space. In Austria, there was a discovery of two Upper Paleolithic burials (ca. 26,000 years BC with three neonates, including twins that died around the time of birth and a three-month-old child (Einwögerer et al. 2006). A much more recent case from the large Siberian Mesolithic cemetery site of Lokomotiv by Irkutsk (Lake Baikal area) dates to 6000 to 5000 BC. In one of the graves, a skeleton of a twenty- to twenty-five-year-old woman was found with two perinatal skeletons (twins) by her pelvis (Lieverse et al. 2015). Geographically closer, from the Late Neolithic cemetery R12, located in the Northern Dongola area (Northern Sudan), there is evidence of neonatal burial treatment similar to that recorded at Gebel Ramlah. Here, perinates (four individuals) and neonates (three individuals) were deposited into richly furnished family graves (Salvatori and Usai 2008). However, before the Gebel Ramlah discovery, there were no other cases of neonate cemeteries known from the Neolithic or older periods. The first case of a separate cemetery for children postdates Gebel Ramlah by approximately a thousand years, dating to 3020 to 2880 BC. This child cemetery at Elkab in the Nile Valley was composed of three circular structures containing forty-one burials. Most graves contained the remains of children, including three neonates (Hendrickx et al. 2002). Some adults and older children were found in a contracted position on their sides, while smaller children were buried in jars. Another, much younger example of an infant cemetery comes from the northern Mediterranean Sea coast, at Cartagina, dated between the eighth and second century BC. At the Carthage site, a cemetery for adults and older children and another for very young children were investigated. The latter is located at the periphery of a settlement and was initially interpreted as a "Tophet"—a place for sacrifices. Most burials were of perinates up to five or six months old. According to Jeffrey Schwartz and colleagues (2010), it was not a place to sacrifice children but primarily a cemetery for children who died before or shortly after birth.

From ethnographic data, it is known that the treatment of children in African societies differs radically (Gottlieb 2004a, 2004b; Pawlik 2004). In many contemporary African societies, the social recognition and status of a child increases with its physiological development (Erny 1968). For example, newborn children might not be considered group members, having no name until the age of two years, after weaning. Such a perspective strongly influences their mortuary practices. (For further discussion, see Scott and Betsinger, chapter 7.) Newborns who die are buried quickly and without specific ceremony. Among the Anyi (Ivory Coast), an old woman who was present at the birth immediately wraps a lifeless newborn in dry banana leaves and places it in a shallow pit far away from the village. If the birth was at home, the deceased newborn is discarded in between kitchen middens, and the woman (mother) is considered to have never been pregnant with that child (Eschliman 1985). In the Basari and the Konkombo (northern Togo), old women bury deceased newborns in shallow graves in remote places (Pawlik 1990). Among the Ashanti (southern and central Ghana), the deceased newborns are beaten and maimed before they are buried. In other cases, infants who died up to several days after birth are buried in a shallow pit at a crossroads where waste is discarded (Froelich 1954; Zimoń 1994). All of these practices are rooted in the spiritual life of the society that concentrates on living members of the group rather than the deceased child. The intention of these practices and their deprivation of respect to the dead is to protect parents of the deceased and other children in the village against the bad spirits of newborns (Pawlik 1990).

In contrast to this perspective, children of the Beng tribe (Ivory Coast) are considered fully developed and are thought to be complex persons who are reincarnations of an ancestor. As such, newborns (and generally small children) have rich social lives like adult members of the society, while simultaneously having complete care and attention. In the Beng world, children have a very important social position (Gottlieb 2004a, 2004b). The Beng do not organize funerals for deceased newborns; they simply consider them a person who was not completely ready to come back to the world of the living (Bielo 2015).

Materials and Methods

The Gebel Ramlah area is located approximately 150 kilometers west of Abu Simbel. The most characteristic landscape feature is a pronounced rocky massif called Gebel Ramlah with remains of a paleolake (Playa) adjacent to the mountain on its southern side. Gebel Ramlah lies within the Egyptian portion of the Western (or Libyan) Desert; the latter covers close to three million square kilometers and extends from the western edge of the Nile Valley on the east to the Libyan border on the west, and from the Mediterranean Sea to the north to the Sudanese border to south (Issawi et al. 1999). In 2001 and 2003, Michał Kobusiewicz, Jacek Kabaciński, and Joel Irish excavated a complex of three final Neolithic cemeteries (Kobusiewicz et al. 2004, 2010; Schild et al. 2002, 2005). In 2009, Agnieszka Czekaj-Zastawny and Jacek Kabaciński (2015) discovered additional Neolithic cemeteries (Czekaj-Zastawny et al. 2017), including the first that contain inhumations of infants. This finding led to a full investigation of the site beginning the following year, as part of a new CPE project that concentrated specifically on burial practices at Gebel Ramlah. To date, six different cemeteries have been recorded, including the infant cemetery, and detailed analyses of burial customs and grave goods suggest cemeteries may be related to different populations visiting the area (Kobusiewicz et al. 2010). Located on the crest of a low-lying hill on the paleolake's south-southwest shore, the infant cemetery appears to have been intentionally set apart from a larger cemetery for adults.

The infant cemetery covered an oval area of six by eight meters (fig. 6.1). Within that spatially limited area, thirty-five burials were recorded, which contained the skeletal remains of forty-two individuals. Grave pits, well discerned in most cases, were found in three different stratigraphic layers (i.e., within the topmost sands, partially within the underlying lake silt, and completely within the lake silt). However, radiocarbon dates precluded any chronological meaning to the pits' vertical sequence. The size of burial pits, usually

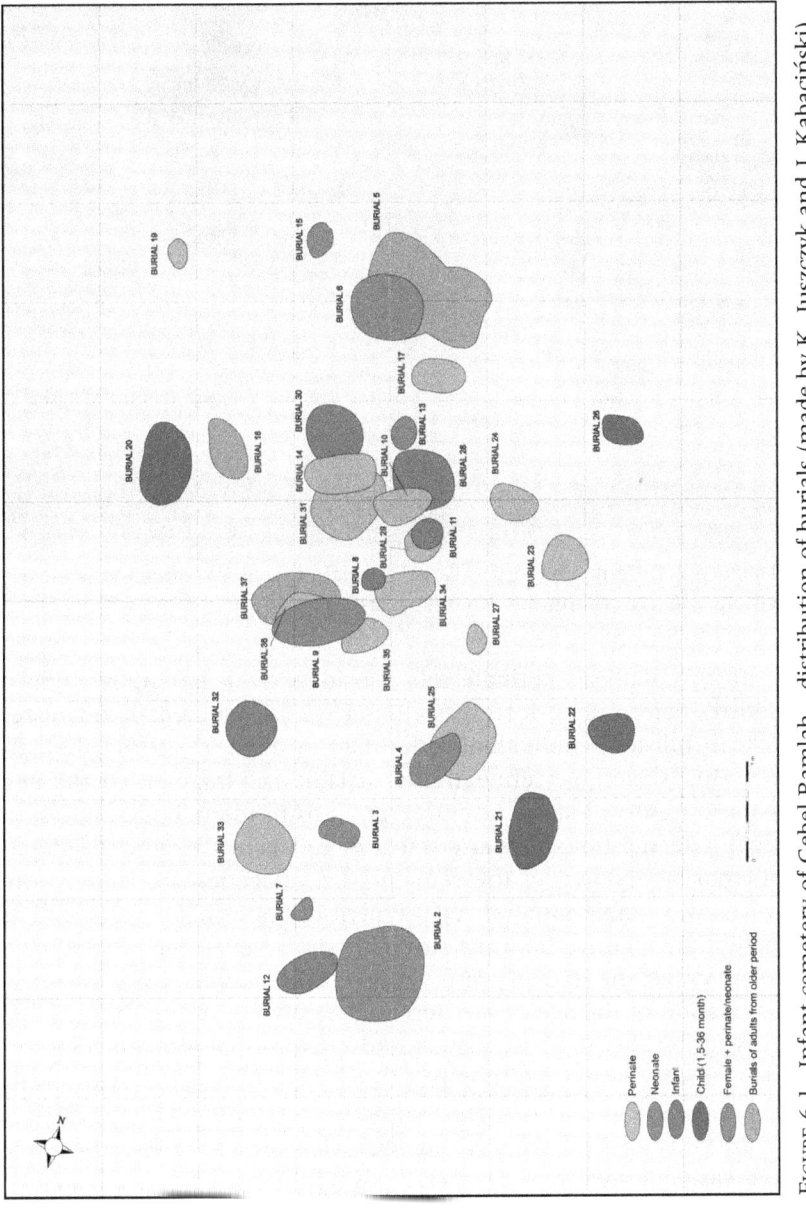

FIGURE 6.1. Infant cemetery of Gebel Ramlah—distribution of burials (made by K. Juszczyk and J. Kabaciński).

appearing as elongated ovals in shape, varied from a maximum of 0.3 to 1 meters along the long axis (fig. 6.2). Because of soil deflation, most pits were quite shallow (i.e., fifteen to twenty centimeters), although in a few cases the depth was measured at thirty to thirty-five centimeters.

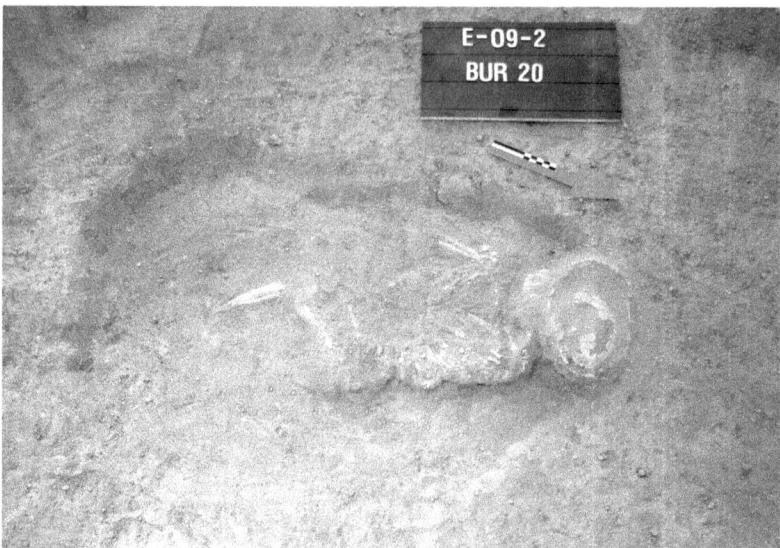

FIGURE 6.2. Infant cemetery of Gebel Ramlah. Burial 20—skeleton in clearly visible grave pit (photo by A. Czekaj-Zastawny)

In this analysis, perinate refers to individuals having been born shortly before or at forty gestational weeks, whereas neonate refers to full-term individuals who died at birth or shortly thereafter. Most skeletons were extremely friable and fragmented and in a poorly preserved state. Chemical analysis of many bone and tooth fragments did not show any traces of collagen. Extreme postdepositional breakdown of the remains (i.e., diagenesis) resulted from a lack of suitable mineral replacement of bone collagen, as well as potential exposure to direct sun and wind. Nevertheless, in almost every grave, the state of preservation was at least adequate to permit estimation of the age of the deceased. It was fortunate that the site was found, because intensive wind erosion had already removed the uppermost parts of the grave pits, and, in several more years, the cemetery and its contents would have essentially disappeared.

In every case where it was detectable, the body within the grave pit was deposited in a contracted position, mostly on the right side. No rules apparently applied to orientation of the body, as they were found in all possible directions. Very few grave goods were found accompanying the skeletal remains. The only item, found in almost every grave, was red ochre in the form of small nodules. In a few graves, shells of mollusks or snails imported from the Nile valley or the Red Sea were present. A unique finding was a bracelet made of hippopotamus (*Hippopotamus amphibius*) tusk put on the right arm of a perinate in grave number 3.

Results

A series of AMS datings (accelerator mass spectrometry is used to measure the concentration of the isotope carbon-14 to determine the age of archaeological samples) from the human remains recovered have a maximal calibrated age range from 4700 to 4350 BC, though most graves are dated between 4500 and 4300 BC. As such, they are related to the Final Neolithic stage of the Nabta Playa Neolithic as defined lately by Schild and Wendorf (2013).

According to site stratigraphy, three of thirty-five burials do not belong to infant cemetery (fig. 6.1). Remaining thirty-two burials contained thirty-nine inhumations. Thirty-six of the total thirty-nine individuals (92.3 percent) comprise the skeletal remains of infants or young children. These results are based on *in situ* measurements of the diagnostic skeletal elements (AlQahtani 2009; Baker et al. 2005; Bass 1995; Brothwell 1981; Buikstra and Ubelaker 1994; Gindhart 1973; Fazekas and Kósa 1978; Holcomb and Konigsberg 1995; Irish 2001; Jeanty 1983; Liversidge and Molleson 2004; Maresh 1970; Molleson and Cox 1993; Schaefer et al. 2009; Scheuer et al. 1980; Scheuer and MacLaughlin-Black 1994; Schutkowski 1993; Sherwood et al. 2000; Ubelaker 1987, 1989; Weaver 1980). In all cases, their young ages at death prevent the assignment of sex. Thirty of these died at the low end of this age range; their state of preservation is mostly poor to very poor so, depending on completeness, age estimates could be classified as near or full-term (perinate) or full-term (neonate), though in some cases they could only be classified as "infant?". With regard to the skeletons themselves, given the poor preservation, an individual skeletal element inventory was supplemented by simple estimates of overall completeness; these values range from 1 percent to a maximum 50 percent. In several cases, there was evidence of burning, based on the charred appearance of skeletal elements (burials 3, 5, 7, 33/I, and II).

In four cases, two children were buried in the same pit (burials 10, 25, 31, and 33). That last one may contain the remains of twins, based on their deposition and similar size. Three adult burials are also present in this cemetery and are specifically unique, as all are likely female and have been buried in association with infants (burial 2, 6, and 9). While it is difficult to distinguish their relationship to the infants buried with them, these women may be biological mothers or symbolic mothers to these children. Burial 2, an older adult female, had a perinate buried in her pelvic area, whereas burial 6 had neonate remains buried over her chest cavity.

Little additional information is apparent based on osteological evidence. No evidence of hard tissue pathology is present, either because of non-bone-affecting illness or because of the poor preservation. The fragmentary bone also prevents other basic observations, including standard osteological measurements; however, despite these limitations, some additional data was able to be collected.

Discussion and Conclusions

Research on fetuses in the field of Stone Age archaeology is limited because of very scarce finds of the kind that can be studied and the difficulty of those studies due to the fragile nature of such young skeletal remains. This is especially apparent in the case of the Sahara, where extreme climatic conditions are very destructive, which radically narrows the analytical possibilities. Bioarchaeological analyses and biophysiochemical analyses may be used to determine health conditions, diseases and causes of death, age-at-death, genetic sex of a child, degree of relatedness within the studied group and population, and migration patterns. Additionally, archaeological observations may be used to reconstruct social relations between adults and fetuses and/or fetuses and society.

Although it might not seem, at the moment, that research on fetuses found in the prehistoric context has any direct influence on present practice and policy, the findings discussed here remind us that that the social significance ascribed to small children are not necessarily a modern invention but are evidenced also in prehistoric times. It also demonstrates cultural differences between past societies, which may make it easier to understand observed differences in attitude toward fetuses in today's societies.

During archeological research on the Gebel Ramlah infant cemetery, substantial information on mortuary behavior was collected. As a rule, a newborn (neonate?) was buried in a specially prepared grave pit, usually very small, closely surrounding the body. Sometimes pit digging required a lot of effort as it reached very hard, silty sediments of the paleolake. The body of the deceased, in every case where preservation of a skeleton was sufficient for such observation, was always in a contracted position on the left or right side. Heads and bodies were orientated in different directions. Grave offerings were very rare, sometimes including a single shell from Red Sea and, in one case, an ivory bracelet. Strikingly, however, in every grave, small nodules of red ochre were found.

Research was also conducted on a neighboring cemetery for adults. There are no differences in mortuary practices observed when we compare both cemeteries: children (including fetuses/perinates/neonates) and adults were buried in the same way, at least based on archaeological data. From the presence of a separate cemetery for newborns (perinates and neonates), it may be suggested that newborns played a significant role in this society with this individual holding special status. Perhaps a separate cemetery expressed special care after their deaths. It is also possible that in a few cases, women (mothers?) were buried together with infants, which could be interpreted as symbolic mothers, whether they were related to the infants of not, taking care of children even after their deaths.

In sum, within the social structure of many indigenous African populations, for the most part in prehistory, perinates and neonates often were not seen as full members of society and are thus treated differently in death. However, that is not the case for the Gebel Ramlah society. Unusual care and respect for dead infants, rarely recorded in prehistory, leads us to assume that children in this population were a very important part of society.

Acknowledgments

The research was sponsored by the Polish National Science Centre (grant no. NCN 2012/05/B/HS3/03928) and by the Combined Prehistoric Expedition Foundation. The authors would like to thank both institutions for that support.

Jacek Kabaciński is a professor of archaeology in the Institute of Archeology and Ethnology, Polish Academy of Sciences. His research concentrates on hunter-gatherers of the North European Plain and pastoralists of northeast Africa. He is the author of more than 150 publications, including 13 books and numerous peer-reviewed articles.

Agnieszka Czekaj-Zastawny is a professor of archaeology at the Institute of Archaeology and Ethnology, Polish Academy of Sciences. She conducts research on the Neolithic (settlements, funerary rituals, populations, material culture) of Europe and Northeastern Africa. She is the author and coauthor of numerous articles and books on the Neolithic of north-central Europe and northeast Africa.

Joel D. Irish is a professor of bioarchaeology in the Research Centre in Evolutionary Anthropology and Palaeoecology, Liverpool John Moores University. He conducts research on topics ranging from hominin origins in Africa to the first Americans, resulting in six co-authored and coedited books and more than seventy peer-reviewed articles.

Notes

1. In the text, we refer to terms like *gebel* and *playa*. *Gebel* means mountain in Arabic. *Playa* is a Spanish term for temporary lake. Names "Gebel Ramlah" and "Gebel Nabta" were initially introduced by archaeologists for two mountains distinct in the landscape and now are used as geographic names. Gebel Ramlah Playa or Gebel Nabta Playa are synonyms of paleolakes located by those mountains.
2. All dates mentioned in this text are given in calibrated years BC.

References

AlQahtani, Sakher Jaber. 2009. "Atlas of Human Tooth Development and Eruption." Queen Mary and Westfield College. www.atlas.dentistry.qmul.ac.uk.

Baker, Brenda J., Tosha L. Dupras, and Matthew Tocheri. 2005. *The Osteology of Infants and Children*. College Station: Texas A&M University Press.

Bass, William M. 1995. *Human Osteology: A Laboratory and Field Manual*. Columbia: Missouri Archaeological Society.

Bielo, James S. 2015. *Anthropology of Religion: The Basic*. New York. Routledge.

Bobrowski, Przemek, Agnieszka Czekaj-Zastawny, and Romuald Schild. 2014. "Gebel El Muqaddas (Site E-06-4): The Early Neolithic Tumuli from Nabta Playa (Western Desert, Egypt)." In *The Fourth Cataract and Beyond: Proceedings of the 12th International Conference for Nubian Studies*, ed. Julie R. Anderson and Derek A. Welsby, 293–301. Peeters: Leuven-Paris-Walpole.

Brothwell, Don R. 1981. *Digging Up Bones*. Ithaca, NY: Cornell University Press.

Buikstra, Jane E., and Douglas H. Ubelaker. 1994. *Standards for Data Collection from Human Skeletal Remains*. Arkansas Archaeological Survey Report No. 44. Fayetteville: Arkansas Archeological Survey.

Chapman, John. 1994. "The Origins of Farming in South East Europe." *Préhistoire Européenne* 6: 133–156.

Czekaj-Zastawny, Agnieszka, and Jacek Kabaciński. 2015. "New Final Neolithic Cemetery E-09-4, Gebel Ramlah Playa, Western Desert of Egypt." In *Hunter-Gatherers and Early Food Producing Societies in the Northeastern Africa*, ed. Jacek Kabaciński, Marek Chłodnicki, and Michał Kobusiewicz, 375–384. Poznań: Poznań Archaeological Museum.

Czekaj-Zastawny, Agnieszka, Joel D. Irish, Jacek Kabaciński, and Jakub Mugaj. 2017. "The Neolithic Settlements by a Paleolake of Gebel Ramlah, Western Desert of Egypt." In *Desert and the Nile: Prehistory of the Nile Basin and the Sahara—Papers in Honor of Fred Wendorf*, ed. Jacek Kabaciński, Marek Chłodnicki, Michał Kobusiewicz, and Małgorzata Winiarska-Kabacińska. Poznań: Poznań Archaeological Museum.

Einwögerer, Thomas, Herwig Friesinger, Marc Handel, Christine Neugebauer-Maresch, Ulrich Simon, and Maria Teschler-Nicola. 2006. "Upper Palaeolithic Infant Burials." *Nature* 444 (285): 285.

Erny, Pierre. 1968. *L'Enfant dans la pensée traditionnelle de l'Afrique Noire* (The child in the traditional thought of Black Africa) Paris: Editions le Livre Africain.

Eschliman, Jean-Paul. 1985. *Les Agni devant la mort (Côte d'Ivoire)* (The Agni before death (Ivory Coast). Paris: Karthala.

Fazekas, István Gyula, and F. Kósa. 1978. *Forensic Fetal Osteology*. Budapest: Akademiai Kiado.

Froelich, Jean Claude. 1954. *La tribu konkomba du Nord Togo* (The Konkomba tribe of North Togo). Dakar: Cahors.

Gindhart, Patricia S. 1973. "Growth Standards for the Tibia and Radius in Children Aged One Month through Eighteen Years." *American Journal of Physical Anthropology* 39 (1): 41–48.

Gottlieb, Alma. 2004a. "Babies as Ancestors, Babies as Spirits: The Culture on Infancy in West Africa." *Expedition: The Magazine of the University of Pennsylvania* 46 (3): 13–21.

Gottlieb, Alma. 2004b. *The Afterlife Is Where We Come From: The Culture of Infancy in West Africa*. Chicago: University of Chicago Press.

Gronenborn, Detlef. 2003. "Migration, Acculturation and Cultural Change in Western Temperate Europe, 6500–5000 cal BC." *Documenta Praehistorica* 30: 79–91.

Hendrickx, Stan, Dirk Huyge, and Eugène Warmenbol. 2002. "Un cimetière particulier de la deuxième dynastie à Elkab (A special cemetery of the Second Dynasty in Elkab)" *Archèo-Nil* 12: 47–54.

Holcomb, Susan M.C., and Lyle Konigsberg. 1995. "Statistical Study of Sexual Dimorphism in the Human Fetal Sciatic Notch." *American Journal of Physical Anthropology* 97 (2): 113–125.

Irish, Joel D. 2001. "Human Skeletal Remains from Three Nabta Playa Sites." In Wendorf and Schild 2001a, 521–528.

Issawii, Bahay, Mohamed El Hinnawi, Maher Francis, and Ali Mazhar. 1999. *The Phanerozoic Geology of Egypt: A Geodynamic Approach*. Special Publication No. 76. Cairo: The Egyptian Geological Survey.

Jeanty, Philippe M. D. 1983. "Fetal Limb Biometry." *Radiology* 147 (2): 601–602.

Jórdeczka, Maciej, Halina Królik, Mirosław Masojć, and Romuald Schild. 2011. "Early Holocene pottery in the Western Desert of Egypt: new data from Nabta Playa." *Antiquity* 85 (327): 99–115.

Kobusiewicz, Michał, Jacek Kabaciński, Romuald Schild, Joel D. Irish, and Fred Wendorf. 2004. "Discovery of the First Neolithic Cemetery in Egypt's Western Desert." *Antiquity* 78 (301): 566–579.

Kobusiewicz, Michał, Jacek Kabaciński, Romuald Schild, Joel D. Irish, Maria Gatto, and Fred Wendorf. 2010. *Gebel Ramlah: Final Neolithic Cemeteries from the Western Desert of Egypt*. Poznań: Polish Academy of Sciences.

Lieverse, Angela R., Vladimir I. Bazaliiskii, and Andrzej W. Weber. 2015. "Death by Twins: A Remarkable Case of Dystocic Childbirth in Early Neolithic Siberia." *Antiquity* 89 (343): 23–38.

Liversidge, Hellen M., and T. Molleson. 2004. "Variation in Crown and Root Formation and Eruption of Human Deciduous Teeth." *American Journal of Physical Anthropology* 123 (2): 172–180.

Maresh, Marion M. 1970. "Measurements from Roentgenograms." In *Human Growth and Development*, ed. Robert W. McCammon, 157–200. Springfield, IL: C.C. Thomas.

Molleson, Theya, and Margaret Cox. 1993. *The Spitalfields Project, Volume 2: The Anthropology—The Middling Sort*. CBA Research Report 86. London: Council for British Archaeology.

Lubbock, John. 1865. *Prehistoric Times, as Illustrated by Ancient Remains, the Manners and Customs of Modern Savages*. London: Williams & Norgate.

Pawlik, Jacek J. 1990. *Experience sociale de la mort: Etude des rites funeraires des Bassar du Nord-Togo* (Social experience of death: Study of the funerary rites of the Bassar of north Togo). Fribourg: Editions Universitaires Fribourg Suisse.

Pawlik, Jacek J. 2004. "Śmierć dziecka w Afryce." In *Dusza maluczka a strata ogromna* (A little soul and the huge loss): *Funeralia Lednickie, 6*, ed. Wojciech Dzieduszycki and Jacek Wrzesiński, 35–75. Poznań: Stowarzyszenie Naukowe Archaeologów Polskich.

Salvatori, Sandro, and Donatella Usai, eds. 2008. *A Neolithic Cemetery in the Northern Dongola Reach: Excavations at Site R12*. BAR International Series 1814 and Sudan Archaeological Research Society 16. Oxford: Sudan Archaeological Research Society.

Schaefer, Maureen, Sue Black, and Louise Scheuer. 2009. *Juvenile Osteology: A Laboratory and Field Manual*. San Diego: Academic Press.

Scheuer, J.L., J.H. Musgrave, and S.P. Evans. 1980. "The Estimation of Late Fetal and Perinatal Age from Limb Bone Length by Linear and Logarithmic Regression." *Annals of Human Biology* 7 (3): 257–265.

Scheuer, Louise, and Sue MacLaughlin-Black. 1994. "Age Estimation from the Pars Basilaris of the Fetal and Juvenile Occipital Bone." *International Journal of Osteoarchaeology* 4 (3): 377–380.

Schild, Romuald, Michał Kobusiewicz, Fred Wendorf, Joel D. Irish, Jacek Kabaciński, and Halina Królik. 2002. "Gebel Ramlah Playa (Egypt)." In *Tides of the Desert: Contributions to the Archaeology and Environmental History of Africa in Honor of Rudolf Kuper,* Africa Praehistorica 14, ed. Jennerstrasse 8, 117–124. Cologne: Heinrich Barth Institut.

Schild, Romuald, Michał Kobusiewicz, Fred Wendorf, Joel D. Irish, Jacek Kabaciński, Halina Królik, and Gilberto Calderoni. 2005. "A New Important Area of Neolithic Occupation in the Southwestern Desert of Egypt." In *Hunters vs. Pastoralists in the Sahara: Material Culture and Symbolic Aspects,* BAR International Series 1338, ed. Barbara E. Barich, Thierry Tillet, and Karl Heinz Striedter, 51–56. Oxford: Archaeopress.

Schild, Romuald, and Fred Wendorf. 2002. "Forty Years of the Combined Prehistoric Expedition." *Archaeologia Polona* 40: 5–22.

———. 2012. "The New Age Reuse of Nabta Playa's Neolithic Sanctuary." In *Prehistory of Northeastern Africa,* Studies in African Archaeology 10, ed. Jacek Kabaciński, Marek Chłodnicki, and Michał Kobusiewicz, 421–439. Poznań: Poznań Archaeological Museum and Institute of Archaeology and Ethnology, Polish Academy of Sciences.

———. 2013. "Early and Middle Holocene Paleoclimates in the South Western Desert of Egypt: The World before Unification." *Studia Quaternaria* 30 (2): 125–133.

Schutkowski, Holger. 1993. "Sex Determination of Infant and Juvenile Skeletons: I. Morphognostic Features." *American Journal of Physical Anthropology* 90 (2): 199–205.

Schwartz, Jeffery H., Frank Houghton, Roberto Macchiarelli, and Luca Bondioli. 2010. "Skeletal Remains from Punic Carthage Do Not Support Systematic Sacrifice of Infants." *PLOS ONE* 5 (2): e9177. doi:10.1371/journal.pone.0009177.

Sherwood, Richard J., R. S. Meindl, H. B. Robinson, and R. L. May. 2000. "Fetal Age: Methods of Estimation and Effects of Pathology." *American Journal of Physical Anthropology* 113 (3): 305–315.

Ubelaker, Douglas H. 1987. "Estimating Age at Death from Immature Human Skeletons: An Overview." *Journal of Forensic Sciences* 32 (5): 1254–1263.

———. 1989. *Human Skeletal Remains: Excavation, Analysis, Interpretation.* 2nd ed. Washington, DC: Taraxacum.

Weaver, David S. 1980. "Sex Differences in the Ilia of a Known Sex and Age Sample of Fetal and Infant Skeletons." *American Journal of Physical Anthropology* 52 (2): 191–195.

Wendorf, Fred, and Romuald Schild, eds. 2001a. *Holocene Settlement of the Egyptian Sahara, Volume 1: The Archaeology of Nabta Playa.* New York: Kluwer Academic/Plenum Publishers.

Wendorf, Fred, and Romuald Schild. 2001b. "Conclusions." In Wendorf and Schild 2001a: 648–675.

Zimoń, Henryk 1994. "Dziecko w kontekście kultury afrykańskiej (Child in the context of African culture)." *Collectanea Theologica* 64 (4): 69–80.

Zvelebil, Marek. 2001. "The Agricultural Transition and the Origins of Neolithic Society in Europe." *Documenta Praehistorica* 28: 1–26.

Chapter 7

EXCAVATING IDENTITY
BURIAL CONTEXT AND FETAL IDENTITY
IN POSTMEDIEVAL POLAND

Amy B. Scott and Tracy K. Betsinger

As gender archaeology has moved the study of children and childhood to the forefront of bioarchaeological investigation (e.g., Kamp 2001; Lillehammer 1989; see review in Baxter 2005), the scarcity of research on fetal and neonatal remains, and their place and role in society has been highlighted. Whether attributed to poor preservation or purposeful exclusion from communal cemeteries, less is known and understood about the social treatment of perinates (twenty-eight weeks in utero to seven postnatal days), including stillborn infants. The goal of this chapter is to examine and situate fetal identity—that is, the social significance a particular culture group might have ascribed to fetuses—by examining and comparing the funerary treatment of perinates with those of older infants and young children (within the context of typical adult mortuary treatment). Examining patterns in funerary treatment suggests important cultural and social distinctions that may or may not have been made between not only children and adults but also between perinates and older children.

Some of the initial investigations of perinatal infants have focused on assessment and interpretation of infant mortality (e.g., see discussion in Kinaston et al. 2009; Lewis 2007; Lewis and Gowland

2007) and infanticide (e.g., Scott 2001; see discussion in Lewis 2007; Mays and Eyers 2011). More recently, research has begun to include social identity and childhood social theory (e.g., Halcrow and Tayles 2011; Tocheri et al. 2005) to "explore the role of the child as an independent agent in the past" (Lewis 2007: 3). However, such studies rarely address perinates in particular.

A fundamental reason fetal remains have been excluded from such discourse may be attributed to their frequent absences from cemeteries (Lewis 2007; Saunders 2008; Scott 1999, 2001). One explanation for their absence is poor preservation due to a variety of factors including soil acidity, bone density, bone mineral content, and physiochemical properties (e.g., Bello and Andrews 2006; Djurić et al. 2011; Gordon and Buikstra 1981; Guy et al. 1997; Walker et al. 1988). Other possible factors include biases in excavation, failure to recognize small fetal bones, or the misidentification of fetal remains (Sundick 1978; Tocheri et al. 2005).

Beyond causes related to preservation and excavation, many researchers have considered the lack of perinatal remains a function of culture, especially as it relates to selective burial practices. There are numerous cultural reasons perinates may not be included in a general community cemetery, including issues of "personhood" or ontology (Scott 1999; Sofaer Derevenski 1997a), belief systems (Orme 2001; Saunders 2008), economic factors (Scheuer and Black 2000), social policies (Scott 1999), or infanticide (Gowing 1997; Scott 1999). However, precisely these cultural factors, ontology in particular, may also account for a perinate burial's inclusion in a community cemetery. From an anthropological perspective, personhood is understood to be a socially ascribed status (i.e., not taken for granted at birth). Personhood is also accrued over time, as children, especially infants, are not necessarily understood to occupy the same kind of status as adults. Eleanor Scott has found that "societies distinguished infancy from childhood to some degree, and, moreover ... the perinatal/neonatal period was frequently recognized as a particular and different stage of infancy" (1999: 4). Whether this acknowledged period was viewed synonymously with older infants and young children vis-à-vis "personhood" depends on the culture in question and may be determined from the mortuary context. In some situations, infants were excluded from and "considered a stranger" to society, requiring specific admission and recognition to the larger social group at culturally predetermined ages or stages, including naming the child (Scott 1999, 2001).

Ethnographic accounts describe the significance of rites of passage in marking the social recognition of a child as a member of the human community (see DeLoache and Gottlieb 2000). As Lynn Morgan discusses, there are sharp contrasts between societies with and without clear demarcations of personhood particularly when the "contents of the womb are considered ambiguous and uncertain" (1997: 324). Of course, the idea of incomplete personhood is not universal; in some cultures, infants were seen as gods, possessing specific power (Scott 1999), or as individuals "imbued with a spirit ... already conscious and self-aware" (Conklin 2001: 258). In her ethnographic account of Beng infant care in West Africa, Alma Gottlieb (2004) emphasizes that the care of infant bodies is rooted in local understandings about the care of human souls. Beng babies are spirits that must be provided for and thus persuaded to remain in this world rather than return to the afterlife, where all souls make their home; in this context, infant deaths are lamentable but also understandable (see Gottlieb 2004: chapter 8). In contrast, Sally Crawford (2013) discusses the powerful influence perinates may have on the burial process, as an infant death can never be considered natural and as such elicits a nonnormative burial response. It is reasonable, then, to presume that in other cultures, perinates were viewed neither as less than nor more than human; in these situations, fetuses were likely seen as full members of society regardless of their age-at-death. This may be particularly true in societies with Christian belief systems in which the idea of a soul was applicable to all human lives, including perinates (Tocheri et al. 2005), equating these youngest individuals with older infants and young children.

Mortuary archaeology has become a staple of archaeological inquiry, as mortuary data is related to various human behaviors (see reviews in Chapman and Randsborg 1981; Pearson 1999). As Gordon Rakita and Jane Buikstra point out, "since mortuary rites involve manipulations of material culture, social relations, cultural ideals, and the human body, they represent a nexus of anthropological interests" (2005a: 1). Moreover, they suggest that examining how the dead are attended to provides an opportunity to learn about the role of that individual (and their soul) in the living community (Rakita and Buikstra 2005b). The concept that funerary treatment reflects social status, gender, and social roles is not new in mortuary archaeology; however, more recent critiques contend that a person may be "thoroughly misrepresented in death," as emotion and obligation may distort mortuary treatment to fulfill individual or communal expectations. Therefore, these treatments and any observable variations, necessitate a more careful consideration of mortuary practices (Pearson 1999: 32).

Funerary practices, then, may be more reflective of the living than the dead and provide insight into their ideas and beliefs as it relates to the deceased. This may be especially pertinent for perinatal deaths, where their active roles in society during their life would be greatly limited. Funerary rites conducted for perinates would therefore reveal how the family and community viewed the youngest of children and would be suggestive of fetal identity. As Mary Lewis (2007) notes, researchers have often focused on the family and community response to a child's death, providing insight to the attitude of adults toward children. Since adults bury children, we only learn about children as part of the adult world and do not see the world from the children's viewpoint (Lucy 1994; Lewis 2007). Consequently, "we never experience the world of children, only the experiences of adults coming to terms with and attempting to ascribe meaning to their foreshortened lives and premature deaths" (Pearson 1999: 103). Therefore, funerary treatment becomes a valid method by which to examine fetal identity and how it compares to the mortuary traditions associated with older infants and young children.

The intention of this chapter, therefore, is to examine the mortuary context of perinates compared to older infants and young children in order to learn more about fetal identity. We test the hypothesis that if fetal identity in a postmedieval Polish community were distinguished from that of older infants and young children, there would be obvious deviations from the traditional mortuary patterns observed in the remainder of the population. There is limited archaeological knowledge about the identity of fetuses apart from historical accounts, such as Peter Ucko's description of the Ashanti tribe of West Africa that believes those who die before eight postnatal days are "ghost children," and give them separate burial outside communal cemeteries (1969: 271). Mortuary archaeology provides an opportunity to learn more about fetal identity, especially in prehistoric contexts (see Kabaciński et al., chapter 7) or in societies lacking historical documentation.

In this chapter, the term "subadult" designates those individuals collectively who have not reached biological adulthood in terms of skeletal and dental maturation. Based on maturation landmarks, the standards outlined by Lewis (2007) were used for this study with the term "perinate" referring to those between twenty-eight weeks in utero and seven postnatal days, and "postneonate" referring to those between seven postnatal days and one year. While "child" often refers to subadults between one and fourteen years of age, we only include those between one and three years and will designate the category "young child."

Christianity and Burial Traditions in Poland

The settlement of Drawsko is located in the Noteć River Valley of west-central Poland (Wyrwa 2004, 2005). The site was established during the Middle Ages and was continuously occupied through the present era. During the seventeenth and eighteenth centuries, a cemetery was established on the eastern edge of the settlement (designated Drawsko 1). Drawsko 1 was initially excavated in 1929, with follow-up excavations in 2002 and 2003 (Wyrwa 2004, 2005). The cemetery is located beneath agricultural fields, necessitating excavations to remove the burials beginning in 2008. Systematic excavation has continued since then with more than three hundred human skeletal remains recovered up to 2013. The remains, including perinates, postneonates, and young children, are generally well preserved, enabling ongoing bioarchaeological analyses.

With the introduction of Christianity in Poland in the later part of the tenth century, many changes were made to mortuary tradition as this Slavic region moved away from pagan beliefs and rituals. This transition however, did not occur instantaneously within Poland, as bi-ritual cemeteries incorporating both pagan and Christian elements were still present into the twelfth century (Buko 2008). These bi-ritual cemeteries were characterized by both cremation and interment burials and the inclusion of grave goods—a tradition that would eventually be abandoned under Christian doctrine (Buko 2008; Davies 1999). Not until the full adoption of Christianity in Poland did the importance of individual inhumation and the physical preservation of the body become widely recognized and form the foundation of Christian burial practices (Davies 1999; Richardson 2000). In the pre-Christian era in Europe, cemeteries were kept separate from populated settlements, creating a clear delineation between the world of the dead and the world of the living (Buko 2008). However, as Christianity became more entrenched in the early medieval period, cemeteries were established in centralized locations and controlled by the church (Davies 1999). By the mid-thirteenth century, these centralized cemeteries were well established throughout Poland because of the increased number of clergy available to perform Christian burial rites and to consecrate cemetery ground (Buko 2008; Daniell 1997). Typical Christian burials at this time were marked by orientation, body preparation, cemetery location, and grave goods. Orientation was considered a significant aspect of the burial process with individuals oriented on an east-west axis with the head positioned to the west and the face toward the east (Pearson 1999).

In addition to individual orientation, the larger cemetery organization was also significant, as burial location was often reflective of kinship, status, gender, or other social factors (Pearson 1999; Perry 2005). As Megan Perry (2005) discusses, both biological and cultural age of children may have also played a role in burial treatment. In the Anglo-Saxon and medieval periods, children were clustered and/or segregated within the communal cemetery (Lewis 2007). Indeed, even in modern cemeteries, age may play a role in the organization of the cemetery; for example, in the Milton Cemetery (Portsmouth, United Kingdom), deceased infants are clustered along the edge of the cemetery, removed from the general population (Scott 1999). This is also similar to contemporary cemeteries that identify specific regions for infant and children burials. For the majority of these medieval and postmedieval Christian burials, the individual was wrapped naked in a burial shroud and interred without grave goods, whereas pre-Christian traditions generally had rich burials with clothing, food, and items to accompany the individual into the afterlife (Davies 1999; Pearson 1999). While some variation existed in how a body could be buried, it was expected that all Christians were interred on consecrated ground; exceptions were those considered "unfit" for Christian burial, such as thieves, the cursed, accidental death victims, strangers, and the unbaptized (Daniell 1997; Finlay 2000; Hamlin and Foley 1983; Lynch 1998).

Since the introduction of Christianity into Europe, baptism has been a significant sacrament for the devout. Considered a means by which to wash away "original sin," baptism in the early Christian period was generally reserved for adults who were able to verbally declare their faith to their family and communities (Walsh 2005). The inability for children to speak and declare their faith in God placed these youngest members of society in an "ambivalent spiritual state" (Crawford 2013: 57) and eventually led to a shift in the church where baptism became the initial rite of passage after birth (Crawford 2013; Walsh 2005). Once baptized, an infant would be considered spiritually equal to all other members of their community; however, those individuals omitted from the baptism ritual were penalized through exclusionary mortuary traditions (Crawford 2013). This was most common among the youngest members of a community, the stillborn perinates, or those who died before baptism could take place. While early Christian doctrine condemned these individuals to hell, later amendments to Catholic canon saw the creation of a liminal state in which, upon death, these individuals would be trapped for eternity in this state considered neither heaven nor hell (Walsh 2005). This spiritual limbo also led to a change in the mortuary process where these unbaptized, pseudo-Christian individuals were physically separated from their community and buried outside of consecrated cemetery ground (Orme 2001).

However, these rules of faith were pliable, as reactions to death were socially and culturally charged and contributed to the variability in how individuals were treated after death. As Crawford (2013) discusses, infant death is imbued with different emotional and physical reactions than that of adult death, confounded by the challenges of religious obligations and expectations. Similarly, Nancy Scheper-Hughes (1992) discusses different maternal reactions to fetal deaths in northeastern Brazil compared to those of older children and adults and the inability to imbue newborns with the same type of emotional attachment when survival is unlikely in a violent and impoverished environment. Because the body of a child is created, sustained, and influenced in life by adult caregivers, the body of a child after death must be similarly cared for, which can produce differing mortuary traditions from that of adults (Crawford 2013; Scott 1999). Nyree Finlay argues that "the ambiguous personhood of these individuals promotes a diverse repertoire of material responses" (2013: 210) and as such provides an opportunity to explore perinatal burial traditions and the creation of identity and personhood in a posthumous context.

Materials and Methods

A sample of forty-seven subadults was used for this study. Age-at-death and gestational age estimates were calculated using dental formation, eruption, and long bone length measurements (Fazekas and Kósa 1978; Moorrees et al. 1963a, 1963b; Schaefer et al. 2009; Scheuer and MacLaughlin-Black 1994; Ubelaker 1979). These represent the best estimators of subadult age. However, when these elements were either missing or poorly preserved, other postcranial elements were measured and/or the fusion of different elements was evaluated in order to provide an age estimate (Fazekas and Kósa 1978; Schaefer et al. 2009). To ensure consistency in data collection and resultant age estimations, both authors repeated all macroscopic observations and skeletal measurements. Sex was not determined for any of these subadult individuals because of the difficulty in consistently assigning sex to prepubescent skeletal remains (see Cox and Mays 2000).

Mortuary data were collected for each individual at the time of excavation and included observations of coffin use, burial orientation, artifact associations, and burial location. The presence of a coffin was determined by any organic staining that followed a linear pattern around the body (i.e., a coffin outline), if remaining wood fragments were present around or under the body, and/or if any associated coffin hardware was in the grave (e.g., coffin nails). Burial orientation was determined based on head and body position using the cardinal planes. In instances of poor preservation or postmortem disturbance, positional estimation was made; however, this was rare. Burial location was determined from the cemetery site maps created post-excavation with individuals compared within and between age categories. Polish archaeologists identified all artifacts in the field with their physical position in the grave and elevation recorded. In this sample of subadult individuals, metal pins, flint, items of personal adornment, and copper coins were recovered. While still a grave inclusion, the latter artifact is significant for its cultural association with anti-vampire practices and protecting the dead (Barber 1988; Betsinger and Scott 2014; Keyworth 2002; Holly and Cordy 2007; Perkowski 1976; Scott and Betsinger 2012). Numerous adult and subadult burials have been recovered with an associated burial coin (n = 212, 64 percent) (fig. 7.1), including some of the burials designated as "deviant" or vampiristic in nature (for discussion, see Betsinger and Scott 2014; Scott and Betsinger 2012). Because of this association between "deviant" burials and these coins, this particular artifact was analyzed separately.

Nonparametric statistical tests were employed to compare the burial treatments (i.e., coffins, artifacts, coins) among the three age groups (Fisher's exact, p < 0.05). Burial orientation and location was visually examined from the Drawsko 1 excavation site maps. It was determined that a visual examination would provide a better overview of the emerging burial trends at this cemetery site. As excavation is ongoing, using spatial analysis to assess burial location at this stage of research may bias interpretations of cemetery organization, potentially negating the emerging burial patterns; therefore, for this preliminary analysis, burial location and orientation were considered cautiously.

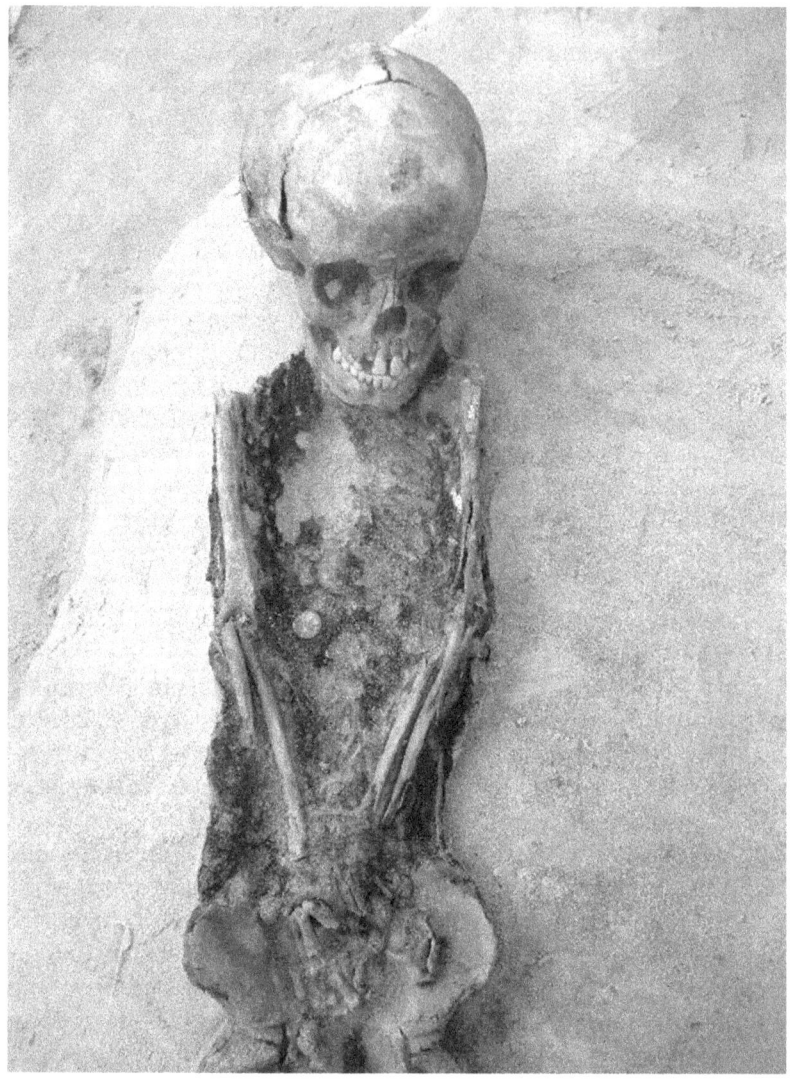

FIGURE 7.1. Example of a copper coin burial positioned over the right thorax in a subadult burial (burial 81/2010) (image courtesy of Amy B. Scott).

Results

Of the forty-seven subadult individuals used for this study, fifteen were classified as perinates, fifteen as postneonates, and seventeen as young children. In comparing the three age groups, no statistical differences existed for the use of coffins (p = 1.0000 for all three comparisons). A greater number of burial goods (excluding copper coins) were found with young children (n = 5) versus perinates (n = 1) and postneonates (n = 1); however, this difference was not statistically significant (p = 0.3696, p = 0.3696, respectively). Additionally, no statistical difference existed between perinates and postneonates (p = 1.0000). The type of burial artifact was also explored in these age groups. Bronze pins were found in all three age groups with additional flint and metal artifacts found in the young child group. In assessing the inclusion of copper burial coins, similar results emerged (table 7.1). In the perinatal category, four inhumations had coins, while five of the postneonates had coins. This difference was not statistically significant (p = 1.0000). In the young child category, ten of the seventeen individuals were buried with coins, but this increase in prevalence was not statistically greater than the number of coins buried with the perinates (p = 0.3352) or postneonates (p = 0.5293).

All individuals in each age category, except one young child, were buried along the east to west cardinal plane, with the cranium to the west and the feet to the east. In observing the overall site map (fig. 7.2), all subadult remains were clearly dispersed over a large area of land, approximately thirty meters east to west and ten meters north to south. There appeared to be some clustering of all age groups in excavation units 1C–D and 3C, but it was not consistent. The Drawsko 1 site is located on a slight topographic rise with its peak in units 1A and 1B; therefore, it was not surprising that fewer remains were recovered in these units as erosion of the sandy soil would have exposed any shallow burials in this region. Further, the northern boundary of units 2B and 1A–B were previously excavated in the 1929 and 2003 field seasons (Wyrwa 2004, 2005) disturbing any burials that had originally been placed in this section of the cemetery. While many of these subadult burials were clustered within larger groupings of adult and older subadult burials, possibly suggestive of family groupings, this clustering effect may also reflect

TABLE 7.1. Distribution of Burial Treatments across Age Categories

Age Category	N	Coin	Coffin	Artifacts
Perinatal	15	4	5	1
Post-neonatal	15	5	5	1
Early Childhood	17	10	7	5
Total	47	19	17	7

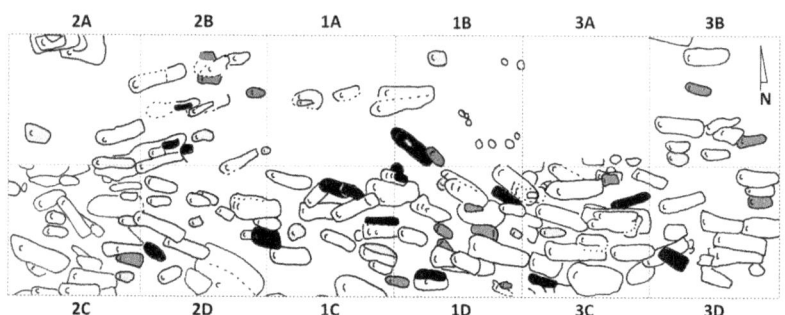

FIGURE 7.2. Drawsko 1 site burial map with perinate burials highlighted in light gray, postneonates in dark gray, and young children in black (modified from Gregoricka et al. 2014).

the temporal range of the cemetery. To date, no evidence of grave markers have been recovered; therefore, overlapping burials would be expected as newer interments encroached on older burials, particularly in the centralized region of the cemetery (i.e., units 1D and 3C).

When looking specifically at the perinates, these youngest individuals were spread across excavation units 1A–D, 2A–D, and 3B–C. Of the fifteen perinatal burials, two were buried in association with other individuals, individual 35/2010 was buried on top of an older subadult, and individual 20/2008 was associated with a thirty-five- to forty-four-year-old adult male, who also had a young child buried in association. All individuals in this burial complex had an associated copper coin. In looking at the distribution of individuals with and without coffins and artifacts, including coins, no clear pattern emerged, as those afforded more burial treatment were scattered throughout the excavation units, similar to those with less funerary treatment. Further, because of the lack of complete excavation data between 2008 and 2009, coffin information was not available for many of these perinate individuals.

The postneonates and young children were similarly distributed throughout the cemetery with postneonate burials present in units 1B–D, 2B–C, and 3B–D and with young children interred in regions 1B–D, 2B, 2D, and 3C–D. Similar to the perinate individuals, there was a similar distribution of coffins and artifacts throughout the excavation units with no distinct clustering for the postneonate and young child categories. As mentioned, the young child burial 40/2008b was found in association with an adult male and perinate individual. There was also a similar distribution of burial coins among the postneonates and young children with no distinct patterns emerging as to the cemetery placement of individuals with a coin.

Discussion

Based on these results, the Drawsko perinates were afforded the same burial treatment as postneonates and young children, suggesting that perinates were viewed synonymously with older children and adults. In other words, fetal identity was created through burial treatment, which demonstrated that the Drawsko community viewed these individuals as regular members of the group worthy of remembrance in the afterlife. This view of perinates and fetal identity surely has roots in this Christian community, for which the life and soul of the fetus was a fundamental belief. Similar to the conclusions of M.W. Tocheri and colleagues (2005), perinates at Drawsko were afforded individualized burial treatment with individualized adornments (e.g., burial shrouds, grave goods) representing an acceptance into community burial traditions where any age was considered worthy of a Christian burial or at least specific aspects of this tradition.

This observation of equal burial treatment is perhaps best explained through the demographic makeup of past populations and the role of each member within that community. Unquestionably, children made up a significant portion of archaeological populations, as much as 50 percent, and were arguably vital contributors to their communities (Lillehammer 2000; Lillie 1997; Mizoguchi 2000; Rega 1997; Sofaer Derevenski 1997b). When considering the burial treatments of children, the formation of identity in death, and why fetal remains were treated similar to older children, focus can be placed on what Grete Lillehammer refers to as the "potentiality of children" (2000: 22). When a child dies at any age, in the womb or shortly after birth, it signifies an interruption of the life cycle that equally interrupts the formation of identity. As children may be considered the foundation of future generations, death within these age categories may be approached from a reconciliation perspective, where the adults within a community establish the identity of the individual based on their "potential" and what their contribution would have been to their larger community if they had lived (Mizoguchi 2000). Historians have argued that the loss of an infant in the Middle Ages, where low life expectancy and childhood mortality was as high as

50 percent (Youngs 2006), would not have been unexpected (Ariès 1973; deMause 1974). However, studies of parental bereavement have clearly demonstrated that while a child may not have been socially active within their community, their premature death is emotionally taxing (Layne 2002; Pollock 1983). If children were highly vulnerable resources, as suggested by Philippe Ariès (1973) and Lloyd deMause (1974), arguably subadult burials would be a rarity in the archaeological record, where the time and effort of a proper burial would be withheld for fully integrated community members. In actuality, subadult burials are extensive at Drawsko with variable mortuary treatment no doubt reflective of intrinsic and extrinsic factors. As Johanna Sofaer Derevenski (1997b) and Linda Layne (2002) argue, material culture, as evidenced in burial treatment, is a means by which to demonstrate social values and tie an individual to their community, making this an important action for the still living community. The evidence at Drawsko suggests this, with the interplay between the creation of identity through burial treatment and a consistent recognition of its importance at any age, whether death was expected or not.

For example, individual 73/2010 is a unique young child burial where all discussed mortuary elements are present. This individual is located in the centralized region of the cemetery and was buried with a piece of flint. Through this relatively "rich" burial treatment, the concept of personhood is demonstrated, as this relatively young individual would have been established in this community through their own physical presence and participation and that of their parents. Even in the perinate age category, this concept of personhood and identity is created. For example, individual 48/2010, aged thirty-two to forty-eight weeks in utero, may have been stillborn or died shortly after birth and would not have been physically or socially established within the Drawsko community. However, the careful interment of this individual in a small coffin wrapped in fabric suggests that despite this lack of established personhood based on physical presence or action within the community, there is recognition of personhood based on this notion of "potentiality." While differences exist between these two burial examples, these differences are arguably reflective of the biological and chronological differences between the two rather than social notions of personhood or identity. The minimal inclusions with the perinate over the young child does not necessarily reflect a lack of care for the youngest members of society but is perhaps distinguished based on what the young child was able to accomplish before death. As Leszek

Gardeła and Paweł Duma (2013) discuss, the young child may have been afforded more grave goods based on what they had acquired in their relatively short life-span, whereas the perinate burials can only reflect what the parents were able to provide at the time of death. Conversely, however, some subadult individuals show no evidence of burial treatment, yet their location may speak to larger notions of identity. For example, individual 35/2010 is a perinate buried with no coffin, no coin, and no other artifacts, but this individual is located within a large feature connected to multiple surrounding graves. Arguably, this close association to others, perhaps members of the same family group, may represent a different type of identity creation for subadults outside the sphere of physical treatment where burial position within the cemetery may speak to personhood and an interconnected community.

In addition to considering the cultural creation of identity and "potentiality" through mortuary treatment, there must also be recognition of the religious obligations of death and how these obligations may or may not interfere with the establishment of personhood. To be buried in a Christian cemetery required a declaration of faith, and those unable to make that declaration (i.e., stillborn or unbaptized) were at risk of being denied a proper burial (Crawford 2013). In looking specifically at the Drawsko 1 cemetery and the identified attritional population pattern (see Gregoricka et al. 2014), the majority of the community was apparently interred at this site, suggesting an inclusive sample. The inclusive nature of Drawsko 1 is further demonstrated through the presence of multiple "deviant" burials (Betsinger and Scott 2014; Gregoricka et al. 2014). These "deviant" burials were likely the individuals considered "outsiders" in the Drawsko community, a consequence of physical malformation or unsavory behavior. Considered dangerous to the living population, these "deviant" individuals were barricaded in their graves after death so they could not reanimate and harm the still living community (Barber 1988). Despite these fears, however, great care was taken in the burial process of these "deviant" individuals, and their final resting place was still within the confines of the community cemetery (Betsinger and Scott 2014). This concept of malevolent spirits after death has also been explored in a contemporary context through the work of Helen Hardacre (1999). *Mizuko kuyo* is a religious ritual performed in Japan to appease the spirits of aborted fetuses that gained in popularity in the 1960s. While arguably driven by entrepreneurial aspirations, the establishment of the *mizuko kuyo* ritual provides women with closure from their aborted fetuses and protection from fetal spirits—arguably similar to the protection sought out by the Drawsko community upon the death of these "deviant" individuals. Further, the complex nature of many Drawsko 1 burials (i.e., multiple arm positions, inconsistent artifact inclusions, variation in body position) suggests this cemetery embraced mortuary adaptations that both adhered to and departed from strict Christian doctrine during this period.

As Eileen Murphy (2011) discusses, unbaptized and stillborn individuals in postmedieval Ireland may have been denied burial in a consecrated cemetery but were still buried in places of remembrance that protected these perinate individuals. Known as *cillíní*, these burial grounds were located in prominent places on the landscape, either culturally or geographically (Donnelly et al. 1999; Donnelly and Murphy 2008; Finlay 2000; Hamlin and Foley 1983; Murphy 2011). These *cillíní*, while separated from the Catholic Church, still provided a landscape in which to protect and honor the dead, as these spaces were deliberately selected by the community and remembered through verbal and written histories. Despite this separation from the church, individuals buried within these *cillíní* were still afforded standard Christian burial treatment and belonged to a unique cemetery assemblage consisting of the "unfit" members of society, arguably appeasing the community's need protect their own eventual burial ground (Donnelly and Murphy 2008; Murphy 2011). While the majority of these *cillíní* were occupied by unbaptized perinates and infants, other members of the community considered "unfit" for proper burial were also included in these cemeteries (Finlay 2000). The Drawsko cemetery cannot necessarily be considered a *cillíní* site, but some of the perinates interred there were likely stillborn or unbaptized based on osteological analyses of age, suggesting that while the community members did not omit these individuals from the Christian burial record, they may have altered the burial treatment of these "unfit" individuals.

While the omission of burial artifacts could be considered a statement of social status or the "potentially" ascribed to that particular individual, it could equally represent the defined burial treatment for those considered spiritually "unfit" due to the circumstances of their death. Just as individual 35/2010 may have been identified by their close proximity to other community members, the lack of any physical mortuary treatment may speak to the larger cultural circumstances of their death. Additionally, the funerary process may have also been altered for these individuals, where ceremonies may have been more private family affairs without the intervention of the church. As much as the physical remains of the mortuary process may guide our interpretations of personhood and identity, unique aspects of the cultural process of interment may have also equally contributed to how these individuals were treated and remembered after death. Even through differential burial where all mortuary treatment appears to be omitted, identity and a sense of personhood were still manufactured for these individuals, as they were purposely acknowledged as those not belonging to the community. While this identity would have been different if they had lived, the circumstances of their death dictate the remembrance of these individuals.

However identity was being constructed within the Drawsko community, this study demonstrates that subadults were considered active members within this postmedieval village, deserving of similar burial treatment to that of their peers and older adults. While the mortuary differences between individuals within and between the three subadult groups may reflect differing levels of identity based on age and community involvement, the physical inclusion of all children in this centralized cemetery suggests social inclusion in the community, yet variable treatment, is based on intrinsic and extrinsic factors. The identification of these factors is difficult to ascertain from archaeological remains, but future studies of neonatal dentition and isotopic analysis may provide vital information identifying those born alive and those not, possibly reflecting the subtle mortuary differences among the perinatal age category. The incorporation of ancient DNA analysis to determine subadult sex may also guide this ongoing analysis, as mortuary treatment guided by biological sex may also be influencing the emerging patterns visible among these subadult burials.

Conclusion

Through this examination of the fetal remains at the Drawsko 1 cemetery, it is obvious that archaeological inquiry must consider the youngest members of society and how they fit within mortuary traditions. In comparing the Drawsko perinates with other subadult age categories, variation clearly exists within the cemetery assemblage, but perinates were generally offered a similar burial to that of older children in the community. Despite low life expectancy and high mortality rates among children in the Middle Ages, the death of a child was an unpredictable occurrence and as such, contributed to differential burial treatments. Because the Drawsko cemetery is comprised of normal and "deviant" inhumations, it is not unreasonable to assume that all perinate individuals from this community were buried in this location, regardless of being stillborn or dying before baptism.

In the *cillini* sites in postmedieval Ireland, stillborn or unbaptized were interred in locations separate from the church but were still provided a proper Christian burial. It is possible that the Drawsko cemetery, removed from the village site and separate from any visible church foundation, may be the final resting place for those considered "unfit" by the community, similar to these *cillini* sites. Further, there is a similarity between Drawsko and the demonstrated care shown to perinates at these sites. The Drawsko community may have adapted their Christian doctrine to include these individuals within the larger cemetery, but distinguished them from their peers through shifts in the mortuary process. While no archaeological evidence suggests an obvious pattern of differential mortuary treatment in the perinate age category, cultural factors such as the funerary process may have been what defined the different burial types in the Drawsko community.

Overall, the Drawsko community shows a clear pattern of care for the youngest members of their community. Whether stillborn, unbaptized, or fully integrated into social life, the children of Drawsko show a similar pattern of interment. Through this demonstration of care, the perinate individuals in particular were provided a sense of identity, personhood, and acceptance within their community. The "potentiality" of these perinates was arguably recognized and embraced by this community, not only because their physical remains were considered equal to well-established, older subadult individuals, but also through the process of burial where they were remembered by their community.

Situating this within a larger context of fetal identity, this archaeological analysis contributes to an historical understanding of fetuses and their physical placement within a community. As Linda Layne discusses, the process of burying the infant is to "assist the deceased in completing the transition" (2002: 62), with the parameters of that transition dictated by the surrounding community. The archaeological context of this transition is unique in that it provides the material evidence of how particular communities felt about fetal death, specifically the physical obligations and requirements associated with death. As changing cultural views dictate how the fetus is considered and integrated into contemporary society, our views are no doubt formulated and negotiated by the historical notions of identity and personhood within this distinct biological and social group.

Acknowledgments

We wish to thank the Slavia Foundation and Dr. Marek Polcyn for access to the collection. Additionally, we thank Site Director Eiżbieta Gajda, Ewa Lobo-Bronowicka, Marta Słonimska, Arek Klimowicz, Michał Rozwadowski, Marta Gwizdała, and the osteological field and lab staff and students. We also thank the three anonymous reviewers who provided useful feedback to improve this chapter.

Amy B. Scott is Assistant Professor of Anthropology at the University of New Brunswick. Her research interests include biochemical analyses of health and stress, skeletal growth and development, and mortuary burial patterns in medieval and postmedieval Europe and eighteenth-century Atlantic Canada.

Tracy K. Betsinger is Associate Professor of Anthropology at State University of New York Oneonta. She conducts bioarchaeological studies of health and mortuary patterns with medieval/postmedieval European populations and prehistoric populations from the Southeastern United States.

References

Ariès, Philippe. 1973. *Centuries of Childhood: A Social History of Family Life.* London: Jonathan Cape.

Barber, Paul. 1988. *Vampires, Burial, and Death: Folklore and Reality.* New Haven, CT: Yale University Press.

Baxter, Jane Eva. 2005. "Introduction: The Archaeology of Childhood in Context." In "Children in Action: Perspectives on the Archaeology of Childhood," ed. Jane Eva Baxter. Special issue, *Archaeological Papers of the American Anthropological Association* 15 (1): 1–9.

Bello, Silvia, and Peter Andrews. 2006. "The Intrinsic Pattern of Preservation of Human Skeletons and its Influence on the Interpretation of Funerary Behaviours." In *Social Archaeology of Funerary Remains*, ed. Rebecca Gowland and Christopher Knüsel, 1–13. Oxford: Oxbow Books.

Betsinger, Tracy K., and Amy B. Scott. 2012. "From Popular Culture to Scientific Inquiry: A Bioarchaeological Analysis of Vampires in Post-medieval Poland." *American Journal of Physical Anthropology* 147 (S54): 99.

Betsinger, Tracy K., and Amy B Scott. 2014. "Governing from the Grave: Vampire Burials and Social Order in Post-medieval Poland." Cambridge Archaeological Journal 24 (3): 467–476.

Buko, Andrezej. 2008. *The Archeology of Early Medieval Poland: Discoveries, Hypotheses, Interpretations.* Leiden: Brill.

Chapman, Robert, and Klavs Randsborg. 1981. "Approaches to the Archaeology of Death." In *The Archaeology of Death,* ed. Robert Chapman, Ian Kinnes, and Klavs Randsborg, 1–24. Cambridge: Cambridge University Press.

Conklin, Beth A. 2001. *Consuming Grief: Compassionate Cannibalism in an Amazonian Society.* Austin: University of Texas Press.

Cox, Margaret, and Simon Mays. 2000. *Human Osteology: In Archaeology and Forensic Science* London: Greenwich Medieval Media.

Crawford, Sally. 2013. "Baptism and Infant Burials in Anglo-Saxon England." In *Medieval Life Cycles: Continuity and Change,* ed. Isabelle Cochelin and Karen Smyth, 55–80. Turnhout: Brepols Publishers.

Daniell, Christopher. 1997. *Death and Burial in Medieval England, 1066–1550.* London: Routledge.

Davies, Jon. 1999. *Death, Burial, and Rebirth in the Religions of Antiquity.* London: Routledge.

DeLoache, Judy S., and Alma Gottlieb, eds. 2000. *A World of Babies: Imagined Child Care Guides for Seven Societies.* Cambridge: Cambridge University Press.

deMause, Lloyd. 1974. "The Evolution of Childhood." In *The History of Childhood,* ed. Lloyd deMause, 1–73. New York: Psychohistory Press.

Djurić, Marija, Ksenija Djukić, Petar Milovanović, Aleksa Janović, and Petar Milenković. 2011. "Representing Children in Excavated Cemeteries: The Intrinsic Preservation Factors." *Antiquity* 85 (327): 250–262.

Donnelly, Colm J., and Eileen M. Murphy. 2008. "The Origins of *Cillíní* in Ireland." In *Deviant Burial in the Archaeological Record,* ed. Eileen M. Murphy, 191–223. Oxford: Oxbow Books.

Donnelly, Seamus, Colm J. Donnelly, and Eileen M. Murphy. 1999. "The Forgotten Dead: The *Cillíní* and Disused Burial Grounds of Ballintoy, County Antrim." *Ulster Journal of Archaeology* 58: 109–113.

Fazekas, István Gyula, and F. Kósa. 1978. *Forensic Fetal Osteology.* Budapest: Akademiai Kiado.

Finlay, Nyree. 2000. "Outside of Life: Traditions of Infant Burial in Ireland from *Cillín* to Cist." *World Archaeology* 31 (3): 407–422.

Finlay, Nyree. 2013. "Archaeologies of the Beginnings of Life." *World Archaeology* 45 (2): 207–214.

Gardeła, Leszek, and Paweł Duma. 2013. "Untimely Death: Atypical Burials of Children in Early and Late Medieval Poland." *World Archaeology* 45 (2): 314–332.

Gottlieb, Alma. 2004. *The Afterlife Is Where We Come From: The Culture of Infancy in West Africa.* Chicago: University of Chicago Press.

Gordon, Claire C., and Jane E. Buikstra. 1981. "Soil pH, Bone Preservation, and Sampling Bias at Mortuary Sites." *American Antiquity* 46 (3): 566–571.

Gowing, Laura. 1997. "Secret Births and Infanticide in Seventeenth-Century England." *Past and Present* 156: 87–115.

Gregoricka, Lesley A., Tracy K. Betsinger, Amy B. Scott, and Marek Polcyn. 2014. "Apotropaic Practices and the Undead: A Biogeochemical Assessment of Deviant Burials in Post-medieval Poland." *PLOS ONE* 9 (11): e113564. doi:10.1371/journal.pone.0113564.

Guy, Hervé, Claude Masset, and Charles-Albert Baud. 1997. "Infant Taphonomy." *International Journal of Osteoarchaeology* 7 (3): 221–299.

Halcrow, Siân E., and Nancy Tayles. 2011. "The Bioarchaeological Investigation of Children and Childhood." In *Social Bioarchaeology*, ed. Sabrina C. Agarwal and Bonnie A. Glencross, 333–360. Chichester: Wiley-Blackwell.

Hamlin, Ann, and Claire Foley. 1983. "A Women's Grave Yard at Carrickmore, County Tyrone, and the Separate Burial of Women." *Ulster Journal of Archaeology* 46: 41–46.

Hardacre, Helen. 1999. *Marketing the Menacing Fetus in Japan.* Berkley: University of California Press.

Holly, Donald H., and Casey E. Cordy. 2007. "What's in a Coin? Reading the Material Culture of Legend Tripping and Other Activities." *Journal of American Folklore* 120 (477): 335–354.

Kamp, Kathryn A. 2001. "Where Have All the Children Gone?: The Archaeology of Childhood." *Journal of Archaeological Method and Theory* 8 (1): 1–34.

Keyworth, David. 2002. "The Socio-Religious Beliefs and Nature of Contemporary Vampire Subculture." *Journal of Contemporary Religion* 17 (3): 355–370.

Kinaston, Rebecca L., Hallie R. Buckley, Siân E. Halcrow, Matthew J.T. Spriggs, Stuart Bedford, Ken Neal, and A. Gray. 2009. "Investigating Foetal and Perinatal Mortality in Prehistoric Skeletal Samples: A Case Study from a 3000-Year-Old Pacific Island Cemetery Site." *Journal of Archaeological Science* 36 (12): 2780–2787.

Layne, Linda L. 2002. *Motherhood Lost: A Feminist Account of Pregnancy Loss in America.* London: Routledge.

Lewis, Mary E. 2007. *The Bioarchaeology of Children: Perspectives from Biological and Forensic Anthropology.* Cambridge: Cambridge University Press.

Lewis, Mary E., and Rebecca Gowland. 2007. "Brief and Precarious Lives: Infant Mortality in Contrasting Sites from Medieval and Post-medieval England (AD 850–1859)." *American Journal of Physical Anthropology* 134 (1): 117–129.

Lillehammer, Grete. 1989. "A Child Is Born: The Child's World in an Archaeological Perspective." *Norwegian Archaeological Review* 22 (2): 89–105.

Lillehammer, Grete. 2000. "The World of Children." In Sofaer Derevenski 2000, 17–26.

Lillie, Malcolm C. 1997. "Women and Children in Prehistory: Resource Sharing and Social Stratification at the Mesolithic-Neolithic Transition in Ukraine." In Moore and Scott 1997, 213–228.

Lucy, Sam. 1994. "Children in Early Medieval Cemeteries." *Archaeological Review of Cambridge* 13 (2): 21–34.

Lynch, Linda G. 1998. "Placeless Souls: Bioarchaeology and Separate Burials in Ireland." MA thesis, Cork: University College Cork.

Mays, Simon, and Jill Eyers. 2011. "Perinatal Infant Death at the Roman Villa Site at Hambleden, Buckinghamshire, England." *Journal of Archaeological Science* 38 (5): 1931–1938.

Mizoguchi, Koji. 2000. "The Child as a Node of Past, Present and Future." In Sofaer Derevenski 2000, 141–150.

Moore, Jenny, and Eleanor Scott. 1997. *Invisible People and Practices: Writing Gender and Children into European Archaeology.* London: Leicester University Press.

Moorrees, Coenraad F.A., Elizabeth A. Fanning, and Edward E. Hunt Jr. 1963a. "Formation and Resorption of Three Deciduous Teeth in Children." *American Journal of Physical Anthropology* 21 (2): 205–213.

Moorrees, Coenraad F.A., Elizabeth Fanning, and Edward E. Hunt Jr. 1963b. "Age Variation of Formation Stages for Ten Permanent Teeth." *Journal of Dental Research* 42: 1490–1502.

Morgan, Lynn M. 1997. "Imagining the Unborn in the Ecuadoran Andes." *Feminist Studies* 23: 322– 350.

Murphy, Eileen M. 2011. "Children's Burial Grounds in Ireland (*Cillíní*) and Parental Emotions toward Infant Death." *International Journal of Historical Archaeology* 15: 409–428.

Orme, Nicholas. 2001. *Medieval Children.* New Haven, CT: Yale University Press.

Pearson, Michael Parker. 1999. *The Archaeology of Death and Burial.* College Station: Texas A&M University Press.

Perkowski, Jan L. 1976. *Vampires of the Slavs.* Cambridge: Slavica Publishers.

Perry, Megan A. 2005. "Redefining Childhood through Bioarchaeology: Toward an Archaeological and Biological Understanding of Children in Antiquity." *Archeological Papers of the American Anthropological Association* 15 (1): 89–111.

Pollock, Linda A. 1983. *Forgotten Children: Parent-Child Relations from 1500 to 1900.* Cambridge: Cambridge University Press.

Rakita, Gordon F.M., and Jane Buikstra. 2005a. "Introduction." In Rakita et al. 2005, 1–14.

Rakita, Gordon F.M., and Jane Buikstra. 2005b. "Bodies and Souls." In Rakita et al. 2005, 93–95.

Rakita, Gordon F.M., Jane Buikstra, Lane A. Beck, and Sloan R. Williams, eds. 2005. *Interacting with the Dead: Perspectives on Mortuary Archaeology for the New Millennium.* Gainesville: University Press of Florida.

Rega, E. 1997. "Age, Gender and Biological Reality in the Early Bronze Age Cemetery at Morkin." In Moore and Scott 1997, 229–247.

Richardson, Ruth. 2000. *Death, Dissection and the Destitute.* Chicago: University of Chicago Press.
Saunders, Shelley R. 2008. "Juvenile Skeletons and Growth-related Studies." In *Biological Anthropology of the Human Skeleton,* ed. M. Anne Katzenberg and Shelley R. Saunders, 117–147. New York: Wiley-Liss.
Schaefer, Maureen, Sue Black, and Louise Scheuer. 2009. *Juvenile Osteology: A Laboratory and Field Manual.* San Diego: Academic Press.
Scheper-Hughes, Nancy. 1992. *Death Without Weeping: The Violence of Everyday Life in Brazil.* Berkley: University of California Press.
Scheuer, Louise, and Sue Black. 2000. *Developmental Juvenile Osteology.* San Diego: Academic Press.
Scheuer, Louise, and Sue MacLaughlin-Black. 1994. "Age Estimation from the Pars Basilaris of Fetal and Juvenile Occipital Bone." *International Journal of Osteoarchaeology* 4 (4): 377–380.
Scott, Amy, and Tracy Betsinger. 2012. "Coins, Kids, and Culture: An Examination of Grave Goods and Health at the Drawsko 1 Cemetery Site (17th–18th centuries). *American Journal of Physical Anthropology* 147 (S54): 264.
Scott, Eleanor. 1999. *The Archaeology of Infancy and Infant Death.* BAR International Series 819. Oxford: Archaeopress.
Scott, Eleanor. 2001. "Killing the Female? Archaeological Narratives of Infanticide." In *Gender and the Archaeology of Death,* ed. Bettina Arnold and Nancy L. Wicker, 3–21. Walnut Creek: AltaMira Press.
Sofaer Derevenski, Johanna. 1997a. "Age and Gender at the Site of Tiszapolgar-Basatanya, Hungary." *Antiquity* 71 (274): 875–889.
Sofaer Derevenski, Johanna. 1997b. "Engendering Children, Engendering Archaeology." In Moore and Scott 1997, 192–202.
Sofaer Derevenski, Johanna, ed. 2000. *Children and Material Culture.* London: Routledge.
Sundick, R. 1978. "Human Skeletal Growth and Age Determination." *Homo* 29 (4): 228–249.
Tocheri, M.W., T.L. Dupras, P. Sheldrick and J.E. Molto. 2005. "Roman Period Fetal Skeletons from the East Cemetery (Kellis 2) of Kellis, Egypt." *International Journal of Osteoarchaeology* 15 (5): 326–341.
Ubelaker, Douglas H. 1979. *Human Skeletal Remains: Excavation, Analysis and Interpretation.* Washington, DC: Smithsonian Institute Press.
Ucko, Peter J. 1969. "Ethnography and Archaeological Interpretation of Funerary Remains." *World Archaeology* 1 (2): 262–280.
Walker, Phillip L., John R. Johnson, and Patricia M. Lambert. 1988. "Age and Sex Biases in the Preservation of Human Skeletal Remains." *American Journal of Physical Anthropology* 76 (2): 183–188.
Walsh, Michael. 2005. *Roman Catholicism: The Basics.* London: Routledge.
Wyrwa, Andrzej Marek. 2004. "Drawsko I jego miejsce w prahistorychnych, średniowiecznych I nowożytnych dziejach osadnictwa ziemi nadnoteckiej." In *Ziemia nadnotecka wczoraj, dziś, jutro: Materiały z sesji naukowej odbytej w dniu 30 maja 2003 roku z okazji nadania Gimnazjum w Drawsku*

imienia "Ziemi Nadnoteckiej, ed. Andrzej Wyrwa and Włodzimierz Gapski, 123–152. Poznań: ZPW M-Druk.

Wyrwa, Andrzej Marek. 2005. "Prahistorycne i średniowieczne ślady osadnictwa w rejonie Drawska, gm. loco, woj. wielkopolskie." *Fontes Archaeologici Posnanienses* 41: 275–298.

Youngs, Deborah. 2006. *The Life-Cycle in Western Europe, c. 1300–c. 1500*. Manchester: Manchester University Press.

Part III

THE ONCE AND FUTURE FETUS
SOCIOCULTURAL ANTHROPOLOGY

Chapter 8

WAITING

THE REDEMPTION OF FROZEN EMBRYOS THROUGH EMBRYO ADOPTION AND STEM CELL RESEARCH IN THE UNITED STATES

Risa D. Cromer

Hallelujah! Finding Places for Frozen Embryos

"Hallelujah, 2PNs!" Wendy announced one afternoon as she flipped through a stack of embryology reports.[1] Such gleeful outbursts about donated embryos are sometimes overheard from Wendy's sun-filled office within Bay University's stem cell institute.[2] The reports accompanied a portable cryopreservation tank filled with frozen embryos that arrived via FedEx that afternoon. "Those have to go to the bottom of the tank," Wendy said, where they would be immersed safely in liquid nitrogen. "I want to prioritize them." Wendy is the lab manager for the REDEEM Biobank, a university-based program that receives, stores, and conducts research on frozen embryos donated by fertility patients from across the United States. After reviewing the donation documents confirming receipt of the rare 2PN stage embryos, Wendy reached for a pair of blue oven-mitt-looking gloves. They protect her skin from the −196 degrees Celsius liquid nitrogen filling the biobank and preserving the hundreds of embryos it contains. Standing atop a footstool with a hooked metal tool in her hand, she reached carefully into the tank's liquid depths in search of a new place for the precious embryos that would serve invaluably in ongoing research projects within Bay University's thriving stem cell institute.

In a Christian adoption agency five hundred miles south of Bay University, efforts to find homes for valuable frozen embryos are also underway. "Yes! Yes! Hallelujah!" Monica exclaimed upon receiving good news. She closed her eyes, said a private prayer, and whispered to herself, "Keep growing, keep growing, keep growing." Monica is the program director of the Blossom Embryo Adoption program, which considers all frozen embryos precious human lives created by God that deserve a chance to be born. The Blossom program strives to match fertility patients possessing leftover embryos with recipients wanting to adopt them for use toward a chance at pregnancy and parenthood. Monica's excitement stemmed from learning that a difficult-to-match embryo donated by the Crown family had finally been thawed and transferred into an adoptive client's uterus. "We prayed for this," she said after detailing the program's tumultuous journey over many years trying to find a place for the Crowns' less desirable embryo. "Risa, we saved this embryo."

Estimates suggest that about a half million human embryos left over from in vitro fertilization (IVF) procedures have accumulated in fertility clinic freezers across the United States since the mid-1980s.[3] The once uncontroversial supply was thrust to the center of public debate prompted by questions about what to do with the growing glut of unwanted embryos considered too precious to discard. REDEEM Biobank and the Blossom Embryo Adoption program represent two putatively opposing solutions: Blossom typifies a pro-life Christian approach to rescuing souls through politics promoting embryo personhood, while REDEEM exemplifies scientists' efforts to procure research materials aimed at curing disease. Each organization has emerged as a leader in the business of saving lives, redeeming value, and converting frozen embryos and their potential into one of today's most timely and contested assets (Ferry and Limbert 2008).

America's frozen embryos have become dually regarded as unwanted and unwastable. This chapter examines how Blossom and REDEEM Biobank give chances, redeem value, and find places for these categorically ambiguous reproductive remainders. A comparative look at two embryo redemption programs reveals that leftover embryos are frozen in multiple forms of limbo, materially and symbolically. They are paused in freezers and positioned in the United States as potential persons and profitable things, undesired and too precious to discard, salvation objects for both scientists and Christians, and other seeming contradictions. This chapter explores how the categorical ambiguity of frozen embryos—as America's unwanted and unwastable remainders—produces the circumstance of *waiting*. Frozen embryos and the ways they wait provide new perspectives on contemporary reproductive politics in the United States and, more broadly, life itself.

Frozen Embryo Subjects, Feminist Positions

My research builds on efforts that began four decades ago when feminist scholars started systematically dragging reproduction to the center of social theory, expanding its definition beyond biological procreation, and demonstrating the invisible centrality of reproduction to social life (cf. Ginsburg and Rapp 1995). Part of this scholarship traced the emergence of embryonic and fetal subjects both cross-culturally and historically, and documented the protean forms they take. Embryos and fetuses have been deemed testable (Rapp 1999), tentative (Rothman 1986), dangerous (Reagan 2010), haunting (Gammeltoft 2014), and born as well as made (Franklin and Roberts 2006). They have been put to work as patients (Casper 1998), consumers (Taylor 2008), citizens (Berlant 1997), icons (Morgan 2009), persons (Hartouni 1999), and kin (Roberts 2012). Sometimes they are lost (Layne 2003) or let go (Scheper-Hughes 1992); other times they tell tales about evolution (Morgan 2003), race (Tsing 2007), and environmental risk (Steingraber 2001). Feminist social scientists also confronted areas of their own underdeveloped positions concerning embryo and fetuses, which are sometimes, though not inherently, regarded as political subjects (Morgan and Michaels 1999).

This project approaches frozen embryos—and the organizations that repurpose them—as neither stable nor naturally occurring but as historically produced entities that are culturally infused with ever-shifting meanings. Further, I chose the Blossom program and REDEEM Biobank as primary field sites with interest in exploring beyond the oversimplified binaries common within contemporary life politics, for example, pro-life versus pro-choice, science versus religion, woman versus conceptus.

This chapter draws from twenty-seven months of comparative ethnographic research from 2008 to 2013 within Pacific Adoptions' Blossom Embryo Adoption program and Bay University's REDEEM Biobank. I also spent four months at Western Fertility Clinic, a private IVF clinic unaffiliated with either redemption program. With frozen embryos situated at the center of analysis, I used three main ethnographic methods—participant observation, interviews, and textual analysis—to trace their "social lives" from IVF clinic freezers to wombs, labs, and waste bins (Appadurai 1986b).[4]

Saving Frozen Embryos

Early scientific curiosities about the survival of life at the frosty extremes reframed the cold from a fearful, threatening mystery to a resource for saving, preserving, and resurrecting life. From these curiosities emerged the field of cryobiology in the early twentieth century, which made possible the preservation of various biological materials for future use (Leibo 2004: 349). Cryobiology figured centrally in the 1940s revolution in domestic animal breeding through the freezing of sperm and embryos (Bavister 2001; Foote 2002), the postwar 1950s eruption of frozen foods and home freezers (Smith 2001), the 1960s cryonics movement to preserve one's brain or body for later revivification (Farman 2013; Sheskin 1979), and the banking of genetic materials from exotic or endangered plant and animal species in the 1970s (Watson and Holt 2001). Freezing human embryos in the United States began in the mid-1980s a few years after the introduction of IVF technology. While the cryopreservation of extra embryos in the United States has been common practice for more than thirty years, saving human embryos is neither customary nor legal in many regions of the world providing IVF (Inhorn and Van Balen 2002; Roberts 2012).

No two IVF patients or clinics are alike, though Sandra's case at Western Fertility Clinic is typical in many ways. Western began as small mom-and-pop fertility clinic in the late 1980s in California and grew to become one of the California's highest volume providers.[5] Sandra was thirty-six when she and her husband first came to Western after two years of multiple miscarriages. After a series of tests, she was diagnosed with "unexplained infertility" and encouraged by her physician to consider IVF. On the day of Sandra's first egg retrieval, she produced thirteen eggs that were coincubated with her partner's sperm for fertilization. A few days later, ten viable embryos remained. Two were selected for transfer into her uterus while the extra eight embryos were frozen for potential later use. Her embryologist immersed the leftover embryos in individual droplets of cryoprotectant media, carefully packaged them in straws, labeled each with Sandra's name and birth date, and plunged the straws into liquid nitrogen for storage. When Sandra and her husband's efforts to build a family through IVF come to an end, they will have a few options for any embryos that remain: they can donate them for research or procreation, discard as medical waste, or keep cryopreserved indefinitely.

Derek, an embryologist at Western, explains why embryos like Sandra's are routinely saved in fertility clinics across the United

States: "The patient has been through high doses of drugs, a needle into her vagina and ovaries, and paid the costs of lab embryology. You've got some good embryos—why waste them? You don't bin them. You want to freeze them for the future." But saving spare embryos for the future has become an increasing burden for fertility clinics and patients alike. Charging each patient anywhere from $400 to $1,000 a year for storage makes freezing embryos seem like easy money for fertility clinics, yet the liabilities may outweigh the revenue. The accrual of frozen embryos in storage tanks represents a growing problem and worry for the nearly five hundred IVF clinics in the United States.

"We have embryos from the 1990s in those tanks," Ken, a senior embryologist at Western Fertility Clinic, explained as he pointed toward the room off the lab where twelve cryopreservation tanks are located. In addition to his daily activities working under the microscope with human eggs, sperm, embryos, Ken manages the physical inventory of frozen embryos preserved at Western. This involves maintaining expensive equipment stored in a room designed to keep precious materials secure. At Western, tanks are strapped to the wall to withstand earthquakes and wired with multiple monitoring systems for security from theft and risk of exposure to unsafe temperatures. Such safeguards are in place to protect embryos as much as fertility clinics for within the litigious US legal environment, fertility patients have successfully sued for damages to their frozen embryos that are legally considered forms of property (Andrews 1986; Litman and Robertson 1993).

Part of Ken's job also involves caring for abandoned embryos that are no longer being paid for by fertility patients. Unlike storage units filled with furniture that can be auctioned off to the highest bidder, determining what to do with the contents of abandoned embryo accounts is more complicated. "We don't discard them after patients stop payment because we are worried about lawsuits," expressed a concerned embryologist to a panel of lawyers convened at the 2011 American Society for Reproductive Medicine (ASRM) annual meeting. In the absence of legal guidance in US law about how to proceed when patients forgo payment, fertility clinics are stuck caring for a growing glut of reproductive remainders that nobody seems to want. Despite profits gained from storage fees, clinics have come to realize that freezing leftover embryos is risky business.

Saving embryos also poses burdens to IVF patients who agree to keep them on ice. Having backup embryos appeals to thousands like Sandra who desire the chance to become a parent, yet bearing the responsibility for leftovers comes with unanticipated costs. Beyond the financial expense of annual storage fees, many patients report feeling differently about their extra embryos at the end of IVF than at the time of freezing (Lyerly et al. 2010; Nachtigall et al. 2005). Some are overwhelmed by the responsibility of deciding what to do with embryos they no longer need but for various reasons are hard to let go. Other difficulties arise for patients in cases of divorce or death that have thrust leftover embryos into the middle of court proceedings that try to determine to whom they belong. Why freeze embryos at all when saving them presents legal, financial, ethical, political, and emotional burdens on the individuals and institutions tasked with their management?

As any computer user today knows each time she clicks to "save" an electronic file, saving is fundamentally about the preservation of something considered valuable. Despite the mainstream practice of freezing embryos in the United States, what to do with leftover embryos became the subject of ethical and political controversy as their perceived potentials came to acquire multiple kinds of value. In 1998, two events thrust the growing US embryo supply into a public debate about the kinds of "returns" that embryos saved today may bring tomorrow. In a university lab that year, scientists established the first human embryonic stem cell line from a donated leftover embryo (Thomson et al. 1998). Researchers began procuring these precious materials for their invaluable promise to revolutionize medicine in the quest for lifesaving cures for diseases like diabetes, Alzheimer's, and cancer (Scott 2006). Meanwhile, a pro-life Christian adoption agency created the first embryo adoption program for turning unwanted embryos into born children. It decried the devaluation of unborn human life by championing embryo personhood and facilitating their chances to be born.

The end of the twentieth century was a pivotal moment when IVF embryos, frozen and awaiting future use, became potential children waiting to be born as well as promissory research material. Hundreds of thousands of embryos in cryopreservation tanks across the United States were newly subjected to simultaneous regimes of value (Appadurai 1986a), which paved the way for organizations like REDEEM Biobank and the Blossom Embryo Adoption program to promise redemption for America's unwanted reproductive leftovers. Blossom and REDEEM emerged as the Christian and scientific vanguard on the new frontier of frozen embryo saving.

Despite their profound differences, I suggest they are both in the business of giving chances and share an orientation toward saving, valuing, and redeeming "life." The next section provides a closer look at each program's origin and mission to put into relief their similarities, laying a foundation for understanding how they approach the challenges of working with entities frozen in categorical ambiguity.

Mission: Redemption

Blossom Embryo Adoption Program: Giving Embryos a Chance at Birth

Topping the backside of a Blossom Embryo Adoption brochure is a statement explaining its mission and name: *Like a tiny seed, each embryo is small but contains everything it needs to blossom into a beautiful flower.* Blossom emphasizes the uniqueness contained within each embryo and its value as a preborn child deserving the chance to be born. The program operates within Pacific Adoptions, a Christian adoption agency that uses adoption as a model to facilitate the placement of remaining IVF embryos from donors to recipients. The first program of its kind, Blossom began in 1998 and is a leading proponent of extending the rights of persons to frozen embryos (George and Tollefsen 2008).[6]

"Embryo adoption comes out of what I consider to be a social need," explained Tim Shoener, the executive director of Pacific Adoptions.

> I think there is a problem. Part of the problem is as a society we are valuing life less, and I think one of the symptoms of that is five hundred thousand embryos frozen that we, as a society, have commodified because it's more economical. There's an attitude that as long as you are doing it, you might as well make it by the dozen. But life isn't cheaper by the dozen. Embryo adoption is a social movement to remind people that life begins at conception.

Blossom developed in response to a society that Tim argues has left some, literally, in the cold. The financial means to forward their mission is supported in part through program fees paid for by adoptive clients, but a sizeable share of their budget comes from federal grant monies. Since 2002, Pacific has received multiple millions of dollars in grants administered by the Office of Population Affairs (OPA) for increasing awareness around embryo donation and adoption.[7]

The Blossom Embryo Adoption program serves clientele across the United States from a modest office building in suburban California. Decorative wooden cutouts of the word "family" and a painting of Jesus walking hand in hand with a child decorate the reception area. Large metal filing cabinets lining the walls contain hundreds of Blossom client files, each brimming with personal letters, family photos, health histories, and matching preferences. A placard stating, "Life is fragile, handle with prayer," rests on a bookshelf alongside angel figurines and artful images of newborns by photographer Anne Geddes. A faux cryotank—used as a prop for Blossom outreach events—is stored in the conference room with an accompanying sign that reads, "Frozen Embryo Nursery: where children wait for their dreams to come true" (fig. 8.1).

Blossom's mission is twofold. Its first goal is to extend the dignity and protections of personhood to frozen embryos. In its official embryo adoption contract, it defines embryos as "preborn children who are endowed by God with unique characteristics and are entitled to the rights and protection accorded to all children, legally and morally." The program's second goal is to recognize all embryos as equally deserving of the opportunity to achieve their full potential through birth. "We consider that every embryo is a potential continued life," said Stacy, one of Pacific's social workers. "We want to give them all a chance." For this reason, they accept all frozen embryos into their program for placement with adoptive families, regardless

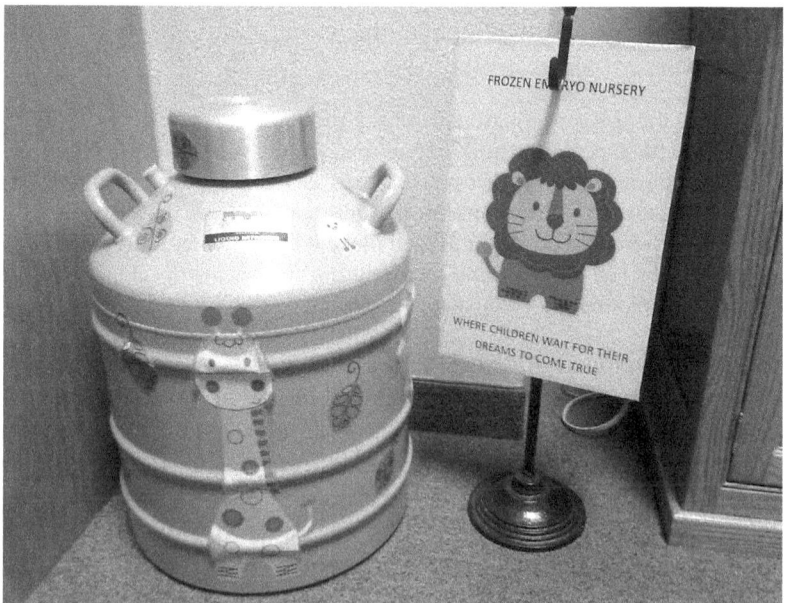

FIGURE 8.1: Faux frozen embryo cryotank in Blossom Embryo Adoption (© Risa D. Cromer).

of how they are ranked clinically. "We're not looking at the embryology report saying, 'Oh this one's not really worth saving.' That's just horrible to us," Stacy explained.

Embryology reports, or what some clinics describe as "baby's first report card," are documents produced by embryologists tasked with evaluating IVF embryos for clinical purposes. Embryo rankings inform decisions about which embryos to transfer for pregnancy, freeze for future use, or discard as medical waste. "Grading embryos is a beauty contest," explained Ken, Western's embryologist, based visually on cell symmetry, clarity, and absence of fragmentation; some get "slam dunk" grades while others barely pass with "cruddy" or "ugly" marks. Blossom challenges the prevailing fertility clinic perspective that embryos are rankable by striving to find homes for all considered equally valuable and awaiting the chance to be born.

REDEEM Biobank: Giving Scientists a Chance to Save Lives

A decade after Blossom began and few hundred miles away, REDEEM Biobank's first coordinator, Tanya, brainstormed a list of names for the new tissue bank she was tasked with launching. Tanya searched for words she hoped would capture the intent of newly funded embryo and oocyte resource center. She tried words like regrow, revive, strive, pluri, and potent, eventually settling on the program name, REDEEM, which she made into an acronym: **RE**generative **M**edicine and **D**iscovery through the **E**thical Procurement of **E**mbryonic **M**aterials.

The precipitating context for REDEEM Biobank was the limiting of US federal funding for research on human embryonic stem cell research (Scott 2006). Then President George W. Bush made an executive order in 2001 limiting federal funding that had a cooling effect on the burgeoning field of embryonic stem cell research in the United States (Korobkin and Munzer 2007). In response to the restriction of federal dollars, California voters passed Proposition 71 in 2004, a $3 billion measure to invest in embryo research over a decade's time. Proposition 71 established the granting agency, the California Institute for Regenerative Medicine (CIRM), that oversees the allocation of funds for human embryonic stem cell research within California. REDEEM Biobank was established through a CIRM grant meant to launch a Bay University–based frozen embryo resource center that would provide expert management of and access to human embryos for scientists around the university and state.[8]

REDEEM's mission, according to the current director, Dr. Pat Dunn, is altruistic and twofold. An early brochure describes the first part of its mission in formal terms: "To optimize the use of precious resources for an increased knowledge of basic science and the future treatment of human diseases." Embryonic stem cells are considered precious because they provide scientists alternatives and complements to traditional disease research based on animal models. Alzheimer's disease, for example, has proven challenging to mimic in animal models. Therapies with promising results in animals exhibiting a form of Alzheimer's have proved ineffective in human trials. Embryonic stem cells allow researchers to develop stem cell models in vitro that illuminate how Alzheimer's develops as a disease, as well as provide the opportunity to test drugs and therapies on actual human cells. Additionally, the pluripotency of embryonic stem cells (or the ability to become nearly any type of the two hundred cells that make up human bodies) has inspired researchers to try developing therapies, such as regenerative tissue transplants, that may replace cells destroyed by degenerative diseases like Alzheimer's. According to Dr. Dunn, the precious resources Bay scientists needed were embryos in order to continue the work of deriving stem cell lines, studying disease development, and testing therapies on human tissues that may lead to cures. "The situation was that we need blastocysts," she explained, "so we decided that we'll take any unwanted embryos."[9] This began a multiyear process of grant writing, legal consulting, and protocol development to be able to begin receiving the first frozen embryo donations from fertility patients in June 2008. REDEEM's welcoming of all leftover IVF embryos without exception to stage or grade made the REDEEM Biobank one of the premier donor sites for fertility patients around the country.

"By the same token," Dr. Dunn explained, "the biobank provides an ethical disposition option for patients who have a difficult decision to make." For IVF patients considering donation options, Dr. Dunn feels that REDEEM "provides a way for people to dispose of their embryos in an honorable way. It is truly a tremendous savior option for people who spent money and effort to get to what they achieved. Now they want to stop paying but don't want to throw embryos away and don't want to give them to someone else." Serving also as Bay University's IVF clinic director, Dr. Dunn is committed to patient satisfaction and maintaining the university's good name. Thus, one of her primary concerns is providing a smooth donation process for patients. For researchers and fertility patients, the dual purpose of REDEEM Biobank is to provide invaluable research opportunities for saving lives and a pleasant donation experience, as well as to serve as a salve for all.

Blossom and REDEEM are both in the business of giving chances and offering redemption for patients tasked with deciding whether to discard, donate, or keep frozen their remaining embryos.[10] Despite promises and best-laid plans, repurposing culturally and categorically ambiguous remainders sometimes involves waiting. The following ethnographic stories about frozen embryos donated through Blossom and REDEEM detail the challenges that result from frozen embryos being suspended in legal, logistical, and moral limbo.

Waiting in Limbo

According to the *Oxford English Dictionary*, "waiting" means active passivity. It connotes a state of lingering readiness for some future purpose, and it involves feelings of anticipation and hopeful preparations for something specific to happen. Frozen IVF embryos wait in multiple senses of the word. As cells, frozen embryos are paused developmentally in a state where they can linger in subzero temperatures for an indefinite period within fertility clinic freezers. They also serve as sites of symbolic activity where forms of value pool simultaneously and their seeming contradictions are vibrantly alive. "Waiting" is a lens that illuminates how frozen embryos are suspended categorically and practically within spaces of ambiguity. The following ethnographic cases of the Bower and Stoll embryos illustrate how frozen embryos slated for reuse through Blossom and REDEEM can remain in multiple forms of limbo, endlessly deferred and subject to various kinds of value extraction.

The Bower Embryo

"Dear Future Mom and Dad," begins the introductory letter in the Blossom Embryo Adoption application packet. Answers to common questions about embryo adoption—from eligibility requirements to agency fees to statistical chances for pregnancy—are outlined in the twenty-nine pages that follow. "We know you've come through a lot to get to this day," the letter continues. "The journey might have been frustrating, sorrowful, and intimidating. Maybe you're not sure if this is the right adoption choice for you." For lingering questions, applicants are encouraged to contact Monica, the Blossom program's friendly manager. It is her job, the signatory explains, to "work diligently on our behalf." The letter is signed, "Sincerely Yours, *The Waiting Embryos*."

"Waiting" is a word common within international adoption efforts that work to find homes for orphaned children. International adoption programs often justify their mission to rescue children "waiting" for homes based on a moral imperative to relieve human suffering.[11] Waiting is a discursive framework that helps to position orphaned children as legitimately deserving of care and resources (Briggs 2003). The Blossom Embryo Adoption program uses the language of waiting to convey that frozen embryos, like orphaned children, are equally at risk and in need of rescue (Joyce 2013).

In testimony to Congress during the 2001 hearings about federal funding for stem cell research, Blossom's first program manager articulates how embryo adoption advocates perceive the problem of leftover embryos as one of waiting:

> These children [embryos] are not a product of some wonderful medical research. They're a product of the fact that a huge problem exists, that too many embryos have been created ... This [embryo adoption] program is not here to provide a new way for families to get children. It's here to eliminate a problem that currently exists, in that *there are children waiting to be born. It's no different than an orphanage,* an orphanage that has never been really looked at as a really neat opportunity for somebody to add children to their families. It's been seen as a travesty that these children are not being parented. (US Congress 2002: 91; emphasis added)

Frozen embryos come through her testimony as imperiled children waiting expectantly, and the Blossom program as providing the moral response to rescuing those left in the cold.

Blossom is one of the most vocal advocates for treating all frozen embryos as awaiting a particular future—the chance to be born—though within their program, some will wait longer than others. In the fall of 2012 I received an e-mail soliciting interest in one of the Blossom program's "special circumstances" embryos. "Dear Risa, *Are you ready to take a leap of faith?* There is a single embryo waiting for its chance at life. Here is its story ..." The e-mail detailed Dennis and Jolene Bower's journey with infertility and their desire to find a willing recipient for their single remaining embryo. The Bower story illuminates how frozen embryos wait in many forms of limbo.

After several years of trying to become pregnant, Dennis and Jolene turned to IVF in 1998 for assistance. By year's end, they had given birth to one son and were paying storage for seventeen extra embryos. Nine years later, the Bowers decided they were done having children but wanted to help another become a parent. They completed the Blossom program application and ranked their top ten preferences for an adoptive family. They desired a religiously moderate, middle- to upper-class family that would be "college minded" for any resulting children. They also preferred a stay-at-home parent but would accept adopters employed full or part time. An ideal family would have no prior marriages and would be parenting fewer than three kids. With respect to contact with the adoptive family, the Bowers requested photos and letters at least once a year about any children born and were open to phone calls, e-mails, and visitations if the adoptive family would be too.

Within months of applying in 2007, the Bowers chose the Daniels to receive their entire batch of seventeen embryos. Once contracts transferring the legal ownership of the embryos were signed and notarized, FedEx delivered the embryos to the Daniels' clinic, where they went quickly to work. The Daniels felt their family complete after giving birth to their only child. They used five adopted embryos so decided to relinquish their rights to the remaining twelve embryos and "return" them to the Bowers. The Daniels continued to pay for storage of the embryos at their clinic while the Bowers reviewed adoptive family profiles in search of another match.

By early 2008, the Millers were selected as the second family to adopt the Bowers' remaining twelve embryos. In their profile letter, the Millers explained how they felt when they first learned about embryo adoption: "The idea that we could share a pregnancy or childbirth was an idea that we had long given up. To think that we could have that together is priceless! We are ready to give birth to twins or triplets. We have the room in our house and our hearts, and if that is what we are blessed with, we believe that God will give us the energy and patience too."

Despite their ready home and hearts, the Millers' prayers to become parents would not be answered. They thawed and transferred eleven embryos without becoming pregnant. At the end of the year, the Millers wrote to the Blossom program expressing their palpable grief and decision to be done:

> Unfortunately, we got the news yesterday that once again, we are not pregnant. We were better prepared for that possibility this time. But, it doesn't hurt any less. The hardest part is that we are finished. We are financially strapped, emotionally worn out, and just plain tired of this battle to become parents. Somehow, this is God's plan for our lives ... I have spent so many years dreaming of being a Mommy, I don't know what else to dream.

After adoption by two families, the Bowers' original group of seventeen embryos had become a batch of one, and the Millers were done trying to become pregnant. Even though they were grieving the end of their embryo adoption journey, the Millers agreed—like the Daniels previous to them—to follow through on their commitment to pay storage for the "returned" embryo until it could be rematched: "Of course we can store the little guy until you find a place for him," the Millers assured in an e-mail to Blossom staff. "We continue to try to tell people about you guys. We have nothing but positive things to say about our experience even though didn't work out for us."

Finding a place for the single Bower embryo proved difficult and slow. The original seventeen embryos matched within months, but as the batch reduced in number and the freeze date grew more distant, the single Bower embryo's "special circumstances" became more pronounced. According to the embryology report, the remaining embryo was frozen in 1998 using a slow-freeze method at day 1 of development, or at the two pronuclear (2PN) stage. In clinical terms, this is not a "slam dunk" embryo but one with questionable potential to establish pregnancy. Coupled with its solo status, the Bower embryo was rejected numerous times by potential adopters uninterested in the low-graded embryo.

When a possible adoptive family applied to the Blossom program in 2011, Kathy, Blossom's social worker, contacted the Millers' clinic for an updated embryology report. She was surprised to learn from the clinic that in the span of a few years, the Millers had divorced, moved out of state, and left an unpaid $1,500 storage bill for the Bower embryo. According to the fertility clinic, the Millers retained legal and financial rights and responsibilities as the intended parents who brought the donated embryo into their clinic for use. Before the clinic would ship the embryo to another adoptive family, the bill needed payment and the Millers needed to sign paperwork approving its release for shipment. Unless Blossom staff could surmount these obstacles, the Bower embryo would wait indefinitely in legal, financial, and frozen limbo.

Blossom staff consulted with Tim, the Pacific Adoptions' executive director, and came up with a plan: the Blossom program would pay the storage fee of $1,500 to release the embryo for relocation to a long-term storage facility where Blossom was prepared to assume legal and financial responsibility as the storage account holders, if the Millers were not willing to, until the embryo could be matched with its "forever family." Assuming such financial and legal burdens is atypical for the Blossom program but, from my field research observations, is also not unique. "Since the embryo was already in the program," Sarah, the Blossom program assistant, explained, "we assumed responsibility. There were seventeen embryos to begin with. We don't want to give up on one."

Overcoming the second obstacle depended on the cooperation of the Millers. Locating and convincing them to help was one challenge; obtaining signatures from a divorced couple on the clinic release forms posed another. Sarah vented to me one afternoon over the fax machine: "It is so frustrating because *they* are the adoptive family. They didn't transfer the embryo, and they are refusing to pay their storage fees. And now they're not getting around to sign a simple form." Kathy surmised that the Millers' behavior might be because they do not consider the Bower embryo to be "theirs," or to belong to them. As Blossom staff worked doggedly to maneuver the ever-changing circumstances that halt, slow, and ensnare the waiting Bower embryo's chance to be born, they felt the pressure of time.

Time, though, is something alleged to be on the side of frozen embryos. When stored at −196 degrees Celsius, one embryologist explained confidently, "You can keep them in there for the next twenty years, and nothing is going to happen." But Blossom staff has learned from experience that time does not stand still for waiting embryos. As the clock ticks outside the freezer, the "viability" of the leftovers suspended within are subject to countless factors that cause frozen embryos to "age" and bear the social marks of time. The adoptability of the Bower embryo was colored by many variables, including being unused by two adoptive families, abandoned in the wake of divorce, and subjected to changing laws, political climates, clinical best practices, and personal financial circumstances. Cryopreservation might pause the effects of time at the cellular level, but waiting embryos are actively subject to social forces at play beyond the cryotank.

The Blossom team expressed "very little hope that the embryo will result in a pregnancy," yet invested time, energy, resources, and prayers into finding a place for it. After nearly a year of effort, Blossom had received and forwarded all of the corresponding paperwork to release the Bower embryo from the Miller clinic. With the bill paid and forms signed, it was shipped from the Millers' clinic to a long-term storage facility, "proving," in Sarah's view, "that miracles do happen." Monica was so personally overjoyed with her staff's efforts that her husband sent her a "Hallelujah" card as a personal congratulation that she displayed on her desk. A week later, I received the

"Dear Risa" e-mail that was sent to all prospective adopters advertising the readiness of this single embryo, waiting expectantly for its chance to be born.

Upon arriving to the new storage facility, the Bower embryo faced a different categorical problem: to whom does the embryo belong? On paper, the storage facility legally recognizes the Millers as the storing clients; meanwhile, Blossom is paying the storage bill, and the Bowers have the ultimate authority to decide when, where, and to whom their remaining embryo may be adopted. Without a clear steward, owner, parent, payer, or categorical place for understanding how frozen embryos relate to their many handlers, the Bower embryo—like all frozen embryos—are forms of matter representing broader cultural matters that remain ambiguous and unresolved. As Blossom seeks to "place" frozen embryos with adoptive families, waiting embryos like the Bowers' provoke fundamental questions about belonging. Where do frozen embryos belong—materially, culturally, categorically?[12]

The Bower embryo, like all frozen embryos, lingers in limbo as an unwanted yet unwastable reproductive remainder. It waits, according to the Blossom program, like an orphaned child in need of a forever family. The Bower story illustrates how it waits for many other reasons, in frozen and categorical limbo, that span the gamut of legal, financial, relational, emotional, geographic, and clinical circumstance. The ambiguities of frozen embryos and the waiting conditions they produce are not unique to embryos regarded as unborn children or slated for adoption. Waiting is also a common circumstance produced around embryos donated for scientific research where categorical ambiguities render some frozen embryos stuck and endlessly deferred. Angela Stoll's donation to REDEEM Biobank illuminates the similarities between the ways frozen embryos wait across the programs striving to redeem them.

The Stoll Embryo

One embryo remained at the end of a series of unsuccessful IVF cycles for Angela Stoll. At the age of forty-four, she was motivated to start fresh and soon. Her fertility clinic would not begin a new IVF cycle for her until she decided what do to with the one embryo in storage that she did not plan to use. Discarding the remaining embryo would have been the quickest and least expensive option, but she decided to donate it for research. Angela's donation involved a logistically intensive coordination effort, a significant expense to her, and an indeterminate period of waiting for her leftover embryo.

Angela called REDEEM Biobank directly and reached the program coordinator, Donna, who explained to her the process for donation. REDEEM follows inquiries like Angela's with an introductory letter, a six-page informed consent form, and a voluntary health questionnaire. Once patients review the materials at home and decide to donate, they mark their preferences for how researchers may use their embryos, sign the donation forms in the presence of a witness, and return the paperwork to REDEEM for review and final consenting over the phone. REDEEM considers embryo donations "complete" when patients submit their consent forms free of error to the biobank office. The "completion" of donation for fertility patients marks merely the next chapter for REDEEM; with a consent on file, Donna orchestrates the retrieval of the embryos from IVF clinics and, once they arrive to the biobank, stores them in preparation for use by research scientists.

REDEEM established their multistage, interactive, and donor-driven consenting process in response to heightened ethical and political concerns in the United States about using embryos for stem cell research (Thompson 2014). Moreover, REDEEM designed its protocol to promote donor choice through the core principles of informed consent. The consenting process builds in extra time for donors to consider their preferences, make a decision, and review it before finalizing their embryo donation. "It's kind of a big decision," explained Tanya, the biobank's first coordinator. "It is an embryo and they did pay $25,000 to get it made and paid storage for however long."

Embryo donation is also a big deal in terms of human research standards. Every word of REDEEM's paperwork and protocols have been approved under the scrutiny of Bay University's Institutional Review Board and the Stem Cell Research Oversight (SCRO) committee to ensure that the biobank's donation process is in compliance with US, state, and CIRM grant guidelines concerning the protection of embryo donors. It took nearly two years for the protocols to be developed and approved by the various oversight committees before the first embryos were accepted. All of these efforts reflect what Charis Thompson (2014) describes as "ethical accounting procedures," or the principled actions and efforts to adhere to standards for utilizing donated embryos in research. Within the bureaucratic procedures of consent protocols and committee approvals, it is possible to see how the "pro-cure" rhetoric of saving lives is translated into "procuratorial" practices that shape the way human embryonic stem cell research is conducted in labs at universities like Bay.[13]

On paper, the process for donating embryos to REDEEM is straightforward, though everyday life within the biobank office revealed that donating embryos is rarely simple and typically involves a lot of waiting. Beyond the intentional design of the donation protocol to give people ample autonomy and time for thoughtful decision making, some facets of the process present unforeseen obstacles that slow, interrupt, and sometimes preclude donations altogether. Pamela, a biobank assistant, shared her impression of the problem: "Eighty percent of the consents are incorrectly filled out, and 50 percent of patients don't respond to requests to correct their forms. The other 50 percent don't have return phone numbers or clinic contact information written on the form to reach them, so that list grows." Pamela and other biobank staff juggle dozens of calls and e-mails each day from prospective donors. Each manages multiple "problem" files containing incomplete consent forms for one reason or another.

While biobank staff reports "drowning in work," potential donors are also sometimes overwhelmed. For example, Jack, a divorced veteran wanting to donate his leftover embryos, sent in his consent forms twice, both times with mistakes. "When I got him on the phone," Pamela said, "he was exasperated, saying that he has mental and physical trauma that makes filling out the form impossible and that he wants me to just fill it out for him. I told him, 'I am so sorry, but we just cannot do that.'" While REDEEM makes every effort to provide a smooth donation process for donors, staff and patients experience many bumps along the way. Strict institutional protocols that strive for the highest ethical standards and legal safeguards within the highly political climate of human embryo research cause precious leftover embryos to wait.

Time was of the essence for Angela, and she could not afford a slow, bumpy donation process. She asked Donna for help, and together they tackled the donation forms at Angela's fertility clinic and the REDEEM Biobank with uncommon precision. Within a week's time, Angela's completed consent materials were in Donna's hands, followed soon after by a thank-you card expressing Angela's gratitude.

Before she could move forward with her next much anticipated IVF cycle, Angela faced an additional hurdle: news that her fertility clinic would continue charging her for storing the embryo still at her clinic and would not cease until the biobank physically retrieved the embryo. As discussed above, fertility clinics that manage frozen embryo inventories assume all of the legal and financial liabilities that responsibility entails. As a result, many clinics consider embryo donations to be complete not merely when paperwork is signed but when embryos are physically removed from cryotanks. As REDEEM builds relationships with fertility clinics around the United States, it explains that the biobank will likely take a few months from the time a patient submits her donation consent forms to when her embryos will be fetched from the clinic because of shipping costs. It is more economical for REDEEM to send a $200 portable shipping container via UPS to a clinic with enough embryos to fill it rather than sending a tank for each individual donor. In order to relieve patients of any financial burdens for donating their embryos, REDEEM asks clinics to discontinue charging patients cryostorage when they complete the biobank paperwork rather than when embryos physically leave the clinic freezer.

Angela's clinic would not budge. "To keep the patient's one embryo is costing them nothing," vented Dr. Dunn, the biobank director, on hearing the news of Angela's storage bill. "I know they run a business, but the patient has already paid them $25,000 and they want to continue charging her?" To avoid the expense of another year's cryostorage bill, Angela took matters into her own hands. She commissioned a private reproductive tissue shipping company to transport her single embryo across the country from her home clinic to REDEEM Biobank. Donna was touched by Angela's determination and was curious how much her donation to REDEEM Biobank cost her. This was the first time the biobank had a donor willing to pay the shipping costs associated with donation.

Before embryos arrive to REDEEM, staff does not know what kinds of materials they are receiving, which means every tank is greeted with some level of anticipation. REDEEM accepts all embryos for donation, from day 1 2PN stage embryos to day 6 hatching blastocysts, in batches of any number, frozen using any variety of method, and made using the sperm and eggs of the intended parents or gamete donors. When Angela's embryo arrived with a shipping receipt of $500, Donna and Wendy were shocked and even more interested to learn what kind of embryo she had given them. They flipped through seventeen pages of paperwork to come to the embryology report describing the leftover treasure that lay inside: one vitrified day 6 blastocyst made with donor egg and sperm.

Clinically, Angela's embryology report showed that her embryo received the highest grades for establishing pregnancy. Had she instead decided to donate it to the Blossom program for adoption, Angela's embryo likely would have been chosen quickly from the matching pool of adopters wanting to become parents. But for RE-

DEEM, this embryo was one in a long line of donated remainders that had little value to researchers at the time. "Ninety percent of the embryos in the bank are blastocysts," Wendy explained to me as she transferred Angela's embryo into the biobank. "And they probably represent 10 percent of the value." The need for blastocysts to derive human embryonic stem cell lines was the original impetus for the grant that established REDEEM Biobank. Since 2008, new developments in stem cell biology shifted research needs in ways that rendered the common blastocyst leftovers from IVF clinics less useful for researchers. Wendy chose a spot for Angela's embryo in the top rack of the biobank because the lower racks, deep in the liquid nitrogen, were reserved for the most precious embryos aligned with the lab's current research needs: day 1 2PNs.[14]

Tanya, the first REDEEM coordinator, believes most donors "are hopeful that someone can learn something and get useful information instead of us putting it in the trash." Even though Angela's embryo made it successfully to the biobank, it remains in limbo slated for neither a researcher's petri dish nor the trash. Similar to the early twentieth-century embryological specimens preserved in formaldehyde that Lynn Morgan discovered in the basement of Mount Holyoke College, frozen blastocysts like Angela's are "specimens [that] proliferated not because anyone necessarily wanted many of them, but because it was awkward to refuse a well-intentioned gift" (Morgan 2009: 61). Instead, chances are good that the donated Stoll embryo will hang out in frozen, categorical, and value-driven limbo for an unforeseeable future.

"You never know down the road what you're going to need," expressed Caitlin, a senior postdoc in the lab that houses the REDEEM Biobank. "Blastocysts are not being used as quickly as they are coming in, but I think things go full circle sometimes. I feel like science is like that." At the same time, she questions if the resources being used to receive and store embryos like Angela's "that aren't going to be used, or at least not going to be used yet" make sense. She and other biobank staff worry about the day their tank reaches capacity—filled with embryos without present-day purpose—and wonder what they would do next: invest in a new tank, limit the kinds of donations they receive, or close the bank altogether, like many other tissue banks in universities around the country have done? Caitlin believes that if REDEEM is going to continue receiving embryos like Angela's, "we should be doing more collaborations with other researchers around the university to get the blastocysts used. I'm not really sure that's being done."

Angela's donation to REDEEM was driven by a clear purpose and mission that helped her surmount many circumstantial snafus common to fertility patients wanting to donate embryos to the biobank for scientific research. Like the Bower embryo, Angela's embryo waits in frozen and categorical limbo for potential change beyond the freezer to render it "viable" material for research. This possibility hinges on various circumstances, such as scientific funding priorities and opportunities, developments in science and technology, and shifting political terrains. Both the Bower and Stoll embryos overcame ensnarement yet still wait, deferred to the future within America's leading organizations that promise redemption for unwanted yet unwastable leftovers.

Redeeming Embryos: A Different View

This chapter has explored how frozen embryos in the contemporary United States wait as a result of categorical ambiguities and within the comparative contexts of two embryo redemption programs. The stories of the Bower and Stoll embryos reveal the challenges of locating leftover embryos in culturally coherent places. Together, each case illustrates various ways that frozen embryos actively linger in material and symbolic ambiguity.

I have argued that waiting embryos are considered burdensome problems for fertility clinics and patients alike. Their inability to be wasted despite no longer being wanted provoked controversy in the United States and begged a solution. Organizations like Blossom and REDEEM emerged to help find a place for categorically ambiguous embryos. Described in this way, frozen embryos may be understood in anthropologist Mary Douglas's (1966) terms as "matter out of place." To resolve the cultural problem of these disruptive matters, Blossom and REDEEM provide frozen embryos opportunities for redemption via the birth of children and promissory scientific outcomes. Moreover, Blossom and REDEEM represent broader desires for redemption in the United States today among its families and communities, its economic possibilities, and its vitality as a nation.

To conclude, I want to suggest an alternative perspective on how and why frozen embryos wait. What if the ambiguous qualities characteristic of today's frozen IVF embryos were not contradictions to be resolved or "matters" needing to find a "place"? What if frozen embryos' seeming contradictions were socially constructed and politically maintained tensions left productively unresolved? Answers to these questions invite a closer consideration of redemption and value.

In the wake of the controversies arising in the United States around the coinciding events of 1998, frozen embryos acquired what anthropologist Sarah Franklin (2006) describes as "double reproductive value": as unborn children and promissory research material. I have argued that waiting is the effect of the categorical ambiguities that keep frozen embryos in various forms of limbo. In light of Franklin's observation, I suggest that waiting in limbo is the means through which frozen embryos' double reproductive value is extracted and made productive. Waiting embryos, like the Bower and Stoll embryos, are not simply waiting to be redeemed (as born children or research materials), but are redeemed by waiting (via extraction of value). Examples from each case hint at how being in limbo and deferred to the future rendered the Bower and Stoll embryos available for the extraction of various forms of value, such as economic value, use value, moral value, and so on. In this light, waiting is an intricate system for converting and extracting forms of value from frozen leftover embryos, or America's intentionally ambiguous reproductive assets. This alternative perspective on the redemption of embryos casts a different light on the value of the Bower and Stoll embryos; rather than being regarded as "problem cases," they become paradigmatic exemplars of how embryos are redeemed in the United States today through the extraction of value, sometimes in the name of distinct, though not dissimilar, values.

Blossom and REDEEM are pioneering entrepreneurs in the business of giving chances, saving lives, and—in complex ways—redeeming value. Caitlin, a Bay University researcher, expressed a sentiment shared by staff of the Blossom program and REDEEM Biobank: "While there might be some logistical problems with the biobank, I think in the end I would prefer to deal with the logistics and have the bank than not." For all of the stressors associated with redeeming leftover frozen IVF embryos, the Blossom Embryo Adoption program and REDEEM Biobank work for redemption because, put simply, it is *worth* it.

Acknowledgments

I would like to thank the Wenner-Gren Foundation for Anthropological Research, the Charlotte W. Newcombe Foundation, and the Graduate Center of the City University of New York for providing generous support for this research.

Risa D. Cromer is a postdoctoral fellow in Stanford University's Thinking Matters program and Center for Biomedical Ethics. Her research investigates reproductive and racial politics within bioethical controversies. Her current book project examines the fates of frozen embryos left over from IVF and saved for the future.

Notes

1. 2PN, or two pronuclear, describes the first developmental stage of a fertilized egg. A 2PN embryo typically emerges eighteen to twenty-four hours after fertilization on the first day of in vitro fertilization. Each pronucleus at the 2PN stage comes from the nucleus of the sperm and the egg.
2. Considerable efforts have been made to protect the confidentiality of individuals and organizations participating in this study. All names in this chapter are pseudonyms.
3. A 2003 RAND study report of four hundred thousand frozen embryos is the most commonly cited estimate (Hoffman et al. 2003). More recent estimates suggest increasing numbers, such as the six hundred thousand figure reported by the Office of Population Affairs (2017).
4. This project is based on two hundred formal, semistructured interviews with two main groups: relevant professionals (e.g. doctors, nurses, social workers, counselors, embryologists, students, study coordinators, lab managers, researchers, lawyers, theologians, bioethicists) and program participants (e.g. donors, recipients, fertility patients). Interviewees included forty embryo adoption professionals, sixty-five embryo adopters, thirty embryo donors, thirty fertility clinic professionals, ten tissue bank professionals, ten stem cell scientists, ten stem cell biology PhD students, and five active fertility patients. Additionally, I attended weekly sessions at Blossom to match embryos between giving and receiving clients, required parenting classes, and counseling sessions to learn how embryos become persons and clients become parents. At Western Fertility Clinic and Bay University's IVF clinic, I shadowed fertility clinic staff to understand how embryos are made, managed, used, stored, and planned for within a clinic. At REDEEM, I followed donated embryos to the biobank where coordinators facilitate their transformation from "excess reproductive waste" into "precious materials" for use by stem cell scientists.
5. California is home to 75 fertility clinics, the largest concentration in the United States, and Western Fertility Clinic is one of the highest volume clinics in the state. Of the 25,500 IVF cycles performed across California in 2012, clinics ranged in performing 16 to 1,737 cycles, for a state average of 331 cycles per clinic. Western, by comparison, completed 1,500 IVF cycles. One reason Western attracts a large number of patients is because they are one of the first clinics on the West Coast to offer cutting-edge and controversial technologies like intracytoplasmic

sperm injection (ICSI), preimplantation genetic diagnosis (PGD) for disease screening and gender selection, and egg freezing. Western supports its large patient load by keeping a full schedule of back-to-back procedures most mornings and maintaining a robust staff of eleven reproductive endocrinologists, ten embryologists, a dozen accounting specialists, and more than thirty nurses and medical assistants. IVF statistics like these are tracked in the United States per mandate by the Fertility and Clinic Success Rate and Fertilization Act of 1992. This congressional law requires clinics in the United States performing IVF to annually submit data to the Centers for Disease Control, which compiles and publishes annual reports of reproductive technology trends and outcomes (see CDC et al. 2014).

6. The Blossom program's attribution of personhood status to embryos echoes a longer tradition in US history in which the embryo, fetus, and child have been mobilized as *"the* plausible innocent in whose name moral claims can be made" (Comaroff 1997). The symbol of the child, for example, served as the linchpin in abolitionist rhetoric, the cornerstone of border politics, and a decoy for transforming social problems into private matters (Levander 2006). The fetus, as historian Sarah Dubow (2010) shows, figured centrally within US politics and imaginaries in "protean" ways from the 1870s to end of the second millennia toward inhabiting what Lynn Morgan (2009) describes as the cultural status of "icons of life."

7. In 2002, the United States Congress began earmarking funds for an Embryo Donation and/or Adoption Awareness campaign authorized under Section 301 of the Public Health Service Act. The grant program was congressionally approved and backed by the George W. Bush administration, though its origins are attributed to Pennsylvania Senator Arlen Specter, a pro-choice, pro–stem cell research Republican. Specter chaired the subcommittee responsible for determining appropriations for the Departments of Labor, Health and Human Services, and Education. In reference to the $123 billion bill his subcommittee helped pass and the $2.9 billion increase in funding to the National Institutes of Health (NIH) that year, he spoke about the allocation of $1 million to a new program for embryo adoption awareness during the 20 December 2001 Senate session: "A controversy has arisen because some object to stem cell research because they are extracted from embryos and embryos can produce life ... If any of those embryos could produce life, I think they ought to produce life and not to be used for stem cell production. But if they're not going to produce life, then why throw them away? Why not use them for saving lives? We put into this bill $1 million, sort of a test program on embryo adoption. Let us try to find people who will adopt embryos and take the necessary next steps on implanting them in a woman to produce a life. If that can be done and use all of the embryos, that would be marvelous to produce life. But where those embryos are going to be discarded, then I think the

sensible thing to do is to use them for saving lives" (US Senate 2001). Since 2002, three to five awardees each year have received hundreds of thousands of public grant dollars to "educate Americans about the existence of frozen embryos (resulting from in vitro fertilization), which may be available for donation/adoption for family building" (OPA 2017). The OPA within the Office of the Assistant Secretary of Health (OASH) administers the grant of which Pacific Adoptions has been a regular recipient for promoting embryo adoption broadly and the Blossom program in particular.

8. In his first presidential address to the nation on 9 August 2001, Bush announced an executive policy that limited federal funding for embryonic stem cell research. This policy restricted federal dollars (though not state or private funds) to research on human embryonic stem cell lines derived before August 2001. Several states responded by filling gaps in federal funding with state tax funds while others went beyond the Bush policy to pass laws prohibiting all human embryonic stem cell research at the state level. California's Proposition 71, the Stem Cell Research and Cures Initiative of 2004, was an ambitious state effort to allocate public funds—more than $3 billion—for human embryonic stem cell research. Bush remained firm in his position and vetoed two bipartisan congressional bills in 2006 and 2007 that would have freed up federal funding for embryonic stem cell research. In March 2009, then President Barack Obama announced an executive order revoking the Bush-era funding restrictions that allowed the NIH to fund research on a wider range of embryonic stem cell lines. (For more substantive analyses of US ethical frameworks concerning on embryonic stem cell advancements in the United States, see Thompson 2014. For comparative case studies in and across other countries, see Bharadwaj and Glasner 2009 on India, Franklin 2006 on the United Kingdom, and Gottweis et al. 2009 for an international perspective.)

9. Blastocyst describes the developmental stage of a fertilized egg typically five to six days after fertilization.

10. Comparable statistics for each site are not available, but here are some data compiled with the assistance of staff from each site. From 1 October 2010 to 31 March 2013, the REDEEM Biobank completed the informed consent process for 841 donors. This is a 208 percent increase in donations from 1 January 2008 to 30 September 2010. Donations to REDEEM were received from 50 clinics across 26 states. At the time of donation, female donor ages ranged from 25 to 53 years old with an average age of 37.8 years; male donors ranged from 28 to 65 years old with an average age of 40.6 years. On average, patients donated 4 embryos and waited 3.8 years before donating. From the program's inception in 1998 to March 2012, Blossom has worked with 300 of the 486 IVF clinics in the United States. More than 500 individuals and couples donated 5,000 embryos through the Blossom program, resulting in the birth of 500 babies.

11. International adoption mobilizes humanitarian frameworks in ways that "salvage" (Anagnost 2000) forgotten kids by transforming them from "needy objects" into "treasured subjects" (Eng 2003).
12. The emergence of programs like Blossom and REDEEM to find a place for categorically ambiguous embryos suggest they may be, in anthropologist Mary Douglas's (1966) terms, "matter out of place." But stories from within each program reveal the challenges of locating leftover embryos in culturally coherent places. Frozen embryos are neither in nor out of place, but operate as entities that remain conveniently in limbo.
13. Within the climate of federal funding restrictions from 2001 to 2009, Thompson observed a "pro-cure" rhetoric emergent in California that reframed embryonic stem cell research as an ethical practice toward saving lives and regenerating the state's economy. The pro-cure, pro-embryonic stem cell framework is premised on "principled procurement and curatorial practices" (Thompson 2014: 45). REDEEM Biobank came about amid high-profile debates about ethical mandates to save lives—of the unborn embryos or people living with debilitating disease. REDEEM espouses its principled procurement efforts by retrieving embryos from particular sources (e.g., repurposing leftover IVF embryos from former fertility patients rather than creating new embryos for research) and adhering to guidelines for their use set out by organizations such as the NIH, CIRM, and Bay University's SCRO committee. (For history about SCRO committees, see Thompson 2014, esp. pp. 35–36 and 106–107.)
14. Recall that at the 2PN stage, the Bower embryo was repeatedly passed over by potential Blossom adopters because its chance of establishing a pregnancy were considered low. Had the Bowers donated their frozen embryo to REDEEM Biobank, it likely would have been received with glee by researchers who consider the rare 2PN stage embryos invaluable for their human development experiments. The Bower and Stoll embryo stories reveal how an embryo's context shapes and informs its perceived value, even in programs like REDEEM and Blossom where embryos are regarded as invaluable.

References

Anagnost, Ann. 2000. "Scenes of Misrecognition: Maternal Citizenship in the Age of Transnational Adoption." *Positions* 8 (2): 389–421.

Andrews, Lori B. 1986. "My Body, My Property." *The Hastings Center Report* 16 (5): 28–38.

Appadurai, Arjun. 1986a. *The Social Life of Things: Commodities in Cultural Perspective*. New York: Cambridge University Press.

———. 1986b. "Introduction: Commodities and the Politics of Value." In Appadurai 1986a, 3–63. Cambridge: Cambridge University Press.

Bavister, Barry D. 2001. "How Animal Embryo Research Led to the First Documented Human IVF." *Reproductive BioMedicine Online* 4 (1): 24–29.

Berlant, Lauren Gail. 1997. *The Queen of America Goes to Washington City: Essays on Sex and Citizenship.* Durham, NC: Duke University Press.

Bharadwaj, Aditya, and Peter E. Glasner. 2009. *Local Cells, Global Science: The Rise of Embryonic Stem Cell Research in India.* New York: Routledge.

Briggs, Laura. 2003. "Mother, Child, Race, Nation: The Visual Iconography of Rescue and the Politics of Transnational and Transracial Adoption." *Gender and History* 15 (2): 179–200.

Casper, Monica J. 1998. *The Making of the Unborn Patient: A Social Anatomy of Fetal Surgery.* New Brunswick, NJ: Rutgers University Press.

Centers for Disease Control and Prevention (CDC), American Society for Reproductive Medicine, and Society for Assisted Reproductive Technology. 2014. *Assisted Reproductive Technology: Fertility Clinic Success Rates Report.* Atlanta: US Department of Health and Human Services.

Comaroff, Jean. 1997. "Consuming Passions: Child Abuse, Fetishism, and 'the New World Order.'" *Culture* 17 (1–2): 1–9.

Douglas, Mary. 1966. *Purity and Danger: An Analysis of Concepts of Pollution and Taboo.* London: Routledge.

Eng, David L. 2003. "Transnational Adoption and Queer Diasporas." *Social Text* 21 (3): 1–37.

Farman, Abou. 2013. "Speculative Matter: Secular Bodies, Minds, and Persons." *Cultural Anthropology* 28 (4): 737–759.

Ferry, Elizabeth Emma, and Mandana E. Limbert. 2008. *Timely Assets: The Politics of Resources and Their Temporalities.* Santa Fe, NM: School for Advanced Research Press.

Foote, R.H. 2002. "The History of Artifical Insemination: Selected Notes and Notables. *Journal of Animal Science* 80: 1–10.

Franklin, Sarah. 2006. "Embryonic Economies: The Double Reproductive Value of Stem Cells." *BioSocieties* 1 (1): 71–90.

Franklin, Sarah, and Celia Roberts. 2006. *Born and Made: An Ethnography of Preimplantation Genetic Diagnosis.* Princeton, NJ: Princeton University Press.

Gammeltoft, Tine. 2014. *Haunting Images: A Cultural Account of Selective Reproduction in Vietnam.* Berkeley: University of California Press.

George, Robert, and Christopher Tollefsen. 2008. *Embryo: A Defense of Human Life.* New York: Doubleday.

Ginsburg, Faye D., and Rayna Rapp. 1995. *Conceiving the New World Order: The Global Politics of Reproduction.* Berkeley: University of California Press.

Gottweis, Herbert, Brian Salter, and Cathy Waldby. 2009. *The Global Politics of Human Embryonic Stem Cell Science: Regenerative Medicine in Transition.* New York: Palgrave Macmillan.

Hartouni, Valerie. 1999. "Reflections on Abortion Politics and the Practices Called Person." In *Fetal Subject, Feminist Positions,* ed. Lynn L. Morgan and Meredith Wilson Michaels, 296–303. Philadelphia: University of Pennsylvania Press.

Hoffman, David I., Gail L. Zellmann, C. Christine Fair, Jacob F. Mayer, Joyce G. Zeitz, William E. Gibbons and Thomas G. Turner. 2003. "Cryopreserved Embryos in the United States and Their Availability for Research." *Fertility and Sterility* 79 (5): 1063–1069.

Inhorn, Marcia Claire, and Frank van Balen. 2002. *Infertility Around the Globe: New Thinking on Childlessness, Gender, and Reproductive Technologies.* Berkeley: University of California Press.

Joyce, Kathryn. 2013. *The Child Catchers: Rescue, Trafficking, and the New Gospel of Adoption.* New York: Public Affairs.

Korobkin, Russell, and Stephen R. Munzer. 2007. *Stem Cell Century: Law and Policy for a Breakthrough Technology.* New Haven, CT: Yale University Press.

Layne, Linda L. 2003. *Motherhood Lost: A Feminist Account of Pregnancy Loss in America.* New York: Routledge.

Leibo, S. P. 2004. "The Early History of Gamete Cryobiology." In *Life in the Frozen State*, ed. Barry J. Fuller, Nick Lane, and Erica E. Benson, 347–370. New York: CRC Press.

Levander, Caroline Field. 2006. *Cradle of Liberty: Race, the Child, and National Belonging from Thomas Jefferson to W.E.B. Du Bois.* Durham, NC: Duke University Press.

Litman, M.M., and G.B. Robertson. 1993. "Reproductive Technology: Is a Property Law Regime Appropriate?" In *Overview of Legal Issues in New Reproductive Technologies*, vol. 3, Royal Commission on New Reproductive Technologies, 201–233. Ottawa: Supply and Services Canada.

Lyerly, Anne Drapkin, Karen Steinhauser, Corrine Voils, Emily Namey, Carolyn Alexander, Brandon Bankowski, Robert Cook-Deegan, et al. 2010. "Fertility Patients' Views about Frozen Embryo Disposition: Results of a Multi-Institutional U.S. Survey." *Fertility and Sterility* 93 (2): 499–509.

Morgan, Lynn M. 2003. "Embryo Tales." In *Remaking Life and Death: Toward an Anthropology of the Biosciences*, ed. Sarah Franklin and Margaret Lock, 261–292. Santa Fe, NM: School of American Research Press.

———. 2009. *Icons of Life: A Cultural History of Human Embryos.* Berkeley: University of California Press.

Morgan, Lynn M., and Meredith Wilson Michaels. 1999. *Fetal Subjects, Feminist Positions.* Phildelphia: University of Pennsylvania Press.

Nachtigall, Robert D., Gay Becker, Carrie Friese, Anneliese Butler, and Kirstin MacDougall. 2005. "Parents' Conceptualization of Their Frozen Embryos Complicates the Disposition Decision." *Fertility and Sterility* 84 (2): 431–434.

Office of Population Affairs (OPA). 2017. "Embryo Adoption." *Health and Human Services,* 3 March. https://www.hhs.gov/opa/about-opa-and-initiatives/embryo-adoption/index.html.

Rapp, Rayna. 1999. *Testing Women, Testing the Fetus: The Social Impact of Amniocentesis in America.* New York: Routledge.

Reagan, Leslie. 2010. *Dangerous Pregnancies: Mothers, Disabilities, and Abortion in Modern America.* Berkeley: University of California Press.

Roberts, Elizabeth F.S. 2012. *God's Laboratory: Assisted Reproduction in the Andes.* Berkeley: University of California Press.

Rothman, Barbara Katz. 1986. *The Tentative Pregnancy: Prenatal Diagnosis and the Future of Motherhood.* New York: Viking.

Scheper-Hughes, Nancy. 1992. *Death Without Weeping: The Violence of Everyday Life in Brazil.* Berkeley: University of California Press.

Scott, Christopher Thomas. 2006. *Stem Cell Now: From the Experiment that Shook the World to the New Politics of Life.* New York: Pi Press.

Sheskin, Arlene. 1979. *Cryonics: A Sociology of Death and Bereavement.* New York: Irvington Publishers.

Smith, Christopher H. 2001. "Freeze Frames: Frozen Foods and Memories of the Postwar American Family." In *Kitchen Culture in America,* ed. Sherrie A. Inness, 157–176. Philadephia: University of Pennsylvania Press.

Steingraber, Sandra. 2001. *Having Faith: An Ecologist's Journey to Motherhood.* Cambridge, MA: Perseus Publishing.

Taylor, Janelle S. 2008. *The Public Life of the Fetal Sonogram: Technology, Consumption, and the Politics of Reproduction.* New Brunswick, NJ: Rutgers University Press.

Thompson, Charis. 2014. *Good Science: The Ethical Choreography of Stem Cell Research.* Boston: MIT Press.

Thomson, James A., Joseph Itskovitz-Eldor, Sander S. Shapiro, Michelle A. Waknitz, Jennifer J. Swiegiel, Vivienne S. Marshall, and Jeffrey M. Jones. 1998. "Embryonic Stem Cell Lines Derived from Human Blastocysts." *Science* 282 (5391): 1145–1147.

Tsing, Anna. 2007. "Meratus Embryology." In *Beyond the Body Proper: Reading the Anthropology of Material Life,* ed. Margaret Lock and Judith Farquhar, 232–237. Durham, NC: Duke University Press.

US Congress. 2002. "Opportunities and Advancements in Stem Cell Research: Hearing before the Subcommittee on Criminal Justice, Drug Policy, and Human Resources of the Committee on Government Reform, House of Representatives, One Hundred Seventh Congress, first session, July 17, 2001." Washington, DC: House Committee on Government Reform. http://purl.access.gpo.gov/GPO/LPS19154.

US Senate. 2001. "Senate Session." *C-SPAN* video, 20 December. https://www.c-span.org/video/?167943-1/senate-sessionandstart=5231.

Watson, Paul F., and William V. Holt. 2001. *Cryobanking the Genetic Resource: Wildlife Conservation for the Future?* New York: Taylor & Francis.

Chapter 9

DEPLOYING THE FETUS
CONSTRUCTING PREGNANCY AND ABORTION IN MOROCCO

Jessica Marie Newman

This chapter explores the role of the fetus (*janīn*) in Morocco as a simultaneously authoritative and pliable entity and source of bodily knowledge. It seeks to understand how various actors deploy the fetus in national discourses on sexuality and abortion. To this end, my analysis builds on fieldwork and interviews conducted primarily in a single mothers' association in Casablanca and a maternity hospital in Rabat, as well as participant observation in various activist organizations and events in these cities. This chapter incorporates close readings of academic texts, national media, and individual narratives to unpack the relationship between ethnogynecological practices (nonbiomedical and/or cultural practices and knowledge surrounding women's reproductive experiences) surrounding pregnancy and abortion, and their conjuncture with biomedical, religious, and legal structures that increasingly govern these embodied experiences.

This chapter begins with readings of Islamic jurisprudential texts concerning fetal development and then brings these to bear on the role of the fetus in Moroccan health cultures. I rehearse various experience-near ways of understanding pregnancies and fetuses to show how these health models encounter biomedical and state power. I argue that although both Islamic bioethical scholarship and Moroccan ethnogynecological practices may privilege women's embodied experiences of pregnancies, the consolidation of state biopolitical power through the proliferation of biomedical institutions increasingly limits women's real-world possibilities to make authoritative claims about their own bodies and pregnancies. Knowledge of the fetus thus becomes contested terrain, and conflicting claims about the fetus structure debates about pregnancy and abortion. More importantly, state and biomedical institutions are able to deploy authoritative knowledge about the fetus, rendering female reproductive bodies increasingly legible and governable.

Legal Background: Abortion and Single Motherhood

Article 453 of the Moroccan penal code states that the only time an abortion is considered legal is to save the life or health of the mother (Ministère de la Justice 1962, art. 453). Such a medical intervention requires the consent of the pregnant woman's spouse, and if the spouse refuses, the consulting physician must obtain the written statement from the chief of medicine of the prefecture or province attesting that the woman's life cannot be saved except through termination of the pregnancy (Ministère de la Justice 1962). Moreover, articles 449, 451, and 454 state that a woman seeking or obtaining an abortion can be imprisoned for six months to two years and face a fine, while anyone who assists her in obtaining an abortion, including any health care practitioner, may be imprisoned from one to five years and face a fine. In cases where the woman dies because of the abortion, implicated persons may be imprisoned for ten to twenty years (Ministère de la Justice 1962, art. 449, 451, 454). Beyond this, premarital sex is criminalized under the penal code and punishable by up to two years in prison (art. 490). Finally, a *circulaire* established in the 1980s by the Minister of the Interior stipulates that hospital personnel must alert the authorities when an unmarried woman comes to a hospital to give birth (Ngrou 2013).

The majority of these laws were adopted from the French Napoleonic Code at the time of independence in 1956. Despite the colonial origin of these restrictive laws, they are consonant with the Maliki school of Sunni Islam's jurisprudential opinions on abortion. This is one of the reasons the articles of the penal code that dealt with abortion and sexuality were not reformed at the time of independence, with only the addition of the "health exception" in article 453. As activists and physicians in Morocco have argued, the language of article 453 is vague with regard to the definition of "health." It is usually so narrowly interpreted that preserving "health" operates more as preserving "life" for physicians seeking to avoid state prosecution. None of the laws pertaining to abortion make reference to the weeks or months of pregnancy (gestational development) that physicians or judges are expected to take into account when determining the permissibility of an abortion. Thus, interpretations of these laws prioritize saving the mother's life over questions of gestational development. In contrast, Islamic jurisprudence goes into great detail discussing the development of a fetus in utero, providing nuanced guidelines regarding abortion, paternity, and fetal personhood.

The Fetus in Islamic Bioethics and Health Cultures

Sunni *fiqh* (jurisprudence; pl. *fuqāha*), the foundation of Sunni bioethical scholarship, spends a great deal of time discussing fetal development. While the four schools of Sunni Islam vary in their opinions regarding the permissibility of abortion, they share an understanding of fetal development rooted in Greek, or *Yunani,* medical knowledge. Indeed, the stages of fetal development set forward by early Islamic bioethicists "agreed perfectly with Galen's scientific account" that featured four stages of fetal formation: "(1) as seminal matter; (2) as bloody form ... (3) the fetus acquires flesh and solidity ... and finally (4) all the organs attain their full perfection and the fetus is quickened" (Musallam 1983: 54). Here, fetal "quickening," or the first time fetal movement can be detected by the pregnant woman, is seen as indicative of "ensoulment."

Ensoulment is an indispensable part of Islamic bioethical understandings of the fetus, as it represents the moment at which the divine spirit enters the fetus, making it a member of the *umma,* or community of believers. Ensoulment represents the final stage of development during which a fetus acquires personhood and becomes a Muslim. Importantly, this stage is connected to the pregnant woman's ability to detect the movement of the fetus inside her body. For this reason, medical historian Basim Musallam emphasizes a view of "the relation between religion and science not as a dichotomy but as an intimate continuum" (1983: 55). Indeed, this continuum is also physical, located in deeply personal and private spaces of the female reproductive body.

Significantly, much of Islamic bioethical debate centering on abortion draws on these conceptions of fetal development. A common misconception is that all schools of Islam regard abortion as *haram* (wrong and forbidden). On the contrary, Muslim scholars view the ethical and moral issues related to abortion as inherently jurisprudential questions open to interpretation and debate varying across the four schools of Sunni Islam. These debates are fraught with epistemological nuances because the Qur'an, on which Islamic jurisprudence is based, does not explicitly discuss abortion. Rather, the Qur'an discusses the sanctity of life, prohibitions against murder and infanticide, and the importance of the human soul. Thus, determining the point at which ensoulment takes places is of the utmost importance in differentiating abortion from murder. Sociologist Jamila Bargach is correct in her assertion that "the fetus has rights in Islam" (2002: 69). However, it is more correct to say, as she does in a note later in her ethnography, that the "Qur'an places high value on life and its preservation" and punishment for abortions can only be decided "after determining the 'quality of personhood' of the fetus." Given this qualitative assessment of fetal life, a "consensus that abortion is homicide" is not at all certain (Bargach 2002: 244n85).

These inconsistencies in claims regarding the permissibility of abortion can in part be attributed to the diversity of Islamic jurisprudential and bioethical opinions regarding the moment of ensoulment. Rather than representing a set of monolithic rules and interdictions, these opinions regarding the moment of ensoulment and the permissibility of abortion should be understood as a dynamic and rigorous scholarly and religious debate (Bowen 1997; Katz 2003; Miller 2007; Musallam 1983). Morocco's official state religion is Sunni Islam, of which there are four schools. In the Maliki school of Islam, the teachings of which form the scaffold of Morocco's state and religious apparatus, a fetus can be considered "ensouled" at conception or within forty days of gestation (Bowen 2003; Brockopp 2003; Musallam 1983). As the malleability of this opinion suggests, there is debate within each school of Sunni Islam as to the moment of ensoulment, and "disagreement about the precise conditions under which abortion is allowed (Katz 2003: 31)," While the Maliki school sets a strict limit within which ensoulment may take place, other schools have more pliable understandings of this phenomenon and connect it to the quickening of the fetus later in pregnancy. These divergent opinions suggest that a physician's, woman's, or religious leader's understandings of a fetus and fetal person may come into conflict in cases of unwanted pregnancies.

The Maliki school represents the most restrictive opinion regarding abortion and does not take into account a woman's detection of fetal movement. Here, ensoulment is a religious phenomenon independent of embodied experiences of pregnancies. All four schools of Sunni Islam require pregnancy termination to take place before 120 days, as such terminations would be "permitted as technically not constitutive of abortion: although the product of conception has been expelled, no soul is being killed as the fetus is not yet animated" (Bowen 2003: 560). As Musallam notes, the division of fetal development into equal forty-day segments is not Qur'anic but rather based in the hadith, or sayings of the Prophet (1983: 54). Thus, Islamic bioethical opinions on abortion should be understood as "a soft 'no but' rather than an adamant 'never'" (Bowen 2003: 51).

Interestingly, sources disagree about the permissibility of abortion in cases of *zinā* (illicit sex). Medical ethicist and historian Kiaresh Aramesh asserts that all four schools of Islam forbid abortion in cases where pregnancy results from "illicit sexual behavior such as an extra-marital relationship" (2007: 30). Conversely, drawing on Muslim theologian and ethicist Abu Hamid al-Ghazali (d. 1111 CE), who stipulated, "if the zygote is the result of adultery, then the allowance of abortion may be envisaged," medical anthropologist Donna Lee Bowen states that "adultery can be considered a valid reason to allow abortion" (2003: 57). It is interesting to note the simultaneously moral and medical character of the fetus in Al-Ghazali's opinion. While he refers to the fetus by the extremely detached and biomedically precise term "zygote," he also links this biological entity to the morally charged term "adultery." Regardless, Al-Ghazali's statement allows us to infer that the illicit nature of the sexual encounter may influence moral valuations of the resultant pregnancy, thereby justifying abortion. It is not clear, however, whether the fetus or zygote itself bears moral markings, or whether the illegitimate pregnancy is the source of social and religious stigma. In the Moroccan case, until the 2004 reform of the Moroccan family code (*Mudawana*), a child of an "adulterous" relationship was considered *wld zina* (child of fornication) or *wld haram* (child from that which is forbidden) and denied legal paternity and a legal last name. Although the legal status of such children has improved, their de facto treatment in society remains highly stigmatized. Despite the divergent opinions regarding abortion in cases of adultery, the very treatment of the issue acknowledges the relationship between abortion and unplanned pregnancies, as well as the relation of the latter to extramarital sex. This will become more important in later sections.

Embodied and Ethnogynecological Knowledge of the Fetus

Some canonical work in anthropology of Morocco focuses on culturally specific, idiomatic explanations for illness and health (Crapanzano 1973, 1985; Mernissi 1977). These health cultures, termed "ethnomedicine," emphasize environmental, spiritual, social, and humoral influences on the body. Subsequent medical anthropological work in Morocco focuses on women's reproduction within these ethnomedical traditions, establishing "ethnogynecology" as a privileged subject of inquiry (Bourqia 1992; Bowen 1998; Kapchan 1993; Laghzaoui 1992; Obermeyer 1993, 2000a, 2000b, 2000c). These studies emphasize a female-centric, experience-near approach to women's bodies and reproduction. Ethnogynecology thus situates the female reproductive body within matrices of natural, supernatural, and social interaction. The ethnogynecological categories discussed below have the potential to provide women and their health care practitioners with a degree of flexibility when reporting and interpreting their pregnancies. Ethnogynecology departs from biomedical understandings of conception and gestation, and may allow women to manage risk and stigma when pregnancies do not conform to the sanctioned model of reproduction within marriage.

The Sleeping Child

Interestingly, religious exegesis and ethnomedicine can coincide to provide flexibility for women's sexual and reproductive experiences. This is true of *fuqāha* that recognize the ethnogynecological phenomenon called "the sleeping child" (*ragued*). The sleeping child is an ethnogynecological model for fetal development that departs from the biomedical model of a normative nine-month pregnancy. Within the logic of the sleeping child, a fetus could "fall asleep"—stop growing at some point during a pregnancy—and remain asleep for a certain period after which it will wake up and continue growing until the woman gives birth. A fetus can be said to have fallen asleep if a woman does not show signs of pregnancy including amenorrhea (cessation of menstruation), a growing belly, morning sickness, or food cravings. A woman could therefore claim to have menstruated during a pregnancy in which the fetus was sleeping, the loss of menstrual blood corresponding with the arrested development of the fetus in utero. A sleeping child could wake up at any point during gestation and develop into a full-term pregnancy. Only a woman's experience of the pregnancy as inactive or undetectable would differentiate it from other normal pregnancies.

Maliki *fuqāha* set the maximum length of fetal gestation at five years, creating the possibility that a fetus could "be 'asleep' in the womb and then wake up after years in a marriage or its dissolution" (Bargach 2002: 162). Importantly, the sleeping child is a way to create a link between a pregnancy and sanctioned martial conjugality, even if the woman becomes visibly pregnant long after she is no longer married. As such, Bargach refers to the sleeping child as "a basic instinct for survival corroborated by the humanist *fiqh* interpretation so that neither mother nor offspring be cast as social pariahs" (2002: 162). We could similarly apply bioethicist Kecia Ali's concept of "legal fictions to legitimize illicit sexual activity" to the sleeping child (2008: 67). The sleeping child allows a woman to claim paternity on behalf of her child years after the dissolution of a marriage or the death of her husband. This ethnogynecological category, combined with Maliki jurisprudential authority, therefore creates an important interstitial space where Moroccan women can "gain certain legitimacy and control in the face of adversity" (Bargach 2002: 162).

Some Islamists and medical professionals with whom I have spoken hotly contest the concept of the sleeping child, basing their opposition to the category either in "pure" Islam or in "modern" medical science. As such, women's assertions about their bodies and pregnancies increasingly clash with patriarchal religious and biomedical institutions and forms of knowledge, competing for authority. The Moroccan state has officially embraced Western biomedicine as the dominant national health model, extending the reach of biomedical authority throughout the country. The advent of prenatal imaging technologies like the ultrasound make it possible for individuals other than the mother to assert knowledge of fetal development, complicating women's recourse to the sleeping child or to quickening in their accounts of their pregnancies. During ultrasounds I observed, doctors took measurements of a fetus's skull and femur, using these measurements rather than information that women provided about their pregnancies to determine gestational age. This is especially significant in institutions with restricted resources like the maternity hospital in which I conducted fieldwork. Ultrasounds were only ordered when preliminary examinations indicated there might be complications in a given pregnancy. In this case, ultrasounds represent the final word on fetal development, consolidating medical authority through diagnostic technology.

It would be theoretically possible for a woman to claim that an ultrasound showed the degree to which a fetus had developed before falling asleep or the point at which a fetus had awakened and continued growing. However, as I discuss in more detail below, the advancement of obstetrics/gynecology and prenatal care seriously undercuts women's abilities to make authoritative claims about the fetus using ethnogynecological categories like *ragued*. Indeed, I have never observed a situation in which a woman's claims about her pregnancy trumped observations made by a doctor during a physical exam or ultrasound, or a case in which a woman claimed to be carrying a sleeping child during a biomedical encounter. Women's imaginative attachment to their fetuses was, however, encouraged within the hospital. One patient even embroidered a representation of her fetus that hospital staff hung in one of the consultation units (Fig. 9.1).

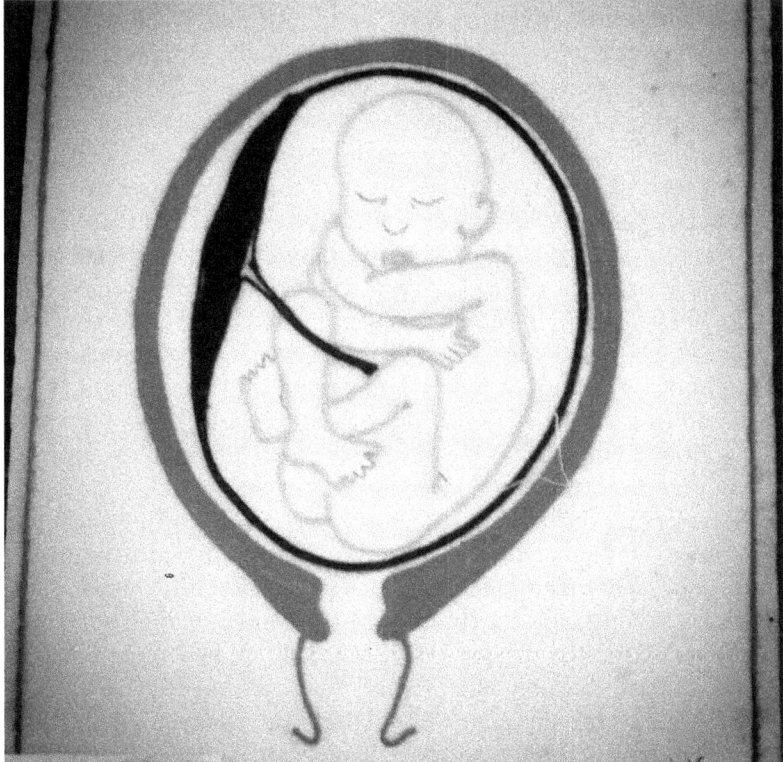

FIGURE 9.1. Embroidery work picturing a fetus, umbilical cord, placenta in blue thread, and uterus in red. A former patient of the maternity hospital in Rabat made the piece and gave it to the head midwife at the family planning unit. Note the level of detail devoted to the fetus's face and its breech positioning (feet-down). One midwife pointed out that the fetus was technically upside down, as fetuses typically turn head-down toward the end of pregnancy (photograph by Jessica Marie Newman).

Cravings and Birthmarks

Another intriguing area of ethnogynecological knowledge centers on cravings during pregnancy and birthmarks attributed to them. Linguistically, the Moroccan Arabic term for a craving during pregnancy (*twahm*) is very closely related to the word for birthmark (*twahima*), and my interlocutors specified that *twahima* refers specifically to a birthmark caused by a craving during pregnancy. Specifically, an unsatisfied craving a woman experiences during pregnancy can result in a birthmark that will be evident on the infant's body after birth. The size, color, and location of the birthmark correspond to the food a woman craved during her pregnancy, from bread to liver. Darker birthmarks are thought to correspond to stronger cravings that went unsatisfied during pregnancy. Birthmarks on the face are considered especially negative, as they are highly visible. Moroccan ethnogynecological understandings of pregnancy cravings and birthmarks thus posit the fetus as a collectively constructed social entity.

Birthmarks become socially legible after the birth of a child, a kind of physical imprinting that corresponds to a woman's treatment during her pregnancy. According to anthropologist Rahma Bourqia (1990), Moroccan woman may instrumentalize birthmark beliefs in order to voice their needs during pregnancy. In fact, recent studies of birthmark beliefs in Morocco have found that the onus is on the pregnant woman to voice her craving. This vocalization functions as a symbolic invitation for family and community members to invest in the fetus and subsequent child. Birthmark beliefs create a moral imperative through which communities become accountable for providing food to pregnant woman (Graves 2011a, 2001b).

Significantly, birthmark beliefs rely on a sense of permeability and mutual influence between the mother and the fetus. While the pregnancy may cause maternal cravings, the woman's experiences during pregnancy and the community's treatment of her have tangible impacts on the fetal body. A birthmark can indicate that a mother either ignored cravings during pregnancy or that her community was unable or unwilling to provide her with necessary food—and support—during her pregnancy. Birthmark beliefs highlight a sense of the woman's body as a substrate for other kinds of social and moral interactions with the fetus.

Ambiguous Bodies, Jinn Possession, and Impregnation

While birthmark beliefs may help pregnant women mobilize social support during pregnancy, other ethnogynecological categories can help protect women against social stigma of pregnancy outside of marriage. Here, women may rely on irregular menstrual cycles or ambiguous bodily signs to refuse pregnancy. Herbalists and religious healers alike may provide treatment for women with "lost" periods caused either by menstrual irregularity or jinn (spirit; pl. *jnun*) possession (Amster 2003; Bargach 2002). Referring to the latter category, herbalists and healers with whom I have spoken detailed the possible conditions under which a woman may become possessed or impregnated by jinn. In most cases, the infiltration of the woman's body is caused by a woman's failure to observe particular norms regarding nudity and the openness of the body, including saying "*bismillah*" (in the name of God) before removing clothes or bathing. This oversight renders her vulnerable to possession by an amorous jinn. In the same way that birthmark beliefs stress the permeability of pregnant bodies—and the interaction between these bodies and the environment—jinn possession indicates an understanding of women's reproductive bodies as susceptible to infiltration and influence.

In one interview about jinn possession, my interlocutor referred to himself as an Islamic healer and denounced many of the herbalists with whom he shared business and clients in the Fez *medina* (old city) as charlatans or even sorcerers. Still, his remedies for lost periods due to impregnation or possession by *jnun* were markedly similar to his less pious counterparts. In my interviews with the Islamic healer and herbalists in Fes, remedies necessarily corresponded to the moment and nature of the possession. *Jnun* can inhabit natural elements including air, water, and earth, and treatments must correspond to these specificities. Thus, in the case of a woman who was possessed by a jinn while undressing, the spirit in question is assumed to be airborne, and the corresponding remedy would include a number of smoke-producing elements, including incense and bundles containing, among other things, papers inscribed with prayers or verses from the Qur'an. Similar remedies include ingesting religious inscriptions on paper mixed with food or water (Amster 2003: 2013).

Herbalists may prescribe other remedies for retrieving lost periods, including vaginal suppositories and physical activities such as kneading the abdomen, ingesting foods with heat-producing properties (following the Galenic tradition, these are foods associated with heat, most commonly spiced with cinnamon, pepper, or *ras al hanout*, a mixture of "hot" spices), and taking trips to the *hammam* (public bath) in conjunction with the ingestion of herbal medicines or the placement of a suppository. Interestingly, physicians warn women in the third trimester of pregnancy not to ingest "hot" foods or visit the *hammam,* indicating an overlap in biomedical and ethnogynecological understandings of pregnancies in this regard.

These remedies are meant to restore equilibrium to the woman's body, which was disrupted by the infiltration of a jinn or other external factors. As medical anthropologist Marybeth MacPhee observes in her study of Saharan ethnomedicine, which she terms "popular health culture," the major sources for health disruptions can be broadly attributed to imbalances in humors, spirit attacks, the evil eye, magical curses, and microbes (2012: 40). Moroccan humoral health models emphasize hot/cold and wet/dry as affecting health.[1] Women are especially susceptible to the first four of these influences, and an imbalance in the humors, the jealousy of a friend, or the whims of a spirit can all infiltrate the body and disrupt normal reproductive functioning. The model of jinn possession therefore represents a way for women to understand menstrual disruptions but does not create a situation in which the existence of a fetus is recognized. On the contrary, potential pregnancies are denied and subsumed under broader categories of menstrual irregularity. Moreover, these herbal methods of menstrual extraction or fertility regulation are available to women who are not pregnant or who employ alternative narratives including jinn possession or menstrual irregularity to account for a "late" or "lost" period.

Conjuncture, Contradiction, and Displacement of Embodied Knowledge

The coexistence of ethnogynecological and biomedical models of pregnancy can provide women with more options in their care-seeking strategies but can also create tensions and contradictory experiences. Furthermore, religious bioethical scholarship on ensoulment derives from but may be at odds with women's experiences of their pregnancies and reproductive bodies. For example, if a Moroccan woman has not sensed the movement of the fetus until after forty days, she could theoretically make the case that the fetus has not yet been ensouled. Indeed, during my observations of patient consultations at a maternity hospital in Rabat, women generally felt unsure whether they sensed fetal movement until the fifteenth or sixteenth week of pregnancy at the earliest. Moreover, in early consultations with pregnant women (within the first twelve weeks), physicians tracked fetal movement in patient charts with the notation "+/−" to indicate uncertainty—for example, when medical exams detected fetal activity but women did not experience it physically.

The ambiguity of this notation is consonant with women's verbal equivocations in response to questions about fetal movement during pregnancies. Women's answers in a single medical consultation early in a pregnancy frequently shifted from saying initially they did not feel the fetus move to then saying they may have felt a little movement (*kayhrk shwiya*) or felt the pregnancy a little (*kanhisbih shwiya*) after being asked the question multiple times throughout an exam. In these cases, women seemed to base their acknowledgment of fetal movement on feedback from physicians that the pregnancies were progressing normally. Women with histories of repeated miscarriages or fetal malformations employed similarly vague language when referring to past pregnancies, sometimes not counting pregnancies that resulted in miscarriages when giving reproductive histories. In one interview, a patient recounted four miscarriages, all of which took place between four and six months of gestation; however, she described a loss that took place during the sixth month of pregnancy as a birth, not a miscarriage. This differentiation between different kinds of pregnancy and loss often caused confusion when doctors and patients discussed women's medical histories. Women attached value to fetuses based on later pregnancy outcomes, while doctors sought to clearly note the number of gestations, viable births, and living children. Medical consultations were thus sites of conjuncture and co-construction of women's reproductive histories.

Despite this flexibility in women's understandings of their pregnancies, the dominance of religious and legal institutions in the country makes it extremely difficult for women to officially advance these sorts of claims. During my fieldwork, I observed that doctors frequently contradicted information that pregnant women provided about their pregnancies, especially concerning date of conception or anticipated date of delivery. Like religious determinations of fetal personhood, medical technologies for calculating fetal development are increasingly independent from women's embodied experiences. Most of the physicians I observed used roundels or smartphone applications for determining gestational age based on information that women provided about the date of their last periods. Doctors asked questions whether women's menstrual cycles were regular and when they had their last period. If women seemed unsure of this information or otherwise equivocated, doctors would write "imprecise" on charts, and gestational age would later be determined based on physical exams and ultrasounds.

In cases where the information on doctors' phones appeared at odds with what women said, they repeatedly asked women, "Are you sure?" (*Nti ekda?*) during the consultation. This was especially true of women who had passed their delivery date. One doctor simply scratched out initial chart notations regarding the gestational age of the fetus, explaining to me that the woman must have told him the wrong date. Another doctor reflected to me that a lot of patients were past their due dates, going on to explain that this was probably because of women's lack of education and understanding of their bodies. He thus assumed patients provided faulty information rather than considering other factors that could cause a woman to carry a pregnancy past her due date.

Doctors' questioning of their patients may simply be a way of making sure they have all of the necessary information. However, it can also be read as the reinforcement of biomedical ideology through patient consultations. During my observations of the family planning unit in the maternity hospital, doctors questioned patients about their birth control methods, advising most patients to use oral contraception or an intrauterine device (IUD). Oral contraceptives, condoms, and emergency contraception (Levonorgestrel), or the "morning after pill," are available over the counter in pharmacies.

Still, some women wish to avoid being seen purchasing these items in a pharmacy and may choose herbal alternatives. One herbalist described a form of menstrual regulation to me, lumping it into early or "emergency" remedies he guaranteed for "up to forty days, but if you take it sooner it works better" (*kaykhadam hatta rbâeen ayam walakin ahssan ila katshrubuh bkree*). It involves taking seven pills, performing reverse somersaults, and inserting a vaginal suppository before going to the *hammam*. Alternatively, herbalists may create herbal suppositories to be inserted vaginally or into the cervix to cause hemorrhage. Interestingly, although the herbalist described almost identical regimens for restoring lost periods and performing herbal abortions, he guaranteed the former treatment for up to forty days, which coincided with the limit placed by Maliki *fiqh* regarding abortion rather than the biomedical guideline of 120 hours for using emergency contraception. Here, ethnogynecological practice aligns itself with religious authority, drawing on these officially sanctioned claims about the fetus.

Consultations with herbalists and traditional midwives (*qabla*; pl. *qablat*) accommodate different embodied experiences and do not force women into potentially stigmatizing interactions based on the admission of pregnancy. Morocco recently restructured its midwifery training programs to conform with biomedical models, and a midwife trained in a hospital and practicing in a medical institution is referred to by the French term *sage-femme*. However, a *qabla* does not necessarily have any biomedical training and would be roughly analogous to the "granny midwives" discussed in work on midwifery in the United States, though with different ethnic and racial connotations (Dawley 2003; Dye 1980; Lee 1996; Simonds et al. 2007). Herbalists, as opposed to pharmacists, similarly have no biomedical training. *Qablat* and herbalists are historically trusted practitioners in their communities, but with the advent of biomedicalization, they have become the archetypal "Others" of highly professionalized pharmacists and *sages-femmes*. The role of *qablat* as a birth attendants and counselors for martial sexuality (including fertility management) makes them ideal resources for women who wish to avoid state and biomedical scrutiny when dealing with ambiguous or stigmatized sexual and reproductive health issues. Both *qablat* and herbalists tend to be integrated into their communities and therefore represent sympathetic sources of advice and treatment.

Yet, precisely by virtue of their connection to "traditional" or "unscientific" practices, these forms of birth control or menstrual regulation are conceived of as threatening to the bodies of women and the body of the nation. Unlike other methods, such as vaginal suppositories or instructions to insert sharp objects such as reeds into the cervix to induce bleeding, the herbal remedies described to me seemed unlikely to induce infection or morbidity. Nevertheless, women seeking pregnancy terminations may combine herbal remedies with other methods including off-label prescription use, physical attempts to rupture placental membranes, and ingestion of toxic chemicals. As such, the national abortion debate subsumes these herbal remedies within the broader category of "unsafe abortions" (*les avortements à risques*), which contributes to maternal mortality rates. Herbalists and *qablat* who may prescribe these herbal remedies thus become equated with public health risks. Thanks to Morocco's commitment to reducing maternal mortality as one of the Millennium Development Goals, persistently high rates of "unsafe abortions" and maternal mortality threaten Morocco's modern image in the eyes of supranational bodies like the World Health Organization.

Unmarried Women and Collective Knowledge of the Fetus

As the foregoing discussion suggests, the advancement of biomedicine has important implications for women who become pregnant outside of the sanctioned marital union. Unmarried teens and women encounter social and structural obstacles to accessing birth control and emergency contraception. Similarly, despite the gestational flexibility *ragued* grants to married or previously married women, it does not provide any protection to never-married women, who supposedly remain virgins until marriage. In this case, the pregnancy of an unwed woman is taken as a de facto confession of *zina* under Maliki jurisprudence (Ali 2008: 63). All extramarital sex in Morocco is not only religiously proscribed but also illegal; individuals found to have engaged in *zina* can be prosecuted and imprisoned for "fornication" under article 490 of the penal code, and children born outside the sanctioned marital contract are highly stigmatized and denied paternity.

In the case of paternity disputes, rather than considering an infant a "child of the [marital] bed," as we saw in the discussion of Sunni *fuqāha* concerning the sleeping child, paternity and DNA testing now make it possible for a man to deny paternity to a child definitively (Ali 2008: 70). However, to demand paternity tests could provide unmarried women with more legal protection for themselves and their children by proving paternity scientifically, though such tests are prohibitively expensive for women of lower socioeconomic backgrounds and must be court ordered. Single mothers I worked with who sought DNA tests to prove the paternity of a child born outside of marriage navigated the court system to provide judges with birth certificates and other official documentation of a birth. This was an extremely arduous process, as many of these women were illiterate and unable to obtain the necessary documents and institutional support to advance their claims in court. Throughout this process, women's assertions about their pregnancies were pitted against biomedical evidence that could either discredit or corroborate these claims.

Furthermore, pregnancies and fetuses take on divergent values when located outside of the heterosexual marital union. The fetuses of single pregnant women become proof of extramarital sexual behavior, taking on shameful connotations. As fetuses grow and pregnancies become more visible, women's bodies, by virtue of their connection to the fetus, become insistent proof of transgressive sexual behaviors. The fetus, located within the maternal body, comes to engulf the woman herself. Unmarried pregnant women I met recounted stories of increased social marginalization and loss of family support. Their individual identities became subsumed within maternal and sexual designations. These isolating experiences contribute to the coalescence of a collective group of single mothers (Fr. *mères célibataires*; Ar. *oumum 'azibat*), known throughout the country by virtue of national media coverage of their plight.

Women who find themselves single and pregnant may seek assistance from associations catering specifically to single mothers. Despite the existence and relative notoriety of these organizations, they are not able to meet the constant demands for assistance. During my fieldwork, organizations turned women away from listening centers (*les salles d'écoute*; spaces provided by nongovernmental organizations for women to tell their stories) on a daily basis because of lack of space. Few of the organizations in Casablanca are able to provide on-site housing to pregnant women, and those that do are not able to accept pregnant women until they are seven or eight months pregnant. One social assistant explained to me that this limit corresponds to the visibility of pregnant women at this point, but also noted that it helped them handle high demands for housing by having an exclusionary criterion like this. Significantly, although friends and family of many women who sought assistance from single mothers' associations knew they were pregnant, they were not pregnant "enough" to qualify for housing. Here, the gestational development of the fetus determines the physical safety of the pregnant woman, who may find herself homeless and additionally vulnerable because she could not find safe housing. The fetus thus serves as a rationale for two different moments of exclusion: the very acknowledgment of the fetus and pregnancy may result in a woman being turned away by her family and community, while organizations may turn women away because their pregnancies are not far enough advanced.

Many women who came to single mothers' associations strategically positioned their knowledge of their bodies and pregnancies in order to gain admission to assistance programs. Based on my observations in the listening center at one these associations, these narratives tended to fall into two categories: asserting immediate knowledge of their pregnancies or describing lack of awareness of their pregnancies until a third party intervened. In cases falling into the first category, women used their bodily knowledge to position themselves as responsible mothers seeking better lives for their children. When asked how they knew they were pregnant, these women tended to say, "I just knew" (*'arft b rasi*), but rarely made references to missed periods or other physical indicators of pregnancy. Alternatively, some women admitted to seeking and attempting abortions but said they came to accept their pregnancies after sensing fetal movement. This experience of the fetus constituted a shift to women's self-identification as mothers.

Single mothers' associations have detailed conduct guidelines for the women in their care, and women whose sexual and reproductive histories indicate a sense of personal responsibility and low risk of recidivism[2] are better candidates for assistance. It was therefore possible for some single mothers and pregnant women to consciously deploy their fetal knowledge in order to appear more desirable to organizations offering assistance. Still, for many women, knowledge of their pregnancies, particularly after they were able to feel the fetus move, was truly transformative. The fetus in these cases became a catalyst for revelatory experiences and new ways of knowing the self.

By contrast, women who claimed to have not known they were pregnant until a friend or relative intervened also tended to rely on third parties and biomedicine for knowledge of the fetus. There are many reasons a woman might not realize she is pregnant. Many women who did not realize they were pregnant described merely feeling sick, not keeping track of their menstrual cycles so not realizing they had missed periods, or being otherwise unwilling or unable to entertain the possibility of pregnancy. In these cases, when women's pregnancies became evident to those around them, a friend or family member typically interceded and either took the woman to a doctor or provided a pregnancy test. In these cases, the pregnant women were not the first people to realize they were pregnant, and tended to characterize their knowledge of the fetus as passive or indirect. They described themselves as being "told" they were pregnant, either by a physician or the person who first approached them. Here, bodily experience was subordinated to biomedical authority, whether in the doctor's office or while taking a pregnancy test. Women's bodies and fetuses were interpreted for them, their self-knowledge externally revised.

For many women in these situations, knowledge of the fetus takes on a communal nature; it is shared between the physician, the pregnant woman, and her friend or family member. It was not uncommon for the friend or family member who first realized the woman was pregnant to accompany her to the listening center. These individuals shared the weight of providing details a woman's pregnancy and/or childbirth. Indeed, in cases where women had no other support besides the person who first noticed their pregnancies, these third-party participants became integral in helping women seek assistance from the available associations. These individuals advocated for the pregnant women or single mothers and sometimes even dominated intake procedures. They became primary sources of knowledge about pregnancies, sexual encounters, menstruation, and family dynamics. In such cases, the would-be beneficiary was mostly silent, deferring to the narrative as told by her advocate, occasionally corroborating or adding details.

Significantly, women who did not realize they were pregnant appeared more reliant on the sympathetic individuals with whom they first shared knowledge of their pregnancies. This not only made them appear more in need of assistance but also perpetuated characterizations of these women as victims unable function without third-party support. However, occupying the role of the victim could undermine the sympathy of individuals being relied on for assistance. I have witnessed numerous conversations between social workers and assistants in which they wondered aloud how a woman could not realize she was pregnant, or speculated on the woman's ability to benefit from the association's assistance in the end.

Deploying the Fetus: Making Authoritative Claims

Biomedical Authority and Knowledge of the Fetus

Thus far, we have seen that the strong tradition of ethnogynecological practices in Morocco have historically allowed women to mobilize a number of sources of knowledge about fetuses and their pregnancies. These forms of knowledge may allow women to claim authoritative knowledge of the fetus while simultaneously asserting morality, paternity, or entitlement to communal care. However, the biomedicalization of the Moroccan health care system makes it increasingly difficult for women to make authoritative claims about their pregnancies. This is especially true in the context of the national abortion debate, where biomedical expertise and state authority dominate ways of speaking about pregnancy and termination.

As discussed, the fetus has become increasingly comprehensible through the proliferation of biomedical technologies like fetal ultrasound. Other kinds of prenatal diagnostic testing have further enhanced doctors' ability to "see" and "know" the fetus. Specifically, the ability to detect fetal anomalies has important implications for the national abortion debate. Although the Moroccan penal code currently only allows for abortions in cases where the mother's life is in danger, activists throughout the country have begun to argue for a liberalization of these laws. Dr. Chafik Chraïbi, former chief of obstetrics and gynecology at Maternité des Orangers in Rabat and arguably the most visible figure in the abortion debate, argues that the state should institute a robust health exception, whereby the "health" of the pregnant woman takes into account physical, mental, and emotional well-being, and that abortion should be allowed in cases of rape, incest, or fetal malformations (*les malformations fœtales*).

The fetus resulting from incest or rape and the fetus developing "abnormally" take on values divergent from those associated with "normal" or normative fetuses. Specifically, arguments for rape and incest exceptions associate the traumatic sexual encounter with the fetus, making the pregnancy and resultant infant physical proof of the transgressive sexual encounter. Until 2014, article 475 of the Moroccan penal code allowed a rapist to avoid prosecution by marrying his victim. With the repeal of this law, rape has become differentiated from other kinds of illicit or extramarital sex (Ministère de la Justice 1962, art. 475). Here, the fetus and/or pregnancy are assumed to cause additional emotional trauma and social exclusion, compounding the original physical violation. Similarly, medical and public health practitioners like Dr. Chraïbi invoke fetal anomalies as sources of distress and hardship for parents. Importantly, in a debate on 16 March 2015, organized by Dr. Chraïbi's Moroccan Association for the Fight against Clandestine Abortions (AMLAC), members of parliament belonging to both majority and opposition parties limited the discourse of fetal malformations warranting abortion to those that are "severe" and life threatening for mother and child.

Interestingly, activists like Dr. Chraïbi base their claims about abortion on their medical expertise and the associated authority that comes with treating patients or being able to detect fetal anomalies and understand their gravity. As such, medical practitioners who have become politicized and active in the abortion debate are able to argue for legal reform without adopting particularly liberal or pro-women positions. Secular, leftist, and feminist movements have historically been at political loggerheads with Islamist groups over social and legal reform, so it is noteworthy that Representative Mustapha Ibrahimi, a deputy of the Islamist Justice and Development Party (PJD), on the one hand disparaged maternal distress during an unwanted pregnancy as "*une petite dépression*" and on the other agreed with and accepted Dr. Chraïbi's definition of "severe" fetal malformations. Here, biomedical evaluations of fetal health and viability serve as common ground for otherwise antagonistic interpretations of rights to abortion. Similarly, the head of the PJD government, Abdelilah Benkirane, initially supported adding rape and incest exceptions to the current laws on abortion when he took office, despite his association with a conservative party (Jay 2012). The PJD's second-in-command, Sâad Eddine El Othmani, also has recently come out in support of abortion in cases of fetal malformations, incest, and rape (Chapon 2015).

Thus, by virtue of their medical authority, many physicians involved in the abortion debate speak for pregnant women—whether victims of sexual assault or those whose infants would be severely disabled—and make authoritative assertions about these experiences. While physicians like Dr. Chraïbi base their claims on decades of experience working in maternity wards and seeing many cases that support their claims, they do not necessarily align themselves with feminist or leftist agendas. Indeed, it was based on his experience as a physician that Ibrahimi dismissed psychosocial factors that pregnant women may face as grounds for abortion. Repeatedly referencing his medical training, he limited the majority of his comments to fetal malformations, the accuracy of these tests, and how early they can be used. Importantly, in my discussions with Dr. Chraïbi, he has actively argued against "abortions on demand" (*les avortements sur command*). Instead, he seeks to reorient public conversations toward public health and medical concerns related to extramedical abortions and unplanned pregnancies.

Politicizing and Personifying the Fetus

Thus far, I have primarily discussed how knowledge claims about the fetus are constructed and asserted by women and health care practitioners, whether biomedical or ethnomedical. In this and the concluding section, I consider how understandings of fetus compete in the public domain and encounter the authority of the state. As I discussed in cases of pregnant women's biomedical encounters, claims about the fetus can sometimes come into conflict, as individual actors may have divergent interpretations of the women's bodies and pregnancies. Significantly, ethnogynecological categories disappear from the discursive field as issues of abortion and single motherhood gain media coverage and are debated in upper echelons of power. Nevertheless, the Moroccan state as a much larger force and collection of actors also has a vested interest in monitoring and directing discussions about pregnancy.

Motherhood (within the marital union) is highly valued in Moroccan society, and women and girls are generally expected to aspire to marriage and motherhood despite women's increased employment and lifestyle changes accompanying rapid urbanization. Beyond this, popular discourse positions Moroccan mothers as reproducers of the nation who give birth to and raise the next generation of Moroccan citizens and Muslims. It is not surprising, therefore, that penal and family codes regulate women's sexuality and the legal status of marriage and offspring through proscription of premarital sex and family code articles—many of which were reformed in 2004—that disadvantage children born to unmarried parents. These laws not only involve the state in structuring the private lives of its citizens but also reproduce and reinscribe national boundaries and identities on the bodies of women and fetuses.

These dynamics are particularly clear when put back into conversation with Sunni *fuqāḥa* regarding fetal development. The concept of ensoulment allows fetuses to be associated with both a community of believers and a national community long before they are born. Concern for the promissory life of the fetus as a future citizen and believer is a powerful narrative that has been marshaled by conservative activists who resisted attempts to liberalize Morocco's abortion laws. During the 16 March debate, Ibrahimi stressed the separateness of the fetus from the maternal body. He parsed the fetal and maternal body by arguing that there were four lungs, four arms, and four legs detectable in a pregnancy, so that fetus should therefore be considered a separate life from the mother. Similarly, the Moroccan Right to Life Association (*Jam'iyyā Maghrebīa ul Haq al Hayat*) plays on these themes in their characterization of abortion as the murder of innocent children. The association uses images and rhetoric that juxtapose life and death, morality and murder. As with arguments about fetal anomalies, this imagery creates a slippage between the fetus and infant, making fetuses seem simultaneously removed from and at the mercy of the pregnant woman. Images of presumably aborted fetuses that have recently accompanied anti-abortion activism in Morocco compound this antagonistic maternal-fetal imagery and depict "prenatal space" as distinct from the female body. This invokes Nathan Stormer's observation that "the spatial rhetoric of the void surrounding the fetus is the synecdochic relation of pure environment to entitlement" (2000: 129).

Although some secular feminist organizations still insist that current abortion laws constitute a form of gender discrimination and argue for unobstructed access to pregnancy terminations, arguments that attempt to balance the entitlement of the fetus with the exigencies of rape or incest and fetal anomalies have the most widespread resonance. This balancing act—and the debate about abortion more broadly—has become of such importance that King Mohammed VI recently promised to intervene (Choukrallah 2015; Kahlaoui 2015). The king's public request for legal reform proposals is markedly similar to his role in the 2004 Mudawana reforms. During that time, the king balanced conservative and liberal agendas, convening an expert review board, and eventually put the royal seal of approval on substantive reforms that moved toward greater women's equality. As with the Mudawana reforms, the king's role as political sovereign and "Commander of the Faithful" will significantly alter the contours of the national abortion debate, circumscribing certain possibilities for dissent.

Significantly, the king's commitment to review proposed reforms to abortion laws reaffirms abortion and pregnancy as central nodes of biopolitical governance. In one of the latest iterations of Dr. Chraïbi's proposed reforms to abortion laws, he suggested that public hospitals establish ethical review boards composed of doctors, clerics, and state officials to review requests for abortions (Choukrallah 2015). Although Chraïbi saw this as a liberalizing step, such a measure would extend rather than diminish biomedical and state control over women's bodies. The review board would deliberate on each case, granting and denying abortions. While such review boards would ostensibly be guided by more liberal guidelines taking women's physical, psychological, and social well-being into account, the ethical review board model recreates the highly bureaucratic and technocratic governance practices that proliferate in Morocco. Following the logic of these review boards, women's bodies would become part of the public domain by virtue of their synecdochic reduction to what we might call "wombscapes," intimate and political spaces that may be by turns nurturing and threatening to the fetus. When reviewing requests for pregnancy terminations, the review boards would be required to weigh the various aspects of a woman's intimate and biological life, parsing and then synthesizing these as factors determining the final approval or denial of an abortion. The intervention of the proposed review committees would thus reduce women's lived experiences to wombscapes, biological and epistemological sites that must be regulated and monitored by the state through the oversight of biomedical institutions. The king's involvement and the precipitous rise in news coverage and public debates about abortion thus position pregnant women and their wombs as biopolitical spaces of ever-increasing interest for politicians and physicians alike.

Conclusions

I have outlined how different understandings of pregnancy and women's bodies influence how individuals assign value to or make claims about the fetus. We saw that women, supported by *fuqāha* and ethnogynecological practices, can deploy different models of pregnancy to mitigate potentially stigmatizing events like single motherhood or abortion. However, as we moved farther away from women's embodied experiences of their pregnancies, biomedicine and state authority increasingly set the terms of the abortion debate unfolding in Morocco. This erasure of women's lived experiences potentiates the ascendance of the fetus as an agentive—if only promissory—member of the body politic and religious community.

As Stormer states, "in the age of biopolitics, prenatal space is a point of articulation for divergent interests in 'life,' with biomedical nomenclature, practices, and imagery in the privileged position of embodying the commonplace of life in the womb" (2000: 136). These divergent interests in "life" are particularly visible in the Moroccan case, especially when abortion and single motherhood come to bear on evaluations of pregnancy, fetuses, and mothers. Single women who become pregnant must balance the realities of their pregnancies and needs of their fetuses against their own social marginalization and precarity. In these cases, it becomes increasingly difficult to speak strictly about fetal or maternal rights in liberal terms. Rather, the fetal-maternal relationship is dynamic, interpersonal, and deeply contingent on socioeconomic status and access to assistance and safe living situations.

Similarly, when activists invoke fetal anomalies or cases of incest and rape in their arguments about abortion, we must be attentive to the shifting rhetorics of life, bodily integrity, and human dignity. The intervention of the king in the ongoing national conversation about these issues demonstrates how the fetus and pregnant woman, beyond representing archetypal biopolitical subjects, are also imbricated in the consolidation of state power. The proliferation of debates and media coverage surrounding abortion and single motherhood may have positive effects for individuals most in need of assistance. At the same time, however, the heightened attention paid to these issues makes them highly visible and thus more vulnerable to redefinition, control, and cooptation. It remains to be seen how, if at all, various deployments of the fetus may change or solidify prevailing social mores and laws governing the fetal and maternal subjects.

Acknowledgments

I would like to thank the editors—Tracy Betsinger, Amy Scott, and Sallie Han—for their generous and careful feedback on earlier versions of this article. I am deeply indebted to the physicians and activists who patiently allowed me to ask questions and observe their daily lives, and whose insight has contributed enormously to my arguments in this chapter. I am especially thankful for Dr. Chafik Chraïbi's advocacy and support of my project. Of course, there would have been no fieldwork at all without the single mothers and hospital patients who shared their lives and stories with me. I am humbled by their honesty and humor, as well as their faith in a *gaouriyya* who showed up one day asking questions. I am grateful for the generous financial support I received for dissertation fieldwork from the Wenner-Gren Foundation for Anthropological Research, the Fulbright-Hays Doctoral Dissertation Research Abroad fellowship, and the Whitney and Betty MacMillan Center for International and Area Studies at Yale University.

Jessica Marie Newman is Lecturer in Yale University's Department of Anthropology. Her dissertation, for which she received the Association for Feminist Anthropology Dissertation Prize, situates the work of Moroccan single mother advocates and NGOs within national political and moral debates.

Notes

1. Moroccan humoral medicine draws on the Galenic tradition that sees imbalances in the four humors (blood, black bile, yellow bile, and phlegm) as influencing physical health and personal temperament. Illnesses derive from an excess or deficiency of a humor, which could in turn be attributed to the diet or environment of the patient. Blood and yellow bile are associated with heat, while black bile and phlegm are colder, damper humors. In practice, hot and cold imbalances in Moroccan ethnomedicine tend to derive primarily from diet (an excess of "hot foods") or environment. For example, women who shower—and particularly wash their hair—while menstruating risk imbalance in the heat of menstruation with the cold associated with exposing the body to water and air while bathing.

2. The French term *récidivisme* appeared in one association's protocols, referring to the possibility of a woman having repeated pregnancies while unmarried.

References

Ali, Kecia. 2008. *Sexual Ethics and Islam: Feminist Reflections on Qur'an, Hadith, and Jurisprudence.* New York: Oneworld Publications.
Amster, Ellen J. 2003. "Medicine and Sainthood: Islamic Science, French Colonialism and the Politics of Healing in Morocco, 1877–1935." PhD diss., University of Pennsylvania.
———. 2013. *Medicine and the Saints Science, Islam, and the Colonial Encounter in Morocco, 1877–1956.* Austin: University of Texas Press.
Aramesh, Kiarash. 2007. "Abortion: An Islamic Ethical View." *Iranian Journal of Allergy, Asthma and Immunology* 6 (S5): 29–34.
Bargach, Jamila. 2002. *Orphans of Islam: Family, Abandonment, and Secret Adoption in Morocco.* Lanham: Rowman & Littlefield.
Bourqia, Rahma. 1992. "The Womans Body: Strategy of Illness and Fertility in Morocco." In *Towards More Efficacy in Women's Health and Child Survival Strategies: Combining Knowledge for Practical Solutions,* ed. Ismail Sirageldin and Robb Davis, 131–144. Baltimore: Johns Hopkins University, School of Hygiene and Public Health.
Bowen, Donna Lee. 1997. "Abortion, Islam, and the 1994 Cairo Population Conference." *International Journal of Middle East Studies* 29: 161–184.
———. 1998. "Changing Contraceptive Mores in Morocco: Population Data, Trends, Gossip and Rumours." *Journal of North African Studies* 3 (4): 68–90.
———. 2003. "Contemporary Muslim Ethics of Abortion." In Brockopp 2003, 51–80.
Brockopp, Jonathan E. 2003. *Islamic Ethics of Life: Abortion, War, and Euthanasia.* Columbia: University of South Carolina Press.
Chapon, Amanda. 2015. "Avortement: Une Réforme, oui mais Laquelle?" [Abortion: A Reform Yes, but Which?] *Telquel,* 17 March. http://telquel.ma/2015/03/17/avortement-reforme-laquelle_1438775.
Choukrallah, Zakaria. 2015. "Débat sur L'avortement: Quand El Ouardi Revient à la Rasions." [Abortion Debate: When Minister El Ouardi Returned to Reason] *Le Figaro,* 18 March. http://www.h24info.ma/maroc/societe/debat-sur-lavortement-quand-el-ouardi-revient-la-raison/31528.
Crapanzano, Vincent. 1973. *The Ḥamadsha: A Study in Moroccan Ethnopsychiatry.* Berkeley: University of California Press.
———. 1985. *Tuhami: Portrait of a Moroccan.* Chicago: University of Chicago Press.
Dawley, Katy. 2003. "Origins of Nurse-Midwifery in the United States and its Expansion in the 1940s." *Journal of Midwifery and Women's Health* 48 (2): 86–95.

Dye, Nancy Schrom. 1980. "History of Childbirth in America." *Signs* 6 (1): 97–108.
Graves, Anna Joy. 2011a. "Birthmark Beliefs Influence Community Food Provisioning for Pregnant Women in Morocco." Paper presented at the American Anthropological Association Annual Meeting, Montreal, Canada, 19 November.
———. 2011b. "Pregnancy Cravings: Biological 'Wisdom of the Body' or Cultural 'Wisdom of Society?'" MA Thesis, Boston University. http://www.worldcat.org/title/pregnancy-cravings-biological-wisdom-of-the-body-or-cultural-wisdom-of-society/oclc/754656642#PublicationEvent/2011
Jay, Martin. 2012. "New Prime Minister Surprises Moroccans with Support for Abortion." *New York Times*, 11 January. http://www.nytimes.com/2012/01/12/world/africa/new-prime-minister-surprises-moroccans-with-support-for-abortion.html?_r=0.
Kahlaoui, Soraya El. 2015. "Maroc: Débat sur L'avortement en Attendant L'arbitrage du Roi." [Morocco: Abortion Debate Waiting for the King's Arbitrage] *Le Monde Afrique*, 17 March. http://www.lemonde.fr/afrique/article/2015/03/17/au-maroc-debat-brulant-sur-l-avortement-en-attendant-l-arbitrage-du-roi_4595280_3212.html.
Kapchan, Deborah. 1993. "Moroccan Women's Body Signs." In *Bodylore*, ed. Katharine Young, 3–34. Knoxville: University of Tennessee Press.
Katz, Marion Holmes. 2003. "The Problem of Abortion in Classica Sunni Fiqh." In Brockopp 2003, 25–50.
Laghzaoui, Latifa. 1992. "Women and Shrines in Urban Morocco: The Case of the Patron-Saint of Sale." Master's thesis, University of London.
Lee, Valerie. 1996. *Granny Midwives and Black Women Writers: Double-Dutched Readings*. New York: Routledge.
MacPhee, Marybeth. 2012. *Vulnerability and the Art of Protection: Embodiment and Health Care in Moroccan Households*. Durham, NC: Carolina Academic Press.
Mernissi, Fatima. 1977. "Women, Saints, and Sanctuaries." *Signs* 3 (1): 101–112.
Miller, Ruth Austin. 2007. *The Limits of Bodily Integrity: Abortion, Adultery, and Rape Legislation in Comparative Perspective*. Farnham: Ashgate Publishing.
Ministère de la Justice du Maroc. 1962. Dahir No. 1-59-413 du 28 Joumada II 1382 (26 novembre 1962) portant l'approbation du texte du Code Pénal. [Moroccan Penal Code] Rabat: Bulletin Officiel No. 2640 bis du mercredi 5 juin 1963.
Musallam, Basim F. 1983. *Sex and Society in Islam: Birth Control Before the Nineteenth Century*. New York: Cambridge University Press.
Ngrou, Imane. 2013. "Bébés abandonnés: Le drame." [Abandoned Babies: The Drama] *Aujourd'hui au Maroc*, 17 December. http://aujourdhui.ma/focus/bebes-abandonnes-le-drame-106696.
Obermeyer, Carla Makhlouf. 1993. "Culture, Maternal Health Care, and Women's Status: A Comparison of Morocco and Tunisia." *Studies in Family Planning* 24 (6): 354–365.

———. 2000a. "Pluralism and Pragmatism: Knowledge and Practice of Birth in Morocco." *Medical Anthropology Quarterly* 14 (2): 180–201.

———. 2000b. "Risk, Uncertainty, and Agency: Culture and Safe Motherhood in Morocco." *Medical Anthropology* 19 (2): 173–201.

———. 2000c. "Sexuality in Morocco: Changing Context and Contested Domain." *Culture, Health and Sexuality* 2(3): 239–254.

Simonds, Wendy, Barbara Katz Rothman, and Bari Meltzer Norman. 2007. *Laboring On: Birth in Transition in the United States.* New York: Routledge.

Stormer, Nathan. 2000. "Prenatal Space." *Signs* 26 (1): 109–144.

Chapter 10

BEYOND LIFE ITSELF
THE EMBEDDED FETUSES OF RUSSIAN ORTHODOX ANTI-ABORTION ACTIVISM

Sonja Luehrmann

In English-language scholarship on the cultural and political lives of the fetus, the ascription of personhood has been a critical focus of analysis. In their edited volume *Fetal Subjects, Feminist Positions*, Lynn Morgan and Meredith Michaels (1999) outline the underlying paradox: new technologies of prenatal visualization, testing, and bonding have made fetuses into increasingly animated subjects with a powerful hold over the imaginations of expectant parents and the larger public. At the same time, pro-life activists mobilize these images to signify powerlessness, defenselessness, and life at its most vulnerable (see also Petchesky 1987; Rapp 2000). In North American pro-life politics, the fetus becomes a kind of *homo sacer:* a figure both sacred and impure because it exists at the limits of collective moral systems, so transgressions against it become transgressions against life itself (Agamben 1998; Arendt 1951). Like stateless refugees who become the motivating center of political action precisely because they represent forms of human life excluded from full political subjecthood, fetal persons are at their most powerful when they embody biological life at its barest.

When cultural anthropologists look at other times and places, however, it becomes clear that the status of "icons of life" does not come naturally to fetuses (Morgan 2009). During research in the Ecuadorian Andes, Morgan (1998) found that her female interviewees universally proclaimed abortion to be a sin while simultaneously relegating the miscarried fetuses they quite routinely handled to the not-quite-human, semi-wild category of *aucas* that deserved no human burial. For these Catholic women, assent to the Church's condemnation of abortion did not depend on the claim that personhood begins at conception but accommodated "a class of quasi and almost persons that happened to include those not-yet, unborn beings who die in the process of becoming" (Morgan 2009: xiv). Historical research in Russia and Japan has shown that before the second half of the twentieth century, when biomedical advances dramatically lowered the rates of infant death, peasants often used terms that encompassed prenatal losses through miscarriage or abortion and perinatal deaths. The Japanese "water children" (*mizuko*) and Russian "not destined to live in this world" (*ne zhilets na belom svete*) designated beings whose process of becoming was interrupted before or after the end of a pregnancy (LaFleur 1992; Ransel 2000: 186).

In many of such contexts, what anthropologists refer to as "social personhood" (i.e., recognition as a full member of a social group) was only achieved some time after birth, through an initiation ritual such as Christian baptism, Jewish and Muslim circumcision, a name-giving ceremony, or other rites of passage. Based on research on Christian anti-abortion activism in contemporary Russia, this chapter investigates the dilemmas caused by the unstable status of the fetus as a being whose biological, social, and theological meanings do not always add up to one coherent whole. As Russian activists attempt to bring together views of the fetus stemming from Eastern Orthodox theology, Soviet science, and international pro-life discourses, they create a visual and verbalized imaginary of the fetus that is quite different from the North American "icon of life." Fetal imagery from post-Soviet Russia shows how scientific views of the fetus as a biological being are culturally inflected, while theological and political formulations grapple with the biological vulnerability of human engendering.

The sociologist Luc Boltanski (2013: 48–49) speaks of "engendering" as a social process, where a being that has arrived "in the flesh" needs to be affirmed ritually and linguistically in order to be "adopted" as a member of a social group. Adoption usually occurs through the affirmation of the new being by the mother and the wider kin group, allowing the new human being to grow into a role that makes it both a singular individual and someone with a place in a social system. Abortion always does more than interrupt a biological process; it also interrupts, or refuses to set in motion, a process of social engendering that produces a socially embedded human person. In North America, the movement to politicize abortion has led to a focus on biological, genetically human life as the minimal trait of a rights-bearing subject. By contrast, insisting on the social embeddedness of processes of engendering has been a feminist countermove designed to shift emphasis from the discourse of fetal rights to a more complex consideration of life circumstances that lead to difficult decisions (Ginsburg 1989; Mensch and Freeman 1993; Parsons 2010).

But not all anti-abortion movements focus on biological life, and not all arguments for embeddedness advance a feminist agenda. Orthodox Christian activists in twenty-first-century Russia willingly adapt materials and approaches from the Western pro-life movement. They even use discourses of human life beginning at conception to counter evolutionist understandings of fetal development that had been prevalent during Soviet times. At the same time, these activists have theological reservations against ascribing individual personhood to unbaptized fetuses. Rather, they value them for their protosocial qualities, embedding them as potential members in kinship and national groups. In their view, the problem with abortion is less that it violates the individual right to life but rather that it prevents a conceived child from assuming full membership in collectives already under siege. In Russian reproductive politics, fetuses do not embody the pure potential of life itself but are akin to the ancestral remains whose reinvigorated role in postsocialist politics was analyzed by Katherine Verdery (1999). Like the remains of adult victims of socialist regimes, aborted fetuses are assumed to have interrupted biographical trajectories (potentials for biological development and social identity) that connect them to kin and national groups. Like dead ancestors, dead offspring can become a relatively risk-free focus for mourning the lost vitality of a social group, unable to criticize or resist attempts to shape its future. They thus become good candidates for animation in the name of particular political projects, lending strength to visions of what makes a morally good society and what endangers it.

Russia and Abortion

Post-Soviet Russia provides a distinctive arena for the study of abortion politics, because it combines a long-term practice and relatively wide acceptance of the procedure with recent attempts to make it more controversial and impose restrictions. Some of the differences between North American and Russian pro-life views of the fetus lie in the fact that direct experience of abortion is far more widespread in Russia than in many other parts of the world. Legalized in 1920 and then again in 1955 (after a period of severe restrictions on elective abortions under Joseph Stalin), abortion was *the* method of fertility control for postwar Soviet generations. Barrier methods of contraception such as condoms and cervical caps were always in short supply and unpopular with the population, while hormonal contraceptives ("the pill") were never produced in the Soviet Union. Importing the pill was prohibited after a brief period in the early 1970s, because of concerns with the side effects of this early generation of the medication. Surgical abortions, by contrast, were available in the gynecological wards of maternity clinics (*roddoma*, literally "birth houses") and quickly became the principal procedure performed there. At their peak in 1965, abortions outnumbered live births almost three to one, and having multiple abortions across a reproductive life-span remained the norm for Soviet women in the 1970s and 80s (Luehrmann 2017; Zdravomyslova 2009).

Though there has been a gradual decrease since the mid-1990s, it was only around 2008 that there were fewer abortions than live births. Hovering at around five hundred per thousand live births, the abortion ratio remains significantly higher than in North America, where it is around three hundred. In Soviet as in post-Soviet times, married and mature women often use abortion as a spacing mechanism. The typical at-risk fetus that becomes an object of activist concern is not necessarily the offspring of a teen mother but rather a second or third sibling whose progenitors think they are not able to increase their family size (Denisov et al. 2012; Sakevich 2009).

In addition to being far more a part of mainstream female experience than in North America, abortion from Soviet times onward was framed more as a problem of demographic responsibility than of sexual morality. These demographic concerns explain why abortion retained the official status of an evil to be fought against although it was legal and widely practiced throughout much of Soviet history. The prohibition of elective abortions between 1936 and 1955 was mainly an attempt to increase the birth rate, shown by the fact that the struggle against illegal abortion intensified in the late 1940s and early 1950s, accompanied by increased attention to preventing infant deaths and supporting unwed mothers. All these measures were framed as means of "replacing the dead" of World War II, in which twenty million Soviet citizens perished (Nakachi 2008; Randall 2011). After restrictions were lifted under Stalin's successor, Nikita Khrushchev, the skyrocketing rates of abortion raised public concern not as part of a discourse on declining sexual mores but in connection with debates about the quantity and quality of the population as well as women's struggles to combine traditional caregiving roles with the expectation that they become part of the socialist work force (Field 2007; cf. Andaya 2014). The rise of the "one-child family" became a publicly debated issue, and scholars and planners voiced civilizationist concerns because birthrates in the Asian parts of the Soviet Union were higher than in the European ones. During the social and political opening of perestroika and after the collapse of the Soviet Union in 1991, these concerns turned into a full-fledged panic about "demographic crisis," because further decreases in births and a dramatic decline in life expectancy especially for men led to negative population growth (Parsons 2014; Rivkin-Fish 2006). Amid fears about the extinction of the Russian nation, aborted fetuses appear less as individuals deprived of their rights and more as large numbers of missing citizens whose lives could have replenished the nation had they not ended in utero.

Post-Soviet Russia has not seen the dramatic changes in abortion legislation of such postsocialist states as Poland, which passed from permissive legislation to almost complete prohibition, and Romania, which lifted the severe and punitive restrictions imposed by the pronatalist socialist state (Kligman 1998; Zielinska 2000). First-trimester abortion remains available on demand and free if performed at a state health clinic. After the first trimester, abortions are performed for medical and a small number of social indications. But since the fall of the Iron Curtain, the strengthening public presence of the Russian Orthodox Church and increased contacts with international Christian activism have led to the emergence of a pro-life movement largely driven by Orthodox Christians. In terms of influencing legislation, the movement's successes have been limited, though not insignificant. Over the years, the list of admissible social indications for a second- or third-trimester abortion has been reduced to just three: rape, incest, and incarceration of the mother. Since the fall of 2011, new legislation requires a mandatory waiting period of one week between the time when a pregnant woman requests an abortion and the earliest date when it can be carried out., During this time, the pregnant woman must attend a counseling session with a psychologist employed by the health clinic (Rivkin-Fish 2013).

More importantly, perhaps, the movement has taken on new institutional contours, influenced by a turn toward state-backed pronatalism under President Vladimir Putin and Patriarch Kirill's policy of standardizing the social outreach activities of the Church (Chandler 2013; Stoeckl 2014). What began in the 1990s and 2000s as small groups formed around individual activist priests who referred to themselves by the Anglicism *prolaif* is turning into a network of "centers for the defense of the family." The work of such centers typically includes counseling services for pregnant women and material help to single mothers and large families, as well as sometimes marriage counseling and classes for parents and children.

Between 2008 and 2014, I visited centers and conducted interviews with lay and ordained Orthodox activists in Moscow, Saint Petersburg, and the regional capitals of Kazan, Nizhnii Novgorod, and Kirov. As an ethnographer who participated in the organizations' day-to-day outreach activities, I was able to see the networks of people and motivations behind policy shifts. Through formally solicited "procreation stories" (Ginsburg 1989) and casual conversations, I realized that many of the activists had themselves experienced abortion and were parents to living offspring. When remembering aborted fetuses, they were often saying as much about their actual and wished-for families as about the abstract rights and wrongs of abortion (Luehrmann 2017). At the same time, they were engaging with the shifting discursive framework provided by church and secular media, which increasingly emphasized the relational and social rather than the individual and biological potentials of fetuses.

Since 2012, every diocese is required to designate a priest who coordinates work to encourage child bearing and family life, and in 2013 a Patriarchal Commission on the Family and the Protection of Motherhood and Childhood was created to collect information on regional activities and offer training and outreach materials while serving as a voice for the moral vision of the Church (Patriarshaia Komissiia 2014). The commission is headed by Archpriest Dmitrii Smirnov, a married parish priest who began raising the issue of abortion and rights of families with many children in the late 1980s and co-founded the Moscow organization Life Center (Tsentr Zhizn') in 1993. His move from director of the Life Center (an organization that still exists but is now run out of the offices of the commission) to chair of a commission that identifies "family," "motherhood," and "childhood" as its key areas of concern is symptomatic of a larger shift away from a focus on biological life. Russian activists who embrace this shift also see it as a move to gain independence from the model of North American pro-life activism. As Sergei, the organizer of a yearly festival of pro-life initiatives that still bears the name "For Life" (*Za Zhizn'*) but increasingly focuses on promoting family-oriented moral frameworks, explained in an interview (February 2012):

> Western pro-life, American pro-life, they consider the highest value to be life from conception to natural death, yes? ... We talked about it and decided that for us, the value is eternal life. That a person is saved in eternal life is more important than that he lives here. So that means that life, well, it can happen the other way round, that we save a child, and he will live in this world, and then a pedophile comes along and kills that child's soul.... So we started the movement as pro-life, defense of children, but we found that we can't do anything without defending the family so that it can protect children from the temptations of the contemporary world.

In Sergei's analysis and that of activist clergy I met at his festival, the Western pro-life movement's focus on biological life as an absolute value was a pragmatic strategy for creating an interreligious coalition in the context of North American multiculturalism. They found that the search for secular and interdenominational partners required Christian organizations to disregard aspects of their traditions in which the value of biological life was subordinate to the eternal fate of the soul, as in ideas about martyrdom, for example. The wish to hold on to a substantive vision of what gave value to human life was a reason the Russian festival welcomed Catholic speakers from Poland, Finland, and other parts of Eastern Europe but did not allow non-Orthodox organizations to compete for festival prizes or participate in joint protest or outreach.

The image of the fetus that emerges from this shift from biological organism to social fabric is complex. As a bearer of "eternal life," a human in utero is less an image of biological perfection whose survival must be promoted at all cost and more a potential that can develop into negative as well as positive directions. Bringing the fetus to a live birth is not enough, because the child that is born is also in need of protection from "pedophiles" (a term widely used in Russia as a derogatory term for homosexuals) and other modern temptations. While Danish in vitro fertilization (IVF) patients asked to donate embryos for stem cell research find it possible to see them as blank figures with potential as biological resources (Svendsen 2011), Russian Orthodox activists insist that an embryo or a fetus is never a biological tabula rasa. Rather, it is a moral entity whose life can take right or wrong turns and who needs a social framing to direct it. Both discourses see the developing human being as a figure of potential, but they have different degrees of openness about how that potential can be realized. If the discarded IVF embryo, "although not yet anything, had the ability to become everything in the future" (Svendsen 2011: 423), the fetus Sergei hoped to save from abortion was already "someone"—a being endowed with a soul. Neither a tabula rasa nor completed at the time of birth, the soul's developmental trajectory connected pre- and postnatal periods and required a specific social environment to unfold in the desired direction.[1] This ideal social environment was imagined in kin and national terms. However, membership in both collectives was not automatic but depended on particular rituals of initiation.

Quasi-Personhood and Protosocial Beings

In this neotraditionalist discourse, fetal personhood mattered but not in the biologist framework familiar from North American debates. One aim of activists in various cities was to establish psychological consultations in the municipal gynecological clinics that gave referrals for surgical abortions. By agreement with the directors of select clinics, Orthodox organizations in Saint Petersburg, Kazan, and other cities paid their own psychologist to hold consultations several times a week to which, ideally, all women presenting for an elective abortion should be referred. While many of these arrangements preceded the legal requirement for a psychological consultation, in some cases the Orthodox psychologist took on the role of providing the mandatory consultations because not every clinic had its own psychologist on staff.

In their approaches, the Orthodox psychologists I spoke to drew on internationally circulating discourses of fetal personhood but gave them specific post-Soviet inflections. They used little plastic models of "preborn" fetuses at various ages of gestation that were originally introduced to Russia by North American pro-life activists but were mainly Russian-made at the time of my fieldwork. Representing life-sized fetuses that can somehow exist and be handled outside of a pregnant woman's body, these models are artifacts of biologistic thinking that can easily be appropriated for relational ends. One psychologist in Kazan told me she encouraged pregnant women to hold one of these dolls during the conversation, wrap it in little swaddling cloths, and put it in a miniature bassinette. She saw these interactions as a natural supplement to showing a brochure with in utero photographs of embryonic and fetal development, both intended to "activate maternal feelings" and make clear that "there is already a person there" (*tam uzhe est' chelovek*). But she and her colleagues also acknowledged that information about human development was not always enough to deter someone from having an abortion. In the 1990s, a longtime Moscow activist explained, one could go into an auditorium and show pictures of fetal development, and people would cry and be shocked. Today's young people know everything, and still have abortions, because "their hearts are hardened."

The idea that "there is a person there" was more surprising in the 1990s because Soviet textbooks taught a theory of embryonic development going back to the German evolutionist Ernst Haeckel (1834–1919), who posited that ontogeny recapitulates phylogeny, and an embryo in utero goes through evolutionary stages resembling various kinds of animals. Haeckel's theories, which popularized Darwinism in much of Central and Eastern Europe, were officially promoted in the Soviet Union because his drawings of fetal development visualized processes of evolution and supported the materialist point of view that no absolute divide existed between human and animal life (Polianski 2012). Several older women reported being influenced by this view in their Soviet-era decisions to abort. "According to Haeckel's teachings, there wasn't a human there, but a fish or a frog—it meant nothing to get rid of it," recalled Valentina (born in 1937), the director of the Saint Petersburg branch of the Life Center. Like many ideas embedded in Soviet-era visual imaginaries, Haeckel's theory of recapitulation still had a place in early twenty-first-century Russian life, for example, in displays at the Saint Petersburg Museum of Zoology that remained unchanged since the fall of the USSR.

For post-Soviet activists, the materialist view of the fetus as a fish or amphibian represented a burden from the past that needed to be overcome, but the more humanist side of socialist discourses about the fetus was less marked as "Soviet." Expressed in medical literature and poetry, socialist humanist discourse on the fetus as a potential member of human collectives serves as one of the sources for how post-Soviet activists frame the harm done by abortion. In Soviet medical literature, the high rates of abortion were treated as a health concern for women, both in terms of physical risks of infection and secondary infertility and mental risks of going against natural maternal feelings. Although the fetus as a rights-bearing individual did not enter Soviet humanism, it was represented as a relational being offering fulfillment to its parents and potential talents that could be of service to society. In a poem I first saw on a sticker distributed by the Life Center, but later found in a Soviet women's health guide from 1965, author Irina Bychenkova asked pregnant woman considering an abortion to "stop to think!" Perhaps, the poem suggests, the one "whose life now hangs on a thin thread, / will turn out to be a scholar or a poet, / and the whole world will speak of him." Although they would deny the implication that only future scholars and poets have a right to survive, post-Soviet anti-abortion activists eagerly embrace the notion of genetic destiny, claiming that everything is already determined (*zalozheno*) in the zygote, from the color of someone's eyes to a love of flowers. The branches of the Life Center and affiliated organizations prominently display a memorandum signed by two embryologists at Moscow State University (Russia's oldest university and one of its most prestigious research institutions). On letterhead depicting the university's distinctive Stalin-era central high-rise, they state that "the life of a human being as a biological individual" begins at conception and that the zygote cannot be considered part of the mother's organism (Golichenkov and Popov n.d.).[2]

Engagement with Soviet discourses thus pushes post-Soviet activists toward biologizing languages of life as an unchanging base of personhood and human worth, both in order to refute particular evolutionist understandings and because they translate an older European discourse of genius and innate talents. But Russian Orthodox theology and practice add complexity by emphasizing social personhood rather than biological engendering. Here, it is baptism, performed forty days after the birth according to Church canons, that confers a name on a newborn and adopts it into the community. By being named after a saint, the infant obtains a spiritual protector and can be included in communal prayers. The infant also receives godparents, aiding in the building of social connections for the family (Herzfeld 1990; Hirschon 2010). The forty days before baptism compose a period when, in rural Russia, both mother and infant were considered in a liminal state in which excessive social contacts could be dangerous for themselves and for visitors. Well into the Soviet period, mothers continued to limit the social exposure of their infants and to seek the cleansing power of the Orthodox churching prayer (*votserkovlenie*) to end their period of relative seclusion (Ransel 2000). Post-Soviet Russian families still practice the celebration of "showing" their infant to neighbors and relatives just before baptism, and

Orthodox families refrain from referring to their infant by name before the baptism, even if they may have picked one.

In this context, a social practice shaped by the theology of baptism and relatively recent experiences of frequent neonatal death[3] stands in tension with the affirmation of life beginning at conception, suggesting a more complex, gradual process of becoming in which neither conception nor birth are decisive events on their own. Both fetuses and newborns are treated as protosocial beings expected to take on a place in a community but who only slowly emerge from relative isolation and ambiguity into full adoption into a socially recognized position.

Fetuses Represented: Unchaste and Chaste Depictions

The sense that focusing on the fetus as a biological entity can be effective but ultimately fails to do justice to its moral status also comes up in visual depictions of fetuses in Russia. Activists were aware of the imagery of "hard pro-life" that comes to mind when thinking of anti-abortion protests internationally: photographs of bloody, aborted fetuses in grotesquely twisted poses. The organization Warriors of Life, made up mainly of university students and other young adults, uses this imagery for signs at demonstrations and "solitary pickets." For a solitary picket, people handing out fliers or displaying signs stand alone or at least fifty meters apart from other activists, and are thereby exempt from requiring a demonstration permit. The photographs of mangled fetuses mainly come from the United States, recognizable by the nickel and dime coins often placed next to the fetal remains to indicate their minuscule size and to hint that they died for somebody's profit. Dmitrii, a leader of the organization in Saint Petersburg, told me that in the United States, "hard pro-life" forced abortion clinics to close and prevented abortion services from being advertised. "It's a proven method," he said. This organization was also vocally opposed to the Russian laws that kept abortion legal, and unwilling to cooperate with any medical institution that offered the procedure. Organizations that focused on collaborating with medical and governmental institutions to set up counseling sessions for pregnant women criticized such confrontational tactics as ineffective, but also had specific concerns about the Warriors' use of bloody imagery.

Iuliia, a psychologist paid by an Orthodox organization to hold consultation sessions in maternity clinics in Kazan, said she would show the "hard" pictures to a male audience but not to pregnant women: "I would show it to young men, to shock them, so that they see what abortion is. Often they cannot hear in any other way." Women, however, were more receptive to positive imagery, which could "activate their maternal instinct" (*podkliuchit' materinskii instinkt*). Pictures of living babies and dolls representing living fetuses were more suitable for that. Svetlana, a counselor who worked with pregnant women in Moscow in face-to-face and telephone consultations, also commented on the ambiguity of fetal pictures, both those taken in utero and post-abortion. In particular, she was against showing such pictures to children: "It is not for nothing that these processes [of fetal development] are hidden from our eyes. Sometimes one could really look and see an animal there. Some kind of chastity (*tselomudrie*) is violated."

The idea of chastity as keeping certain things shrouded in secret speaks to the need to analyze practices of visualizing fetuses in relation to other culturally relevant imagery and related ethical concepts (Harris et al. 2004; Petchesky 1987). A Slavonic calque on the term *sophrosyne* (whole mind) from the Greek New Testament, the term *tselomudrie* refers to the same virtue as the Latin-derived chastity (Latin *castus,* pure). But rather than focusing on sexual restraint, the Greek etymology points to a wider concern with keeping thoughts pure from preoccupations that might be distressing, disturbing, or inappropriate to a particular stage of development. Orthodox educators and media critics often speak of protecting the *tselomudrie* of children, which means limiting their exposure to depictions of sex, nudity, and same-sex relationships but also to violent, frightening, or otherwise distressing content (Medvedeva and Shishova 2012). The frightening or strange-looking fetus disrupts trust in the reliable "humanness" of human beings, and perhaps also in the happy outcomes of pregnancies. During a picket by the Warriors of Life outside a gynecological clinic in Saint Petersburg, several passersby commented that if a pregnant woman saw the photographs of aborted fetuses, she might have a miscarriage.

Misgivings about the efficacy and ethics of some of the standard international pro-life imagery notwithstanding, the Russian antiabortion movement has produced a rich array of visual media. Most notably, fetuses tend to be depicted not as fetuses but as future projections of what they might turn into, depending on the choice their pregnant mother makes. In keeping with the idea of chastity as preserving the mystery of hidden things, Orthodox artists and designers often respect the opacity of a pregnant woman's uterus and attempt instead to see into the postpregnancy future. In such depictions, fetuses appear as growing children, spectral presences, or both at the same time. A church-sponsored advertisement posted on the streets of Nizhnii Novgorod in 2012 (fig. 10.1) featured a black-and-white photograph of a child of three to four years old, shot in profile looking up with a worried expression, with the caption: "Mom, don't have an abortion! I will always do as you say, promise!"

FIGURE 10.1. "Mom, don't have an abortion: I will always do as you say!" Poster commissioned by the Russian Orthodox diocese of Nizhnii Novgorod, 2012 (Photograph by Sonja Luehrmann).

The poster played on the oft-repeated pro-life argument that no one would kill their toddler or preschooler for some of the social reasons given for abortions—lack of time and money, lack of living space, or fear of the difficulty of bringing up another child. The white-on-black writing on the poster evoked the optic of public health warnings against smoking tobacco that were visible elsewhere in Russian cities. The depiction thus deliberately mixed visual codes for referring to children before and after birth and to health concerns relating to unborn and born children and adults, refusing to differentiate between the ethics of caring for a fetus and the ethics of caring for a young child (Casper 1999). At the same time, the white light falling on the child's face evoked the fetal ghost who was one possible outcome of the decision the interpellated pregnant woman was in the process of making.

Less ambiguous depictions of the aborted fetus as a ghost returning to haunt its mother were common in depictions intended for an internal, churched or near-churched audience. A poster hanging in the psychologist's office at the Saint Petersburg crisis pregnancy center represented, "The life of a woman who has a child and one who has an abortion" through a series of graphic-novel style images (Luehrmann 2017: 108). In one image, the aborted child appears to the sleeping woman in a dream, depicted as a baby in white swaddling cloths. More spectral fetuses appear on the digital image *Two Mothers* (*Dve mamy*) by computer artist Boris Zabolotskii, which won the grand prize of the annual pro-life festival in 2010.[4] On the right, a woman in a skirt and headscarf exits the gates of a churchyard accompanied by four children, ranging from a baby in a stroller to a girl of eight to ten years. Behind her we see an Orthodox church and the tower of Moscow State University. On the left, a tall, thin young woman wearing tight jeans and a T-shirt with the English phrase "Sex in the City" stands next to a sports car whose license plate says, also in English, "I ♥ MYSELF." Inside the car are four shadowy silhouettes matching the other woman's four children in size and outline. The car is surrounded by attributes of Western infiltration: post-Soviet steel-and-glass architecture, advertisements for Coca-Cola and Pepsi, a McDonald's restaurant, and a "center of family planning." The graffiti *Proekt Rossiia* (project Russia) on a wall refers to a common claim that birth control and family planning are being promoted in Russia by Western interests intent on reducing Russia's population and gaining control of its natural resources (Leykin 2013; Sperling 2014).

The stakes of reproductive decisions are set high in this image and play out on a national scale rather than as a universal struggle over the sanctity of life as such. The title implies that the woman on the left is also a mother to her aborted fetuses, whose shadows form a group of siblings structured by birth order. Both women are situated in a larger visual field divided between benign Russian culture (framed as a harmonious combination of religion and science) and sinister, threatening "global" or "Western" forces that seek to destroy it. In 2010 when the image was first created, activists still recognized the limited appeal of its explicit brand of Orthodox nationalism. During a discussion at the Saint Petersburg Life Center, staff decided not to use *Two Mothers* on a flyer to hand out during a public event that summer, because "non-Orthodox people won't understand it." However, with the unfolding violence in Ukraine over the course of 2014, the message of the Western threat gained ever more traction in Russia. Putin stated in a December 2014 press conference that the West was not really after Ukraine but Siberian resources (President of Russia 2014), and a New Year's message on the Pro-Life Festival listserv explained that since the outbreak of violence in eastern Ukraine, "we felt that many around us now understand better the meaning of our message, the purpose of our work."

For more general audiences, the Life Center and other organizations continue to avoid imagery of the spectral fetus and of implied enemies, instead focusing on living children and happy families. A series of social advertisements first placed in the Moscow subway in 2008 was designed to convey that having three children is not an excessive burden to be combatted by abortion but rather a good thing. The imagery represents both children and parents through objects, accompanied by the slogan "Congratulations on the addition [to the family]!" (*S popolneniem!*): a third child's toothbrush is added to a cup with two adult' and two children's brushes, number four of a series of Russian nesting dolls opens up to reveal a fifth one, and so on. Similar to the image of the spectral family, these images portray the unborn as always already part of a collective; rather than from life itself, they derive value from "filling up" (the literal meaning of the word *popolnenie*) the existing kin group and strengthening Russia's future.

Compared to the fetal photography that has such a prominent place in Western abortion politics, one could say that the projection into the future of fetal imagery in the Russian Orthodox movement treats fetuses less as pure potential than as bodies subject to political animation, similar to the "lively politics around dead bodies" discussed by Verdery (1999: 23). Like the dead bodies of known and unknown adults, fetuses present the impression of a singular agent but are open to multiple projections of other people's agency. They do not speak for themselves, but words and thoughts can be attributed to them, as in the "Diary of an Unborn Child" or the Nizhnii Novgorod poster. The ambiguity of a dead body comes from the "complex behavior subject to much debate" that is part of actually lived biographies, while the affective power of dead body politics is fueled by notions of kinship obligations and their connections to ideas about cosmic order (Verdery 1999: 28). Living or dead fetuses can be animated through the imaginative work of endowing them with a future biography and inserting them into networks of mutual kin obligations. In these ways, fetal imagery in post-Soviet Russia shows the link between the new reproductive legislation instituted by many postsocialist states (Chandler 2013; Gal and Kligman 2000) and the simultaneous flurry of reburials and posthumous rehabilitations that were part of the reformulation of historical narratives. By focusing simultaneously on ancestors and offspring, the political community rethinks its moral fabric through animating beings on its edges with the qualities desired for its members: loyalty, reliability, and irrepressible vitality.

The visual and liturgical symbol the Orthodox anti-abortion movement has chosen for public commemorations also takes up a narrative of violence in a stylized and aestheticized form, preserving the chastity of viewers. Since the early 1990s, the Life Center in Moscow has marked 11 January, the day the Church commemorates the "14,000 Holy Innocent Infants of Bethlehem in Judah, killed by Herod," as a day to commemorate and express opposition to abortion. Catholic tradition calls this episode the Slaughter of the Innocents, and it refers to the gospel narrative of King Herod ordering the killing of all children under two in the attempt to kill the newborn Jesus.[5]

In the process of creating this ritual commemoration of abortion, the Life Center commissioned an icon depicting the Holy Innocents, a subject previously depicted only as a *kleimo*, a small image in the frame of icons of the Nativity of Christ (fig. 10.2).

In conformity with the classical iconographic style, scenes of murder are relegated to the background, while the small figures in the center stand unharmed, identified as martyrs only by the crosses they hold and the red background. By depicting aborted fetuses as a large group of child victims, the icon once again crosses

FIGURE 10.2. Icon of the Holy Innocent Infants of Bethlehem, Moscow (Photograph by Sonja Luehrmann).

the divide between prenatal and postnatal development. It also puts blame on the state as a perpetrator of abortion, reframing a common Soviet experience—having an abortion in the interest of delaying or spacing childbirth—as a condition of complicity or victimhood in a program of government-sanctioned murder. By stylizing the violence and focusing attention on the inviolate bodies of saintly figures, the icon becomes available for uses that focus less on past abortions than on the present and future vitality of the nation. In many churches, it is used for prayers against infertility and for the support of families. When talking about the decision of the Russian movement not to focus on biological life as an ethical goal, festival organizer Sergei used the infants of Bethlehem as one example of the overriding importance of the eternal life of the soul. He said if they had lived, some of these Jewish children may have participated in the crucifixion of Christ and thereby condemned themselves to eternal damnation.

Reanimating Past Decisions

Focused on Russia's future as their movement appears to be, the specters of past fetuses have a very personal significance for many pro-life activists. Some staff members and volunteers who offer aid to pregnant women and participate in anti-abortion rallies are women of a generation that knows abortion from personal experience. For them, advocating against abortion is a way of expiating their own past reproductive decisions that they now conceptualize as sin. The director of the Saint Petersburg Life Center, for example, was a woman in her seventies who had terminated three pregnancies in the 1960s and '70s as a spacing mechanism between giving birth to three living children. The director of another center had come to church activism through involvement in a voluntary movement that visited children in an orphanage, an oft-recommended penance for abortions. The spectral fetuses of these women were often quite personal and concrete, and showed how the fetus became a field for projection of the family life they might have had. Several interviewees who only had sons speculated that the last pregnancy they terminated might have resulted in the birth of a daughter and wondered what old age might be like with the support of a daughter rather than sons and daughters-in-law. A woman who had only one child because of her job as a railroad conductor speculated what a more settled family life would have been like. Taking dead fetuses and reproductive mishap as objects of speculation about alternative life trajectories is also common in North American narratives of abortion and pregnancy loss (Ginsburg 1989; Layne 2003). But theological reservations against personifying the unbaptized and political discourses of demographic decline posed special problems for these women, pushing them away from their own alternative biographies toward wider social outreach.

Russian Orthodox priests who hear confessions often recommend that the penitent focus on cultivating a counteracting virtue. In the context of demographic anxiety, lay women as well as priests thought that the counterbalancing virtue for ending the life of a fetus was supporting the collective lives of young children and their families: visiting orphanages, giving financial support to a struggling family, or upholding "traditional family values" against perceived threats such as same-sex marriage, LGBT adoption, or government interference with child raising. These activities drew attention away from an aborted fetus to living members of the community that this fetus was not able to join. The book of fictional stories *Pustye Pesochnitsy* (Empty sandboxes) (Fesenko 2011) sold in many church shops and freely distributed by activist groups in print and online linked individual reproductive decisions to the national demographic problem and the traumatic transition period of the 1990s. At that time, the birth rate was so low that many children's playgrounds were allowed to decay and schools and preschools were converted to other uses.

One thing these women could not do was treat their aborted fetuses as persons in the sense of full members of the Church. North American religious groups sometimes allow retroactive namings of children who died in utero or before baptism; the Japanese *mizuko* cult involves couples purchasing a Buddhist mortuary name for their aborted fetus and erecting a small statue of the bodhisattva in the fetus's memory (Hardacre 1997; LaFleur 1992). In Russia, priests categorically denied namings of aborted children and uncanonical rites for their posthumous baptism, although I met women who had engaged in both. One woman claimed that posthumous baptism (according to a rite that the Virgin Mary revealed to a nun in the 1950s) turned the aborted fetus from "a bloody demon" into a full-term, healthy baby waiting for its mother in heaven.

While these clandestine rituals reveal an interest in turning the spectral fetus into a regular dead relative, the main theological objection to such ritual personifications lies in the fact that baptism can only be bestowed on a living person, and only those with baptismal names are members of the Church who can be included in corporate prayers. Officially recommended prayers for aborted fetuses are reserved for "solitary recitation" (*dlia keleinogo chteniia*). The only official rite that can be used to acknowledge abortion or any other kind of prenatal death is one for the "churching" (return into the liturgical community) of a woman after an unintentional miscarriage, which forces her to express repentance for the potential sins that led to the inauspicious outcome of the pregnancy (Kizenko 2013). Not having made it into full Church membership, the fetus as a protosocial being can only be remembered in the privacy of the family. At the same time, church kiosks sell brochures and prayer texts calling for repentance for abortions, presenting the issue as one of collective importance. Precisely because they have no fixed public identity, the spectral presences of aborted fetuses can animate projects that connect very personal doubts, regrets, and speculations to wider diagnoses of where society took a wrong turn.

Conclusion: Fetuses and Life Courses

In Russia and elsewhere, politicizations of abortion show the intimate connection between the problems posed by birth and death for maintaining and reconstituting social orders: the capacity of aborted fetuses to combine future potential, social relatedness, and death and destruction in one symbol with deeply private as well as public appeal makes them the ultimate dead bodies of a postsocialist politics of restoration. For a comparative anthropology of fetuses, the Russian example points to the cultural construction of boundaries and continuities between fetuses, neonates, and stages of human life cycles as a crucial area of inquiry.

Religious traditions play crucial roles in determining points of transition, necessary rites of passage, and what counts as a human life worth living (Inhorn and Tremayne 2012). But demographic histories and political traditions are no less important, as are standards of medical care and experiences of lived (in)security. As Morgan (1999) found out, Ecuadorian Catholics and North American Catholics differ in the weight they place on issues of fetal personhood for determining the moral status of abortion. Russian Orthodox anti-abortion activists, for their part, tend to be respectful observers of the North American movement, which they perceive to be far more powerful and influential than their own. They take assertions of fetal personhood seriously and use them as correctives to Soviet views of fetuses as representing prior stages of human evolution. At the same time, they see post-Soviet Russia as a place where the fabric of the social is threatened by economic and moral decline and threats from outside. In this context, fetuses are not so much embodiments of universal and individualizable biological life, but rather represent society's smallest building blocks, whose vulnerability magnifies the vulnerability of the whole edifice. In a political setting where discourses of individual rights are contested and far from hegemonic even when applied to adults, ideas of the social embeddedness of unborn children become a dominant discourse that imposes its own normative goals on pregnant women, postmenopausal women, and actual and imagined children (Rivkin-Fish 2013).

One may see this emphasis on fetal embeddedness as residual collectivism, left over from socialism or the peasant village. Or one may see it as a reinterpretation of authoritative bioscientific knowledge in a context where "the politics of life itself," conceptualized by Nikolas Rose (2006) as an increasing focus on the quality rather than quantity of human organisms, competes with the legacy of immense population losses through Russia's twentieth century. Fetuses become objects of public concern because of their insufficient numbers, and rather than improving biological organisms, the goal of reproductive activism is to improve the family units that are supposed to raise morally healthy and plentiful offspring. Anthropologists of the fetus will find themselves sympathizing with the Russian activists' insistence on the social contexts without which there can be no human reproduction in either a biological or a cultural sense. Where activists seek to construct the one moral framework in which they claim all fetuses could thrive, anthropologists do well to note how fetuses trouble cultural and political projects at the same time as they can be mobilized to support them. Human and not quite human, disturbing at the same time as appealing, standing for life in a way that emphasizes its close neighborhood to death, fetuses are creatures whose images, however carefully managed, continually undermine the causes to which they summon their viewers.

Acknowledgments

Research for this chapter was made possible by financial support from the University of British Columbia Killam Trust, the International Research and Exchanges Board (IREX), the Social Science Research Council's New Directions in the Study of Prayer initiative, and Simon Fraser University. I am grateful to Anastasia Rogova for research assistance and to participants of the panel on "Four-Field Perspectives on the Fetus" at the 2013 American Anthropological Association meeting for helpful comments.

Sonja Luehrmann is an associate professor of anthropology at Simon Fraser University. She is the author of *Secularism Soviet Style: Teaching Atheism and Religion in a Volga Republic* (Indiana University

Press, 2011) and *Religion in Secular Archives: Soviet Atheism and Historical Knowledge* (Oxford University Press, 2015).

Notes

1. On the continuous malleability of souls across the life course in the Russian imagination, see Pesmen 2000.
2. The idea of a genetically determined love of flowers is expressed in the text "Diary of an Unborn Child" originating in the North American pro-life culture of the 1980s, a Russian translation of which circulates on fliers and in the Russian blogosphere.
3. Cf. Nancy Scheper-Hughes's (1992) descriptions of deferred emotional investment in infants among residents of Brazilian favelas who cannot take survival of their children for granted.
4. The image can be viewed on the artist's website at http://www.bzab.ru/tvorchestvo/za-zhizn-i-semyu/nggallery/image/11-3 (accessed 23 April 2017).
5. The link between the Holy Innocents and abortion has precedents in mid-twentieth-century Catholicism (Stycos 1965), but an iconographer and a priest I interviewed separately at the Life Center recalled no knowledge of this parallel but said they "naturally" settled on the story of murdered children as symbols of aborted fetuses.

References

Agamben, Giorgio. 1998. *Homo sacer: Sovereign Power and Bare Life.* Trans. Daniel Heller-Roazen. Stanford, CA: Stanford University Press.
Andaya, Elise. 2014. *Conceiving Cuba: Reproduction, Women, and the State in the Post-Soviet Era.* New Brunswick, NJ: Rutgers University Press.
Arendt, Hannah. 1951. *The Origins of Totalitarianism.* New York: Harcourt Brace Jovanovich.
Boltanski, Luc. 2013. *The Foetal Condition: A Sociology of Engendering and Abortion.* Translated by Catherine Porter. Oxford: Polity.
Casper, Monica. 1999. "Operation to the Rescue: Feminist Encounters with Fetal Surgery." In Morgan and Michaels 1999, 101–112.
Chandler, Andrea. 2013. *Democracy, Gender and Social Policy in Russia.* Basingstoke: Palgrave Macmillan.
Denisov, Boris P., Victora I. Sakevich, and Aiva Jasilioniene. 2012. "Divergent Trends in Abortion and Birth Control Practices in Belarus, Russia and Ukraine." *PLOS ONE* 7 (11): e49986. doi:10.1371/journal.pone.0049986.
Fesenko, Denis Olegovich. 2011. *Pustye pesochnitsy* [Empty sandboxes]. Moscow.

Field, Deborah. 2007. *Private Life and Communist Morality in Khrushchev's Russia*. New York: Peter Lang.
Gal, Susan, and Gail Kligman. 2000. *The Politics of Gender after Socialism: A Comparative-Historical Essay*. Princeton, NJ: Princeton University Press.
Ginsburg, Faye. 1989. *Contested Lives: The Abortion Debate in an American Community*. Berkeley: University of California Press.
Golichenkov, V.A, and D.V. Popov. n.d. Memorandum, Faculty of Biology, Moscow State University. Retrieved from http://theme.orthodoxy.ru/abort/page01.html. Accessed 23 April 2017.
Hardacre, Helen. 1997. *Marketing the Menacing Fetus in Japan*. Berkeley: University of California Press.
Harris, Gillian, Linda Connor, Andrew Bisits, and Nick Higginbotham. 2004. "'Seeing the Baby': Pleasures and Dilemmas of Ultrasound Technology for Primiparous Australian Women." *Medical Anthropology Quarterly* 18 (1): 23–47.
Herzfeld, Michael. 1990. "Icons and Identity: Religious Orthodoxy and Social Practice in Rural Crete." *Anthropological Quarterly* 63 (3): 109–121.
Hirschon, Renée. 2010. "Indigenous Persons and Imported Individuals: Changing Paradigms of Personal Identity in Contemporary Greece." In *Eastern Christians in Anthropological Perspective*, ed. Chris Hann and Hermann Goltz, 289–310. Berkeley: University of California Press.
Inhorn, Marcia, and Soraya Tremayne, eds. 2012. *Islam and Assisted Reproductive Technologies: Sunni and Shia Perspectives*. New York: Berghahn Books.
Kizenko, Nadieszda. 2013. "Feminized Patriarchy? Orthodoxy and Gender in Post-Soviet Russia." *Signs* 38 (3): 595–621.
Kligman, Gail. 1998. *The Politics of Duplicity: Controlling Reproduction in Ceausescu's Romania*. Berkeley: University of California Press.
LaFleur, William R. 1992. *Liquid Life: Abortion and Buddhism in Japan*. Princeton, NJ: Princeton University Press.
Layne, Linda L. 2003. *Motherhood Lost: A Feminist Account of Pregnancy Loss in America*. New York: Routledge.
Leykin, Inna. 2013. "Population Prescriptions: State, Morality, and Population Politics in Contemporary Russia." PhD diss., Brown University.
Luehrmann, Sonja. 2017. "Innocence and Demographic Crisis: Transposing Post-Abortion Syndrome into a Russian Orthodox Key." In *A Fragmented Landscape: Abortion Governance and Protest Logics in Postwar Europe*, ed. Silvia de Zordo, Joanna Mishtal, and Lorena Anton, 103–122. New York: Berghahn Books.
Medvedeva, Irina, and Tat'iana Shishova. 2012. *Bomby v Sakharnoi Glazuri: Tckhnologii Obmana* [Bombs under Sugary Icing: Technologies of Deceit]. Moscow: Zerna-Slovo.
Mensch, Elizabeth, and Alan Freeman. 1993. *The Politics of Virtue: Is Abortion Debatable?* Durham, NC: Duke University Press.
Morgan, Lynn. 1998. "Ambiguities Lost: Fashioning the Fetus into a Child in Ecuador and the United States." In *Small Wars: The Cultural Politics*

of Childhood, ed. Nancy Scheper-Hughes and Carolyn Sargent, 58–74. Berkeley: University of California Press.

———. 2009. *Icons of Life: A Cultural History of Human Embryos.* Berkeley: University of California Press.

Morgan, Lynn, and Meredith Michaels, eds. 1999. *Fetal Subjects, Feminist Positions.* Philadelphia: University of Pennsylvania Press.

Nakachi, Mie. 2008. "Replacing the Dead: The Politics of Reproduction in the Post-War Soviet Union, 1944–1955," PhD diss., University of Chicago.

Parsons, Kate. 2010. "Feminist Reflections on Miscarriage, in Light of Abortion." *International Journal of Feminist Approaches to Bioethics* 3 (1): 1–22.

Parsons, Michelle. 2014. *Dying Unneeded: The Cultural Context of the Russian Mortality Crisis.* Nashville, TN: Vanderbilt University Press.

Patriarshaia Komissiia po voprosam sem'i, zashchity materinstva i detstva. 2014. "Komissiia" [The Commission] http://www.pk-semya.ru/komis siya.html.

Pesmen, Dale. 2000. *Russia and Soul: An Exploration.* Ithaca, NY: Cornell University Press.

Petchesky, Rosalind. 1987. "Fetal Images: The Power of Visual Culture in the Politics of Reproduction." *Feminist Studies* 13(2): 263–292.

Polianski, Igor J. 2012. "Between Hegel and Haeckel: Monistic Worldview, Marxist Philosophy, and Biomedicine in Russia and the Soviet Union." In *Monism: Science, Philosophy, Religion and the History of a Worldview,* ed. Todd H. Weir, 197–222. New York: Palgrave Macmillan.

Randall, Amy. 2011. "'Abortion Will Deprive You of Happiness!' Soviet Reproductive Politics in the Post-Stalin Era." *Journal of Women's History* 23 (3): 13–38.

Ransel, David L. 2000. *Village Mothers: Three Generations of Change in Russia and Tataria.* Bloomington: Indiana University Press.

Rapp, Rayna. 2000. *Testing Women, Testing the Fetus: The Social Impact of Amniocentesis in America.* New York: Routledge.

Rivkin-Fish, Michele. 2006. "From 'Demographic Crisis' to 'Dying Nation': The Politics of Language and Reproduction in Russia." In *Gender and National Identity in Twentieth-Century Russian Culture,* ed. Helena Goscilo and Andrea Lanoux, 151–173. DeKalb: University of Northern Illinois Press.

———. 2013. "Conceptualizing Feminist Strategies for Russian Reproductive Politics: Abortion, Surrogate Motherhood, and Family Support after Socialism." *Signs* 38 (3): 569–593.

Rose, Nikolas. 2006. *The Politics of Life Itself: Biomedicine, Power, and Subjectivity in the Twenty-First Century.* Princeton, NJ: Princeton University Press.

Sakevich, Viktoriia. 2009. "Problema aborta v sovremennoi Rossii." [The Problem of Abortion in Contemporary Russia] In Zdravomyslova and Temkina, 136–152.

Scheper-Hughes, Nancy. 1992. *Death Without Weeping: The Violence of Everyday Life in Brazil.* Berkeley: University of California Press.

Sperling, Valerie. 2014. *Sex, Politics, and Putin: Political Legitimacy in Russia.* New York: Oxford University Press.

Stoeckl, Kristina. 2014. *The Russian Orthodox Church and Human Rights.* London: Routledge.

Stycos, J. Mayone. 1965. "Opinions of Latin-American Intellectuals on Population Problems and Birth Control." *Annals of the American Academy of Political and Social Science* 360: 11–26.

Svendsen, Mette. 2011. "Articulating Potentiality: Notes on the Delineation of the Blank Figure in Human Embryonic Stem Cell Research." *Cultural Anthropology* 26 (3): 414–437.

Verdery, Katherine. 1999. *The Political Lives of Dead Bodies: Reburial and Postsocialist Change.* New York: Columbia University Press.

Zdravomyslova, Elena. 2009. "Gendernoe grazhdanstvo i abortnaia kul'tura" [Gendered Citizenship and Abortion Culture]. In Zdravomyslova and Temkina, 2009, 108–135.

Zdravomyslova, Elena, and Anna Temkina, eds. 2009. *Zdorov'e i doverie: Gendernyi podkhod k reproduktivnoi meditsine* [Health and Trust: A Gender Approach to Reproductive Medicine]. Saint Petersburg: Izdatel'stvo Evropeiskogo Universiteta.

Zielinska, Eleonora. 2000. "Between Ideology, Politics, and Common Sense: The Discourse of Reproductive Rights in Poland." In *Reproducing Gender: Politics, Publics, and Everyday Life after Socialism,* ed. Susan Gal and Gail Kligman, 23–57. Princeton, NJ: Princeton University Press.

Chapter 11

THE "SOUND" OF LIFE
OR, HOW SHOULD WE HEAR A FETAL "VOICE"?

Rebecca Howes-Mischel

What is the "sound" of life? Why, and when, is it necessary that we hear life in fetal form? How do our conceptions about that life—and its sensing—constitute a perhaps nascent biopolitics (the intertwined scientific and political regimes of knowing through which bodies mediate between state and population)? And what kinds of methodological and theoretical approaches help us consider the nature of such fetal claims?

In this chapter I draw on public encounters with fetuses in two seemingly disparate locales—the theatricality of US anti-abortion legislative activism and everyday interactions between doctors and patients in a southern Mexican public hospital—in which audiences were asked to recognize a fetal subject as social, prompted by the sound of its amplified heartbeat. Rather than offer them as comparative cases within equivalent reproductive politics, I query how they might together illuminate implicit propositions about fetal biosocial existence and the proof thereof. As such, this chapter offers a speculative set of methodological and analytic approaches for glimpsing the production of fetal personhood as it emerges through the mediation of diagnostic technologies. Here, I name such productions "fetal propositions" to highlight their not-yet-settled and suggestive nature, while arguing that anthropology offers a kind of tool kit for identifying and analyzing their emergent social claims about fetal beings. Ultimately, such approaches necessitate a consideration of the entanglements between reproductive politics and diagnostic technologies, especially as the spread of routine public health logics may open space for political mobilization of an already alive fetus to globalization. Considering fetal propositions through cross-cultural and cross-linguistic analyses illuminates the underlying logics that make biosocial personhood claims plausible, if not always persuasive.

Nascent Propositions

On 2 March 2011, two unusual "expert witnesses" were presented to the Ohio Health and Aging Committee to provide legislative "testimony" for House Bill 125. The first in a series of a new legislative campaigns to restrict abortions once a heartbeat could be "medically detected," the bill framed this *sound* as both the origin of legal life and as a form of primal and intentional "voice."[1] In the words of Oklahoma State Senator Dan Newberry when he introduced his own bill: "the heartbeat is the only way for a fetus to *communicate* that it wants to live" (Olafson 2012, emphasis added). To facilitate this communiqué, two women at fifteen and nine weeks gestation volunteered for public vaginal ultrasounds—projecting the sight and sound of "the state's youngest legislative witnesses'" live beating hearts, in the words of activists' press releases.

As I followed the narrative framing of this Ohio bill and activists' rhetoric about a heartbeat's "self-evident" claim to a kind of social and legal status, I heard eerie echoes from my ethnographic fieldwork in public health spaces of Oaxaca, Mexico. There, in everyday and unremarkable encounters rather than amid charged politics, heartbeats served as symbolic tools through which rural doctors could intertwine social and medical discourses about the status of the fetus in the interest of public health. Working in underresourced conditions, completing mandatory six- to twelve-month service requirements, and under pressure to improve maternal health outcomes while working efficiently, these doctors grappled with how to make their medical knowledge socially and culturally meaningful so women would *cuidarse mejor* (care for themselves better). In response to the potential anxieties provoked by hearing that their pregnancies were "high risk," one gynecological resident, Dr. Celia,[2] tried to reassure her patients by emphasizing the importance of both affective (or emotional) and diagnostic information. After finishing a routine examination, she held up a small pink plastic Doppler device and explained to one patient: "Don't worry about your baby; see, with this, everyone can hear, not just me. With this, you can meet your baby and know that everything is OK. (*Con esto, se puedes conocer a tu bebé y saber que todo está bien.*)" While she had earlier drawn on her medical training and professional knowledge to frame the diagnostic sequence, here she suggested that her patient, Maria Elena, did not need this expertise to recognize her fetus as a social presence. This proposition rests on a subtle distinction between forms of knowing that Spanish recognizes and English does not: knowing things objectively (*saber*, to have knowledge about) and knowing things subjectively (*conocer*, to be acquainted with). As she suggested, the sound of her fetus's heartbeat should initiate a new acquaintance, and this would in turn reassure Maria Elena about its biological status.

I contrast the rhetorical propositions in these two cases to argue that this epistemological presumption about forms of "knowing" also implicitly undergirds expectations about the power of fetal heartbeats in contemporary US anti-abortion politics. Activists who grant fetuses the ability to "testify" and communicate a form of intentionality—as in the Ohio State House—are similarly proposing that the corporeal body is a biosocial site of *conocimiento* (acquaintance with a person or subject). In effect, social and political claims that draw on diagnostic presentations of fetal biology rely on this slippage between these forms of knowing: sensing the body (*saber*) and recognizing the social subject (*conocer*). As a sensing practice, fetal biological materiality appears only through social and technological mediation, offering us a glimpse into the implicit cultural presumptions through which fleshy persons come into being (in line with Conklin and Morgan 1996).

Taking a cue from a rich tradition in feminist anthropology, I approach fetuses as simultaneously biological and cultural. Indeed, the biological and cultural are mutually constituted in the process of making "persons"—or, alternatively, the invading wandering spirits or fleshly evidence of human and non-human relations as fetuses may be variously considered (Conklin and Morgan 1996; Morgan 1989)—through ideas about developmental potential and social recognition (Kaufman and Morgan 2005). That is, fetuses become socially present as particular kinds of objects or subjects only through cultural ideas about bodies, spirits, and people. This happens as evidence from pregnant women's bodies, their haptic testimonies, and (increasingly) diagnostic technologies is socially materialized through extant narratives and expectations about who or what a person is

(Erikson 2007; Mitchell 2001; Mitchell and Georges 1998; Morgan 2009; Petchesky 1987; Roberts 2012; Taylor 1998, 2008). Thus, the query "what is a fetus" is itself a question both prompted by and resolved within specific cultural contexts and cosmologies. It is at once a question of epistemology (how to know), phenomenology (how to sense), and ontology (how to be).

Further, ideas about fetal personhood are deeply intertwined with contemporary reproductive politics that use the biological status of the fetus as evidence of an already social subject to argue for its right to protection (and care). Arguably initiated by the development and widespread implementation of diagnostic ultrasound in routine prenatal care, Americans have learned to recognize—and fetishize (Taylor 1998)—visible fetal bodies as a particular category of social subjects. This is a historically recent accomplishment, and "we" the viewing public had to learn to see fetuses as such by selectively ignoring the mediating work of both diagnostic technologies and epistemological expectations about how to recognize embodied persons.[3] Now, the making of fetuses into plausible "persons" relies on not only technologies that produce visual images of fetal bodies but also diagnostic technologies that amplify and allow us to publicly hear what is supposed to be the barest proof of life: the heart's beat. Attending to the underlying claims that undergird the presentation of fetal "voice" through amplified heartbeats offers a model of speculative analysis through which we may glimpse the development of still emergent fetal claims: social, political, and medical propositions made about and through fetal bodies.

In addition to expanding the literature on fetal propositions to include consideration of the power of sound as well as sight, I suggest that an anthropological analysis of the rhetorical deployment of fetal heartbeats sheds light on the underlying assumptions made about biosocial essence and the power of diagnostic technologies that together animate contemporary US reproductive politics. The significance of fetal heartbeats to denote proof of aliveness is not new, but the symbolic meanings attached to this sound as an embodied fetal "voice" are—especially as they are promoted and deployed in politically consequent settings. Between 2011 and 2014 in the United States, nine states introduced heartbeat bills that prohibited abortion after a heartbeat could be medically detected (somewhere between six and eight weeks gestation).[4] While only Arkansas and North Dakota have (at the time of writing) successfully passed laws (both eventually rejected by Federal Courts), heartbeat bills continue to be powerful symbols for anti-abortion activists.[5] In turn, reproductive justice activists have contested these bills on two main grounds: that they (1) violate the landmark *Roe v. Wade* (1973) decision's ruling that viability (approximately twenty-four weeks) is the limit for state bans and (2) rely on assumptions of a standard (and therefore universally enforceable) moment when a fetal heartbeat will be undeniably heard. Instead, opponents of these bills point out that heartbeats are generally initially detectable in a broad window that spans six to eight weeks and that hearing them depends on variables ranging from skill and experience of the medical practitioner to the technical modality selected (e.g., a stethoscope or a fetal Doppler).[6] Thus, even as activists on both sides debate the impact and implications of such legislation, there has been scant attention to their ontological presumptions; a presumption I argue is best described as an unmarked slide from *saber* to *conocer,* implicitly shifting from biological to social forms of being. What is it about a heartbeat (instead of other bodily cues) that seems to reveal "something" about not just biological potential but social as well? In untangling possible answers to this question, anthropological and speculative analyses of emergent life-forms (Franklin 2005) contribute to contemporary reproductive politics by attending to the cultural logics that make such claims possible and plausible, even if not (yet) persuasive.

Constructing Fetal Claims

As this collection itself demonstrates, fetal claims are tied to extant ontological expectations about the constitutive elements of social and biological existence as well as epistemological expectations about how to sense such existence; questions about how to be biosocial and how to know about such biosocial being. That is, fetuses are slippery phenomena that are simultaneously biological and social (although domains are far from distinct); their public emergence is culturally contextual (Gameltoft 2014; Morgan 1989; Morgan and Conklin 1996; Mitchell and Georges 1998; Ivry 2006, 2009), and their personhood status depends on social practices of recognition or "placing the unborn" (James 2000).

As feminist scholars have attended to the intertwined interests that ethnographically shape the politics of reproduction (Ginsburg and Rapp 1991), they have closely tracked the emergence and implications of new social, moral, and medical deployments of diagnostic technology to frame such claims. These fetal constructions are both deeply culturally embedded and globally circulated alongside claims about technologies' acultural objectivity (see, e.g., Oaks 1999).[7] Specifically, claims made in North American reproductive politics over the past three decades have relied on "placing" (James 2000) social subjectivity within the biological materiality mediated by diagnostic technologies. Tied to the increasing routinization of fetal sonography and other diagnostic imaging technologies, public fetal personhood claims are increasingly predicated on *seeing* the fetal body as a plausible child (Petchesky 1987; Rapp 1997; Taylor 1998). By ethnographically attending to ongoing cultural and technological shifts in prenatal practice, much of the scholarship on the emergence of the social, cultural, and political lives of new fetal subjects has targeted propositions about the centrality of visual evidence of anatomical form to delimit the transformative line between base biological and social existence.

Placing new diagnostic artifacts within long-standing narratives about the objective nature of sight and the "North American tendency to see fixed, *structural* markers of personhood" (Conklin and Morgan 1996: 660, emphasis in original), the fertile literature on fetal personhood stresses the cultural ease of imaginatively "disembodying" the fetus from the maternal body. In the words of Lennart Nilsson's (1965) classic, and still used, book of medical photography within US reproductive cultures, "a child is born" when it can visually float outside the womb as a recognizable human "person." Thus, as the routinization of fetal ultrasounds has accelerated over the past two decades, parents, families, and publics have learned to interact with visibly mediated fetuses as social subjects already enmeshed in family contexts and public politics. Indeed, through these visualizing practices, parents and families construct narratives about "knowing" their fetus by identifying heritable features, such as ears and noses, and animate them as (often gendered) subjects through intrauterine activity (Han 2013). In these interactions, the fetus is "placed" as an intersubjective person through the integration of visual mediation, cultural expectations about embodied personhood, and recognition of its position within familial and affective relationships.

While the visual fetus has become a routine feature in personhood claims, it is a historically recent one. Rosalind Petchesky's (1987) analysis in "Fetal Images: The Power of Visual Culture in the Politics of Reproduction" established a framework for scholars to consider the salience and complexity of visual diagnostics' contribution to social and moral fetal claims. By tying claims made about ultrasonic images in both clinical experiences and politicized media spectacles to the influence of broad reproductive politics on women's embodied experiences in pregnancy, she identified a nascent deployment of medical and diagnostic images to make political and cultural statements. Centrally, her analysis of fetal visual culture as a not-yet-solidified phenomenon rests on juxtaposing politicized and banal sites and tracing out their subtle connective lines. That is, fetal encounters became a routine and unmarked part of contemporary pregnancy through the public's accumulated exposure to fetal propositions—both explicitly political and explicitly not—in varied contexts. Methodologically, she suggests identifying and exposing the underlying shared expectations through which we learn to experience fetuses as the same kind of beings across these contexts. Following her model, we can understand still emergent (and thus provisional) fetal claims by attending to the diverse contexts in which aural cues are similarly deployed to make claims about the animation of fetal biology.

Methodological Approaches

If we take seriously the proposition that fetuses are simultaneously social and biological and that their material presence is always culturally contextual, they present interesting methodological challenges—especially for a cross-cultural analysis. The enrollment of diagnostic technologies to both materialize and place fetuses requires methodological consideration of their use in clinical encounters and their contextualization within broader politics. Here, I rely on Annemarie Mol's (2002) concept of "enactment" to show that emergent propositions about embodied subjects and the clinical, political, and technological conditions that make them possible are constantly co-constituted. This approach argues that rather than existing outside of diagnostic and sociopolitical context in which we encounter them, the plausibility of ideas about fetal persons relies on a dynamic relationship with the contexts in which they are materialized—as such, they are assemblages of technologies, bodies, and clinical protocols. With this focus on materializing practices, my analysis considers how new fetal claims may be precarious and contested but also granted plausible status because of their connection to extant and naturalized assumptions about biosocial life. In tracing out these fetal assemblages, I draw on two very different kinds of data and research methods—one classically ethnographic and the other a mix of rhetorical and content analysis of journalistic and activist accounts. I situate both within an analysis of the shifting landscape of public health policies and reproductive health legislation in each country.

The first source of data I draw on is thirteen months of ethnographic fieldwork I conducted in Oaxaca between 2005 and 2013, including nine months of clinic-based ethnography in 2008. Primarily based in a regional hospital and two small community clinics (one a satellite of the hospital) in Oaxaca's Central Valley (southern Mexico), I shadowed community health educators who served as initial sites of contact between these institutions and the surrounding communities, family practitioners as they socialized women through medical practices of diagnosis, surveillance, and self-accounting, and obstetricians as they cared for those patients whose condition was labeled "higher risk." Within the hospital, I traced the institutional circuits pregnant women moved through—from nurse to doctor, from social worker to lab—conducting informal interviews with women and their families about their care-seeking deliberations. Using snowball and self-selection sampling techniques in which medical professionals and patients I had previously interviewed suggested further participants, I observed sociomedical encounters between sixty pregnant women and their doctors, fifty of which I recorded, all augmented with detailed notes. After exams, I conducted short informal interviews with the doctors, asking them to elaborate on the previous exam and place their concerns within larger demographic, reproductive, and health politics.

To contextualize these clinical encounters—to consider how the practices of medicine and diagnosis always reflect and reproduce broader politics—I interviewed the program directors of the six major local and gender focused nonprofit organizations, including an urban health clinic. Finally, to complement this institutionally centered research, I conducted community-based research in a local Zapotec village where I learned about quotidian caregiving practices and gathered reproductive life histories. Drawing on my eventual imbrication within a transnational extended kinship community, I spent another nine months with Oaxacan immigrants in Los Angeles, researching immigrant homelife and reproductive health practices. This extended ethnographic research allowed me to approach the materialization and placing of reproductive subjects within ever expanding broad contexts of Mexican reproductive politics.

I juxtapose clinic-based ethnographic data with textual sources and network analyses gathered from academic and activist involvement in North American reproductive justice movements. Spurred by the political theatricality in Ohio, I began to track narratives about heartbeat bills across different media sources, analyzed activist materials and interviews by proponents of the bill, and contextualized these claims within a decade of research tracking the sociopolitical construction of fetuses in contemporary US reproductive politics. Through content coding and rhetorical analysis, I tie heartbeat bill initiatives to broader (and more successful) "informed consent" restrictions. Throughout these accounts, activists propose that increasing women's access to social knowledge of their fetuses—mediated by medical technologies—is crucial for shifting their assumed affective relationship—that is, hearing a heartbeat will persuade women to not abort because of the power of this fetal "voice."

By welding ethnographic and rhetorical analysis together cross-culturally, I am not suggesting that Oaxacan and US reproductive health politics rely on and reproduce a universal model of the interconnections between fetal personhood, embodied subjectivity, and technological diagnostics. Instead, I bring these cases together to consider how we may learn to see fetal propositions in their underlying and shared epistemologies. I emphasize the importance of the constitutive power ascribed to the sound of the fetal heartbeats as a way for anthropological scholarship to speak to public concerns about reproductive politics. Yet, if fetal heartbeats are marshaled to bridge biological materiality and social presence, the stakes of this knowing are not the same in Ohio and Oaxaca.

Materializing Contexts

Within the Mexican national imaginary, Oaxaca is prefigured as an "indigenous state" marked by the statistics of poverty—not only high rates of maternal mortality but also illiteracy and child malnutrition— and indigenous women's reproduction has been a key site of state-driven modernization projects (Smith-Oka 2013). Public health professionals in rural Oaxaca are thus faced with a challenging mandate: improve the region's epidemiological profile in a context of high patient loads, clinics staffed by doctors rotating through their national service obligations, and deeply entrenched narratives about the cultural differences within their largely indigenous catchment area. Oaxaca's current high rates of maternal mortality not only reveal lapses in the state's ability to care for its most marginalized populations but also make pregnancy and prenatal care into an overdetermined domain of risk and care. Indigenous women face social and medical narratives about the imperative to "modernize" their reproduction by applying scientific principles to "care better." Women's responsibility to follow biomedical prenatal care is encouraged by public welfare programs and a way to demonstrate proper maternal disposition (Smith-Oka 2013). Thus, the logic of care that structures Mexican public health institutions' approach to pre- and postnatal health already displays an entanglement between the forms of knowing marked by *conocer* and *saber*—or affective relation and expert knowledge.

This relationship between feeling and knowing takes a different form in US reproductive politics yet is also entangled with techniques of materializing fetal embodiment. As Petchesky (1987) anticipated, and Celeste Condit (1994) argued, through the ongoing routinization of fetal ultrasound visual, fetal artifacts quickly transformed from fodder for highly politicized claims to banal keepsakes of a normative pregnancy (Han 2013).[8] This normalization parallels (and cannot escape) a shift in anti-abortion activism since the mid-1970s that has heavily relied on photographic "proof" of its material form and substantiality to anchor their claims that a fetus is concretely and already human—that is, the ability of lay audiences to see a developed rather than developing person (Condit 1994).

Concurrently in the United States, as sonograms have achieved routine status and "entertainment" ultrasound studios have developed a niche market (Taylor 2008), we can see an entangling of medical, cultural, and political discourses in the rise of "informed-consent" abortion restrictions. Taking fetal personhood as a settled (rather than contingent) claim, anti-abortion activism since the 1992 *Planned Parenthood of Southeastern Pennsylvania v. Casey* decision has pursued an effective and affective strategy to incarnate the fetus as a social person (Halva-Neubauer and Zeigler 2010).[9] While the language of informed consent evokes an apolitical and ethical imperative of voluntary participation in a medical procedure, in practice, activists in the thirty-five states that mandate pre-abortion counseling emphasize the importance of women's emotional response to this information (Gold and Nash 2007). Informed consent to a medical procedure in these campaigns is deeply tied to normative ideas about the relationship between maternal bonding and the symbolically dense sight of the fetal body (Hopkins et al. 2005 argue that this is also central to British abortion politics). Following the (still contested) success of abortion laws mandating ultrasounds, American activists have turned to fetal heartbeats to amplify their claims that social and legal personhood is tied to the energetic and material body—that is, materializing this body prompts relationships. Ultrasound images and audible heartbeats are thus both employed to provide "better information" through already emotionally laden forms of proof. And "knowing" and "feeling," as in Mexico, are tied together in these propositions.

While the quotidian politics of public health practice in rural Mexico and the theatrical politics of abortion activists in the United States are far from parallel in terms of scope or aim, their juxtaposition illuminates an implicit set of presumptions about the nature of (fetal) biosocial life and the entanglement of ontological claims and diagnostic technologies—even as "personhood bills" remain heavily resisted and face strong legal challenges. Rooted in both diagnostic technologies and their deployment with discourses about maternal affect, together these cases shed light on the increasing entanglement of these claims to make plausible (and purportedly transparent) propositions about the social and political significance of fetal materiality.

Conocimiento Fetal Voice in Mexico

Caring for fetuses, or "bebès," is central to Mexican public prenatal health. While this may seem like an obvious statement, caring (*cuidado*) as an affective orientation and set of bodily practices is central to the accomplishment of rural public health agendas. Oportunidades (the popular public assistance program) targets the maternal-child relationship as central to regional social and economic development (Smith-Oka 2013), and the rural hospital where I primarily conducted research described its community health mission as "assistance to mothers and children." In an effort to fulfill the social welfare promises of the constitutional right to health in a historic context of lagging resources for infrastructure development and minimal culturally responsive initiatives, public health campaigns articulate a neoliberal vision that links mothers' affective demeanor to medically responsible practices (Howes-Mischel 2012). Brightly painted murals echo community health education workshops' message that "giving the breast is the best way to show love" and that to "attend your clinic consultation" is the best way for women to show personal and maternal care. These rhetorics of emotional connection and medical compliance were often used to bring women into the clinic's domain, yet inside the exam room, women's ability to *conocimiento* their baby was filtered and reframed through a doctor's ability to know (*saber*) through the mediation of diagnostic technologies (fig 11.1).

While medical personnel institutionally emphasized the importance of education and building community support for the clinics' health initiatives (and many individually sought to build social rapport with their patients), in practice their activities were limited by

FIGURE 11.1. Public health mural in Oaxaca: "Think of your child, choose a family planning method at the clinic" (photograph © Rebecca Howes-Mischel).

time and resource constraints. Family doctors at the end of their service year talked about how they did the best they could to learn on-the-job cultural competence, even as they reminded patients about the importance of following "scientific things" (*cosas de ciencias*) instead of cultural "beliefs" (*creencias*). From behind the tall stacks of patient charts the nurses (the most visible representatives of the clinic's permanent staff) delivered throughout the day, they found quick and subtle ways to extend human connections into the medical encounter. Often, patients refused these intersubjective gestures because of breached cultural norms (as in the case of the prior gynecologist at the hospital who jokingly suggested to a tense patient that she imagine her husband in order to relax her thighs) or in the mismatch between broader expectations about clinical institutions as the disciplining sites in which indigenous women directly experience state-driven health agendas. Otherwise, in their brief exams, doctors emphasized the strength of their professional knowing, using forms of the verb *saber*—that is, fetuses were the subjects of clinical assessment and description. While doctors acknowledged that women already had social and affective ties to their fetuses, these ties were not the subjects of the clinical encounter.

Doctors followed a standard language practice of using "baby" in ways that acknowledged that these fetuses were already enmeshed in relational and affective ties, using "fetus" only in their discussions of hypothetical standards. As an example from my field notes illustrates: "When a fetus is fifteen weeks old, it is small, like this size [he shows her the space between his hands], kind of like a small tortilla, your baby is this big [gesturing with his hands wider], so the dating must be off." Without direct access to the forms of diagnostic technology they were accustomed to in urban hospitals (i.e., routine ultrasound or electronic fetal monitoring), rural doctors relied on haptic physical exams, dented aluminum fetal stethoscopic "horns," and secondhand analysis of external lab reports.

In the prenatal encounter that opens this chapter, Dr. Celia, the new gynecological resident, proposed something very different in her work with Maria Elena. Instead of delivering knowledge *about* fetal status, she would offer her patient a direct experience *with* her fetus. Earlier in the morning of her second day in the hospital's outpatient clinic, I had asked her how she was going to approach this year of service—especially given the increased surveillance the region's maternal mortality rate draws to reproductive health workers. In response, she had emphasized the importance of her affective as well as effective medical work:

> They come to me because of concerns, you know, that they have to be referred from the other doctors [and classified as at higher risk]. So I think that's important for them to really understand what's going on, to make it real, and to *show* them that it's all OK. So I help them meet the baby. Then, after this, they can understand the process better, and maybe it helps them care for themselves better.

The other doctors could pronounce "your baby's heartbeat sounds good, everything is OK," using the verb form *saber* to present a professional diagnostic assessment based on what only they had heard. Using her personal fetal Doppler, Dr. Celia would instead offer direct (and presumably unmediated) proof as to its social presence—rather than only make an assessment based in her own expert listening (*saber*), she would offer Maria Elena reassurance rooted in an expectation of lay competence of *conocimiento* (personal knowing).

Maria Elena, eighteen years old and about twenty-six weeks' pregnant, had just been referred by one of the hospital's general family practitioners to the specialist because of a concern about her possibility for preeclampsia.[10] While the first half of the exam had been structured by her medical discussion of Maria Elena's risk status that stressed the importance of what Dr. Celia "knew" (*saber*) about fetal health broadly, she now shifted the emotional register of the clinical encounter toward one of sociability marked by her shift to *conocer*. While previously she had listened in silence with her ear pressed firmly into the fetoscope placed on Maria Elena's lower abdomen and timed the heartbeat with her digital watch—noting it on the chart with a brusque "good"—now she moved from addressing Maria Elena's fetus as a medical subject to presenting it a social and intersubjective one. Not only did she shift from contextualizing her observations within standardized norms of fetal development and maternal symptoms to enacting the fetus as a social person, but she also marshaled a particular diagnostic technology to make this claim.

While Dr. Celia had earlier drawn on her medical training and professional knowledge to frame the diagnostic sequence, here she suggested that Maria Elena did not need this expertise to recognize her fetus as a social person: the broadcast heartbeat itself should facilitate *conocimiento* of this in utero presence. This presumption that Maria Elena would recognize her fetus as a familiar and social figure through the mere presence of its auditory heartbeat suggests that the amplified sound contained incontrovertible proof of life and by extension health—a seemingly more authoritative and socially recognizable kind of proof than haptic signals such as movement and pressure that are only mediated by a woman's bodily experience.[11] Dr. Celia was implicitly making an argument about fetal presence bolstered by a kind of public declaration—the fetus could, with her assistance and through its heartbeat, be "met" as a social presence.

This medical encounter was in many ways an example of the quotidian experience of delivering public health amid complicated expectations about medicine, pregnancy, and development. The selection of the heartbeat as a particular proof of biosocial life and "aliveness" is not new. What is new are the attachments made between this sound and claims to affective recognition—exemplified by the distinction between the ability to *saber* a sound and to *conocer* a subject through the sound. Further, by tying this *conocimiento* to "better understanding" and ultimately to better "self-care," Dr. Celia simultaneously made a public health and moral claim about Maria Elena's need to recognize her fetus (see Howes-Mischel 2016 for an expanded discussion of the Oaxacan context).

This bridge between an amplified heartbeat and maternal-fetal recognition exemplifies complicated negotiations over the claims made through diagnostic technologies—and their enactments of biosocial personhood— that exceed the Oaxacan (or Mexican) context. In emphasizing the epistemological distinction between the actions of knowing things objectively and knowing things intersubjectively, this routine case in Oaxaca helps us understand some of the underlying ontological claims made about and through fetal diagnostics in US anti-abortion activism—the implicit power of fetal heartbeats as technologies of truth. Further, one of the great contributions an anthropological approach to fetuses can offer reproductive health policy and agenda is how cross-cultural and cross-linguistic analysis can illuminate the underlying logics that make activists claims plausible, if not persuasive. Thus, I finish this analysis by returning to the political theater that began this chapter.

Conocimiento Fetal Voice in the United States

In 2011, amid more than a thousand state-level legislative proposals targeting reproductive health access (Guttmacher Institute 2012), Ohio Representative Lynn Wachtmann introduced legislation that would ban abortions, except in the case of medical emergency, once a heartbeat could be detected. Drawing mixed responses from anti-abortion organizations and prompting questions about its implementation, the bill nonetheless proposed that "there is something almost magical about a heartbeat," in the words of International Right to Life Federation founder John Wilke (Sanner 2011). To speak to the "obviousness" of this proposition, the House hearing featured the sight and sound of "the state's youngest legislative witnesses'" live beating hearts. Characterizing them as "witnesses," advocates suggested that these fetuses would publically announce themselves as living persons rather than as cellular entities. In a rhetorical synecdoche, Ducia Hamm, executive director of the pregnancy center that supervised the ultrasound, linked this original testimony to a later Senate hearing that featured one of the babies post-utero: "The House heard her heart.... You get a chance to see her face and look into her eyes" (Sanner 2011). Underlying this self-conscious act of political theater was the premise that any reasonable American should recognize the sight and sound of a heartbeat as uncontroversial "proof" of intentional life—and that this "voice" is a kind of intention assertion of intersubjective personhood. Or, in other words, to *saber* a heartbeat is to *conocer* the person.

While anti-abortion materials have long used fetal "voice" as a kind of authentic (Ingold 2000) and affective prompt through which women are encouraged to acknowledge their fetuses as already social subjects, it has typically been represented in writing that emphasizes the mediating work of transcription. For example, a pink "pre-abortion diary" stuck on a bulletin board in my office hallway features a series of entries in which a presumptive fetus writes directly to her mother from in utero. Here, the politics of fetal communication are clearly encoded in its form. Other product-related claims have relied on ideas about fetal bodies and communication tied to parental instinct as in Volvo's advertisement that queried, "Is something inside you telling you to buy a Volvo?" above a visual fetal sonogram (Taylor 1992). Or, the subsequent AT&T advertisement in which a fetus responds to (and bonds with) his father by kicking when the phone is held up to his mother's belly—ironically selling a product called True Voice (Taylor 2008). Each of these examples visually and rhetorically relies on an imputation of fetal *voice* as central and persuasive to prenatal experience. Yet, this new emphasis on diagnostic heartbeats as a mode of fetal voice relies on an expectation of biosocial evidence that appears less mediated and more self-evident.

In these exemplars, it is not the fetus qua fully formed human subject but rather a single body part, the heart, that is mobilized to make this claim of agentive communication. Further, while the visual fetus may be used to make communicative claims, the proposition underlying the alignment of the heart's sound and the social knowing of *conocimiento* is that this technological mediation only amplifies an already existing voice. Heartbeats' "obvious" ontological status as a kind of social proof relies on the entangling of this particular sociomedical discourse about the heart as the site of both biomechanical and emotional life with expectations about the objective and affective transparency of sound as a kind of shared and public sensing of this life. In this equation, a person with a beating heart is an obviously alive feeling person, which is proved in a sound recognized by nonexpert witnesses. This selection of a symbolically laden sound as a gesture of intentional and intersubjective communication firmly locates personhood in the biosocial body.[12] Beyond Ohio's legislative chambers, the proposition that fetuses could offer legislative testimony on the basis of their embodied presence relies on expectations about how to sense the body that are often understood as transparent, neutral, and empirically objective rather than political and cultural. As the activists in Ohio illustrate, new movements to tie "personhood" to abortion restrictions emphasize the symbolic and emotional importance of the heart to signify social "aliveness."

Here, I find it helpful to return to Petchesky's critique of the increasing routinization of visual ultrasound technologies. Writing in the early stages of ultrasounds' unmarked incorporation into US pregnancy culture, she highlights late capitalism's "politics of style" in which the visual fetus's political and familial power comes from placing medical practices within extant contexts of mass culture—reflecting a conscious political strategy to rely on "medicotechnical" discourses and authorities to "win over the courts, the legislatures, and popular hearts and minds" (1987: 264–265). Notably, within her pointed critique of the pernicious implications of granting ultrasound's voyeuresque "window into the womb" objective evidentiary status, Petchesky also notes that this technology works through affective desire too. Her ultimate discussion of women's pleasure at seeing "baby's first picture" points to the persuasive power of collaborations between visual and medical cultures but also to the process by which highly politicized propositions may be transformed into unmarked routines (Mitchell 2001 further nuances this). Considering fetal heartbeat claims in conversation with this trajectory suggests we must attend to the contextualizing logics through which these claims may be considered plausible—logics that ever rely on the medicalization of prenatal affect and the displacement of maternal authoritative knowledge in favor of the evidence produced by technomedical diagnostics. Far from treating these fetal claims about heartbeats and voice as settled, Petchesky's model suggests we may attend to the interweaving of clinical and political discourses about bodies, persons, and technology that together present still nascent fetal propositions.

Toward an Anthropology of Fetal Voice

What ties a political claim in the United States to a routine prenatal exam in Oaxaca is a reliance on an overlapping and implicit set of narratives about the epistemological and ontological proof communicated by the sound of fetal heartbeats. That is, these diagnostics used to provide proof of an already existing personhood collapse multiple social and medical claims about the fetal body. Further, these narratives are enrolled within existing public health discourses that intertwine medical and moral imperatives for women to affectively relate to their fetuses through diagnostic technologies. Thus, implicit in both of these transformations is a moral claim made about the need to make women aware of their fetus's existential presence by offering proof that does not rely on her bodily experience—and thus is ostensibly more "objectively" transparent. Tracing the contours of these claims that displays of fetal biomateriality will—and should—compel a newly "knowing" maternal affective response illustrates a kind of unease about maternally directed fetal instantiations and a quasi-moral imperative to "care."

Dr. Celia and the Ohioan activists draw on a deep expectation about the ability of certain technologies to make objective truth claims—expectations grounded in a Western phenomenology that prioritizes sight and sound over more "subjective" senses like touch. Sound is synesthetic—a sensation that evokes another—in the way it triggers an embodied sense of involvement with another embodied subject. While publics had to consciously learn to see fetuses, their auditory form appears ostensibly simple and transparent, as the heart's beats have long stood in for a fundamental sound of bare life. This proposition rests on a series of nested expectations: that the biomechanical heart is a prime indicator of bodily vitality and that this vitality is somehow associated with affective connection (either symbolically or linguistically). Yet, even as this notion of voice is often hailed as a "true" and "authentic" form of intersubjective and embodied communication, hearing is a deeply cultural and social action (Ingold 2000). We learn to hear the fetal heart by first selecting among meaningful sounds and then attaching embodied significance to this sensation in much the same way we had to learn to see.

Why now are we asked to hear, as well as see, fetuses as embodied plausible persons? In US anti-abortion activist propositions, part of the imperative appears to come from the success of visual claims—much as Petchesky anticipated—as well as their limitations. While a six-week embryo may have a recognizable heartbeat, the humanness of its form is far from legible. Framing heartbeats as a form of preverbal, and yet public, communication locates "personness" in fetal biology in a way that elides the temporal and developmental differences between embryo, fetus, and baby. In Oaxacan public health spaces, something very different is at work. There, rather than an extension of fetal propositions made through ultrasonic visual form, amplified heartbeats are an early form of constructing personhood as nascently embodied. Without either routinely available ultrasound machines or an expectation that diagnostic ultrasounds are "baby pictures," Dr. Celia's small personal plastic Doppler machine—the same marketed to women for at-home use and "belly talk" (Han 2013)—is the only portable and accessible way to materialize the fetal body beyond its mother's body. The need to hear the fetus in that setting is one driven by the intertwining of available technologies and public health expectations about the power of *conocimiento* on maternal practices.

While separated by region, language, and purpose, both these fetal propositions rely on a vision of fetal materiality embedded in the intertwining of global biomedical practices and public health logics. It is crucial to attend to both claims about fetal materiality and the discursive expectations about public or maternal reception. In each of these cases, the "voice" of fetal personhood is predicated on women's need to know—and thus to care—for their fetus differently. Targeting maternal affect—in Oaxaca to *cuidarse mejor* (to care for yourself better) and in the United States to "make more informed choices"—these new fetal propositions reinscribe the same long-standing discourses about gendered responsibility that early feminist critiques of ultrasound indicted. In a similar way, the rapid routinization of fetal ultrasound to make both diagnostic and politicized claims about fetal materiality illustrate the powerful traction of this convergence of knowledge practices and health technologies.

Juxtaposing the discursive uses of the sound of a fetal heart for public health professionals and political activists enables us to query what makes a heartbeat plausible as a personned "voice." Methodologically, this approach relies on a mix of research approaches to show that it is in the overlapping spaces of implicit expectations about knowing and being that we might glimpse nascent propositions about biosocial life that are far from settled or stable. Examining the rhetorical propositions made about the ability to *conocer* a fetus facilitates an ethnographic analysis of the plausible that enables us to trace their emergence across domains without assuming a shared politics. This close focus on discursive propositions supports a careful approach that does not presume technological or linguistic determinism. That is, it aids us in avoiding the cross-cultural conundrum that claims made by and about fetuses, and by and about diagnostic technologies, are always deeply embedded within existing expectations about bodies, persons, and technologies' truth claims. Anthropologies of fetuses thus enable us to expand our expectations about how we recognize the possibility of fetal materiality, as well as to critically interrogate the kinds of medical, cultural, and political claims enrolled in to make these materializations plausible.

This cross-cultural and cross-field attention to fetuses illustrates that their personhood is not only culturally constructed but also enacted through diagnostic, social, and political practices. While the "sound of life" is not a settled claim to fetal personhood, this approach to emergent propositions calls for close attention to the contextualizing conditions in which social claims are made (or not) through and about the biological body. It also suggests that we must consider how the enrollment of diagnostic technologies in globalized public health logics may invite new forms of "truth" claims and open new spaces for the political mobilization of an already alive fetus. Ultimately, close attention to the "sound of life" enables us to query the circumstances of fetal "voice" as well as who (or what) has been authorized to mediate its speech.

Acknowledgments

I am deeply grateful the editors for this volume for their vision and enthusiasm bringing this volume into being. This chapter benefitted greatly from their editorial comments and suggestions. Thanks as well to the Wenner-Gren Foundation (grant no. 7615) for funding for the Oaxaca-based research.

Rebecca Howes-Mischel, PhD, is an assistant professor of anthropology at James Madison University. She conducted ethnographic research in Oaxaca, Mexico, and in the United States from 2005 to 2013 about how public clinics, activist agendas, and policy initiatives shape the material and symbolic care of indigenous women's reproductive bodies and health.

Notes

1. "Voice" draws attention to how the heart's sound is ascribed the ability to symbolically communicate a subject's presence qua subject position. Publically presented as separate from the maternal body and its intermediation, the proposition that an amplified heartbeat signals a "social" presence draws on an association between this voice and personhood claims.
2. All Oaxacan informants have been given pseudonyms.
3. While it is beyond the scope of this particular chapter, my experience tracking Oaxacan women's unscripted responses to ultrasounds as first "medical artifacts" and then "baby pictures" as they migrated from Oaxaca to the United States illustrates that this recognition is the product of both learning and contextual expectations.
4. States vary in the degree to which they mandate a method for detecting the fetal heartbeat; the medical literature also shows that heartbeats may be detected as early as six weeks and as late as eight weeks depending on ultrasonic method, embryo placement, and the woman's body fat distribution. The 2013 Arkansas law required a traditional transabdominal ultrasound, while North Dakota specified "by any technology available," which could include the much more invasive transvaginal ultrasound. There is some consensus that a heartbeat can only be detected by transvaginal ultrasound until eight to twelve weeks gestation.
5. While many mainstream anti-abortion organizations such as statewide Right to Life chapters have vocally opposed the promotion of such legislation, they have been careful to do so out of a concern for tactics rather than out of ideological opposition (Eckholm 2011).
6. This tact also frames opposition to this legislation in terms of technological specificity, which sidesteps the kinds of ontological social claims made about "life" as ultimately rooted in biological development.
7. As other contributions to this volume demonstrate, we must be cautious about extrapolating such claims based on expectations of the intertwining of technological determinism and reproductive politics. However, my ongoing research finds that rural Oaxacan women are increasingly exposed to these forms of routinized diagnostic technologies and that one impact of Oaxacan immigration to the United States is the increasingly adoption of this narrative about technological revelation.
8. And yet, as Lauren Fordyce noted at a roundtable on advocacy at the 2014 American Anthropological Association annual meeting, we have

to find a way of talking about these experiences of the ultrasonic fetus as materializing different kinds of subjects and make space to theorize the wide array of emotional responses they elicit.
9. This Supreme Court decision granted states the ability to regulate abortion as long as they did not present an "undue burden."
10. Preeclampsia is a common life-threatening yet etiologically confusing condition indicated by high blood pressure and protein in the urine; it is also highly correlated with poverty. Preeclampsia is the leading cause of maternal mortality within Oaxaca and the one most commonly referenced by the doctors I observed.
11. It is worth noting that Zapotec women already had a well-elaborated discourse for linking bodily cues to fetal subjective agency or willfulness. While outside clinical awareness, at home and in extended-kinship settings, women would speculate at great length about the relationship between their pregnancy cravings and child's disposition (Howes-Mischel 2012). Yet, this understanding of fetal subjectivity is still mediated by the mother's body and her interpretation of their significance.
12. Further, while both English and Spanish use the same word for the emotional and biomechanical heart, this is not cross-culturally universal. This raises interesting questions and potential research directions about the alignment of technological and linguistic determinism in the further expansion of fetal claims.

References

Condit, Celeste. 1994. *Decoding Abortion Rhetoric: Communicating Social Change.* Champaign: University of Illinois Press.

Conklin, Beth A., and Lynn M. Morgan. 1996. "Babies, Bodies, and the Production of Personhood in North America and a Native Amazonian Society." *Ethos* 24 (4): 657–694.

Dubow, Sara. 2011. *Ourselves Unborn: Fetal Meanings in Modern America.* Oxford: Oxford University Press.

Eckholm, Erik. 2011. "Anti-Abortion Groups Are Split on Legal Tactics." *New York Times,* 4 December, A1.

Erikson, Susan L. 2007. "Fetal Views: Histories and Habits of Prenatal Technology in Germany." *Journal of Medical Humanities* 28 (4): 187–212.

Franklin, Sarah. 2005. "Stem Cells R Us: Emergent Life Forms and the Global Biological." In *Global Assemblages: Technology, Politics, and Ethics as Anthropological Problems,* ed. Aihwa Ong and Stephen Collier, 59–78. Malden, MA: Blackwell Publishing.

Gammeltoft, Tine. 2014. *Haunting Images: A Cultural Account of Selective Reproduction in Vietnam.* Berkeley: University of California Press.

Ginsburg, Faye, and Rayna Rapp. 1991. "The Politics of Reproduction." *Annual Review of Anthropology* 20: 311–343.

Gold, Rachel B., and Elizabeth Nash. 2007. "State Abortion Counseling Policies and the Fundamental Principles of Informed Consent." *Guttmacher Policy Review* 10 (4): 6–13.

Guttmacher Institute. 2012. "States Enact Record Number of Abortion Restrictions in 2011." *Guttmacher,* 5 January. http://www.guttmacher.org/media/inthenews/2012/01/05/endofyear.html.

Halva-Neubauer, Glen, and Sara Zeigler. 2010. "Promoting Fetal Personhood: The Rhetorical and Legislative Strategies of the Pro-Life Movement after Planned Parenthood v. Casey." *Feminist Formations* 22 (2): 101–123.

Han, Sallie. 2013. *Pregnancy in Practice: Expectation and Experience in the Contemporary US.* New York: Berghahn Books.

Hopkins, Nick, Suzanne Zeedyk, and Fiona Raitt. 2005. "Visualising Abortion: Emotion Discourse and Fetal Imagery in a Contemporary Abortion Debate." *Social Science and Medicine* 61 (2): 393–403.

Howes-Mischel, Rebecca. 2012. "Local Contours of Reproductive Risk and Responsibility in Rural Oaxaca." In *Risk, Reproduction and Narratives of Experience,* ed. Lauren Fordyce and Aminata Maraesa, 123–140. Nashville, TN: Vanderbilt University Press.

———. 2016. "'With This You Can Meet Your Baby': Fetal Personhood and Audible Heartbeats in Oaxacan Public Health." *Medical Anthropology Quarterly* 30 (2): 186–202

Ingold, Tim. 2000. *The Perception of the Environment: Essays on Livelihood, Dwelling, and Skill.* London: Routledge.

Ivry, Tipsy. 2006. "At the Back Stage of Prenatal Care: Japanese Ob-Gyns Negotiating Prenatal Diagnosis." *Medical Anthropology Quarterly* 20 (4): 441–468.

———. 2009. "The Ultrasonic Picture Show and the Politics of Threatened Life." *Medical Anthropology Quarterly* 23 (3): 189–211.

James, Wendy R. 2000. "Placing the Unborn: on the Social Recognition of New Life." *Anthropology and Medicine* 7 (2): 169–189.

Kaufman, Sharon R., and Lynn M. Morgan. 2005. "The Anthropology of the Beginnings and Ends of Life." *Annual Review Anthropology* 34: 317–341.

Mitchell, Lisa. 2001. *Baby's First Picture: Ultrasound and the Politics of Fetal Subjects.* Toronto: University of Toronto Press.

Mitchell, Lisa, and Eugenia Georges.1998. "Baby's First Picture: The Cyborg Fetus of Ultrasound Imaging." In *Cyborg Babies: From Techno-Sex to Techno-Tots,* ed. Robbie Davis-Floyd and Joseph Dumit, 105–124. New York: Rutledge.

Mol, Annemarie. 2002. *The Body Multiple: Ontology in Medical Practice.* Durham, NC: Duke University Press.

Morgan, Lynn. 1989. "When Does Life Begin? A Cross-Cultural Perspective on the Personhood of Fetuses and Young Children." In *Abortion Rights and Fetal Personhood,* ed. Edd Doerr and James W. Prescott, 89–107. Long Beach, CA: Centerline Press.

———. 2009. *Icons of Life: a Cultural History of Human Embryos.* Berkeley: University of California Press.

Nilsson, Lennart. 1965. *A Child Is Born.* New York: Delcorte Press.

Oaks, Laurie. 1999. "Irish Trans/National Politics and Locating Fetuses." In *Fetal Subjects, Feminist Positions,* ed. Lynn Morgan and Meredith Michaels, 175–195. Philadelphia: University of Pennsylvania Press.

Olafson, Steve. 2012. "Oklahoma State Senate Passes 'Heartbeat' Abortion Bill." *Reuters,* 6 March. http://www.reuters.com/article/us-oklahoma-abortion-heartbeat-idUSTRE82609J20120307.

Palmer, Julie. 2009. "Seeing and Knowing Ultrasound Images in the Contemporary Abortion Debate." *Feminist Theory* 10 (2): 173–189.

Petchesky, Rosalind. 1987. "Fetal Images: The Power of Visual Culture in the Politics of Reproduction." *Feminist Studies* 3 (2): 263–292.

Rapp, Rayna. 1997. "Real-Time Fetus: The Role of the Sonogram in the Age of Monitored Reproduction." In *Cyborgs and Citadels: Anthropological Interventions in Emerging Sciences and Technologies,* ed. Gary L. Downey and Joseph Dumit, 31–48. Santa Fe, NM: School of American Research Press.

Roberts, Elizabeth. 2011. "Abandonment and Accumulation: Embryonic Futures in the United States and Ecuador." *Medical Anthropology Quarterly* 25 (2): 232–253.

Sanner, Ann. 2011. "Ohio Senators Hear 'Heartbeat' Abortion Bill." *Associated Press,* 8 December.

Smith-Oka, Vania. 2013. *Shaping the Motherhood of Indigenous Mexico.* Nashville, TN: Vanderbilt University Press.

Taylor, Janelle. 1992. "The Public Fetus and the Family Car: From Abortion Politics to a Volvo Advertisement." *Public Culture* 4 (2): 67–80.

———. 1998. "Image of Contradiction: Obstetrical Ultrasound in American Culture." In *Reproducing Reproduction: Kinship, Power, and Technological Innovation,* ed. Sarah Franklin and Helena Ragoné, 15–45. Philadelphia: University of Pennsylvania Press.

———. 2008. *The Public Life of the Fetal Sonogram: Technology, Consumption, and the Politics of Reproduction.* New Brunswick, NJ: Rutgers University Press.

CONCLUSION

Tracy K. Betsinger, Amy B. Scott, and Sallie Han

The four-fields approach of this volume demonstrates the efficacy of encouraging cross-discipline discussion of a topic such as the human fetus and embryo, which is, of course, the focus of substantial discourse today. Anthropology strives to be holistic, and the present volume achieves this aim, bringing together seemingly disparate research on the human fetus and embryo from the anthropological subdisciplines that, at first, seem to have limited interconnectivity to one another. However, on closer inspection, the various chapters of this book are unified in their concerns with the fetus as biological, cultural, and social entity in and across time and place.

Studies of fetal development, which are found not only in biological anthropology but also in bioarchaeology, call attention to the fetus *as* biology and the fetus *in* time. Julienne Rutherford (chapter 1) reminds us that a fetus is simultaneously an embodiment of previous generations and a significant influence on future generations, both in terms of genetics and maternal environment. Kathleen Blake (chapter 2) and Mary Lewis (chapter 5) consider the skeletal remains of fetuses as sources of multigenerational information, as the remains provide insight to the health and well-being of not only the fetus but the mother as well. Siân Halcrow, Nancy Tayles, and Gail Elliott (chapter 4) build on this, outlining how fetal remains can increase our understanding of growth and development and patterns of health and disease for entire populations over several generations.

Studies from archaeological sites and from ethnographic settings provide us with historical and cross-cultural insights into the social lives (and deaths) of fetuses. Jessica Newman (chapter 9) examines their social role and identity in contemporary Morocco, while Jacek Kabacinski, Agnieszka Czekaj-Zastawny, and Joel Irish (chapter 6) investigate the value and social role of fetuses and infants from a 4,700 to 4,350 BCE cemetery in Egypt. Sallie Han (chapter 3), Amy Scott and Tracy Betsinger (chapter 7), Risa Cromer (chapter 8), Sonja Luehrmann (chapter 10), and Rebecca Howes-Mischel (chapter 11) all explore issues of fetal personhood, although the methods employed and populations investigated vary. Han traces the characterization of the fetus in the United States in recent history, while Scott and Betsinger examine funerary treatment of fetal remains from a postmedieval Polish population. Luehrmann studies how value and personhood is ascribed to fetuses from the perspective of modern Russian Orthodox anti-abortion activists, while Cromer's research, focused on the embryos, outlines issues of personhood for contemporary populations in California. Howes-Mischel considers the fetal "voice" and its association with fetal personhood in two distinct settings in the United States (Ohio) and Mexico.

Each chapter addresses the questions: What is a fetus? How is it defined and conceptualized in a particular field of study? Addressing these, some authors drew from understandings more firmly grounded in biology. From this perspective, the developmental age or stage determines what a fetus is. Yet, the authors also acknowledged the limitations of such categorization. Historically and cross-culturally, the question of what a fetus is also becomes what a fetus signifies. In other words, it is what it means, from the beginnings of life to its potentialities. Overall, there seems to be no simple or single agreed definition of a fetus. Rather, there is the recognition that how we, as scholars and researchers, conceptualize the fetus or embryo itself reveals the kinds of questions and answers we pursue. In that sense, none of our concepts can be described as "neutral." Moreover, the definition of a fetus becomes especially charged given the current context in which all of us conduct our research. The human fetus or embryo is the focus not only of scientific and scholarly inquiry but also of political, legal, and legislative action. In the current climate, we face particular challenges in furthering conversations that engage the complexities of the human fetus as it falls at the nexus of biology and culture.

Also addressed here: What does a study of fetuses in a given field contribute to public concerns, such as reproductive policies and practices? Answers to this question generally fall into three categories. First, research outlined in this volume demonstrates how fetal bodies (including ancient and historical remains) might inform our understanding of the health of not only individuals but also communities and populations, providing invaluable insight for fields such as public health. Next, chapters here demonstrate how research on human fetuses offers insight to fetal identity and how it is culturally constructed and defined through any number of ways. These definitions of personhood and identity become imperative in the evolution of our social and cultural ideologies surrounding the fetus and embryo. Finally, the volume points to the complexities associated with embryos and fetuses, as well as the fetal-maternal relationship, which are affected by sociocultural factors, such as socioeconomic status and political practices. Taken together, this collection illustrates that the human fetus, while undeniably a biological organism, is significantly a cultural being with a definition that constantly shifts between these various frameworks constructed both within and outside the womb.

Glossary

achondroplasia: a type of dwarfism in which cartilage does not properly convert into bone, resulting in short stature and short limbs.

ameloblast: the cells responsible for the secretion and mineralization of enamel tissue; found only in the teeth.

anemia: a condition resulting from an iron deficiency in the body, due to either dietary or genetic conditions.

anencephaly: a congenital condition in which part or all of the brain is absent.

aneuploidy: a condition in which the chromosome number differs from what is typical for that organism, usually varying by only one chromosome.

biopolitics: intertwined scientific and political regimes of knowing through which bodies mediate between state and population.

blastocyst: one of the first stages of embryonic differentiation, ca. day 6 post conception in humans. At this stage, the ball of dividing cells develops a cavity that contains an inner cell mass. The outer cell mass gives rise to the major components of the eventual placenta, while the inner cell mass becomes the embryo proper.

conceptus: the products of conception, including the embryo and related structures like the placenta.

congenital: present from the time of birth.

diaphysis: the shaft of the long bone, also known as the primary ossification center.

dysplasia: an abnormal development of cells and tissues.

embryotomy: the dissection of a fetus or embryo.

endochondral growth: a type of ossification that is the most prevalent type of growth in the skeleton, where bones are preceded by a cartilage matrix and eventually mineralize.

endogenous: relating to internal influences or stressors.

epiphysis: the secondary ossification center(s) of the long bone(s) that are separated from the diaphysis during growth by the cartilaginous growth plate.

exogenous: relating to external influence or stressors.

fibroblast: the cells responsible for the creation of the soft tissue matrix on which skeleton mineralization can occur.

fontanelle: the cartilage tissues present at birth that connect the various bones of the cranial vault, allowing brain growth and flexibility in the cranium. All six fontanelles are closed by two years of age.

gynecology: translated as "the science of woman," the field of medicine that is concerned with female reproductive health.

holoprosencephaly: a condition in which the forebrain of the embryo does not develop two distinct hemispheres, potentially leading to abnormal development of the brain and face.

hydrocephaly: a condition involving the accumulation of cerebrospinal fluid within the cranium, leading to enlargement of the head and compression of the brain.

hypoxia: a condition of low oxygen, which can arise because of environmental conditions (e.g., high altitude) or physiological conditions (e.g., poor blood supply, anemia, etc.). However, early embryological development takes place in a normally hypoxic environment through plugging of maternal vessels, because too much oxygen can be damaging.

iniencephaly: a condition in which there is extreme backward bending of the head and defects of the spine, leading to an enlarged head size that is disproportionate to the rest of the body.

infantile cortical hyperostosis: a condition of infants that causes excessive new bone formation, leading to widening of the bones of the arms, legs, and shoulders and swelling of the joints; also known as Caffey disease.

intramembranous ossification: a type of ossification that begins in utero and continues throughout life during the skeletal remodeling process. Once cell differentiation begins (i.e., creation of **osteoblast** cells), mineralization begins immediately without a cartilage matrix (e.g., cranial vault bones).

lytic: bone tissue that is dying due to a disruption of normal cell function (e.g., **osteoblasts** or **osteoclasts**), which may be caused by traumatic injury or disease.

macrosomia: high birth weight, typically defined as greater than 8 pounds 13 ounces (4,000 grams) at birth.

meningocele: a protrusion (or a sac containing cerebrospinal fluid) of the membranes covering the brain through a gap in the skull or vertebral column.

metaphysis: either end of the diaphyseal shaft where the growth plate connects the primary and secondary ossification centers.

mizuko: from Japanese, literally, "water child"; term for aborted and miscarried fetuses and stillborn infants. Since the 1970s, couples in Japan can commemorate their mizuko by purchasing a small stone statue of the bodhisattva Jizō to be erected at memorial sites near Buddhist temples. These statues are often treated as personifications of the dead fetus, dressed in items of baby clothing and given offerings of toys or candy.

neonatal line: a microscopic line found on teeth as the process of amelogenesis, or enamel formation, is halted at the time of birth.

neonate: a term used to refer to a newborn around the time of birth (i.e., forty weeks to seven postnatal days).

neonatology: a subspecialty of **pediatrics** that is focused on newborns.

nutrient foramem (pl. foramina): an opening in the cortical bone through which blood vessels pass in order to nourish skeletal tissues.

obstetrics: the field of medicine and surgery associated with pregnancy, childbirth, and the postpartum period; typically practiced in tandem with **gynecology**.

ossification: the process of bone formation, often developing from a cartilaginous precursor.

osteobiography: skeletal changes that reflect biological and cultural influences that mirror a group or population as a whole.

osteoblast: the bone cells responsible for the secretion and mineralization of osteoid, the precursor to mineralized bone tissue.

osteochondritis: an inflammation of bony cartilage.

osteochondroma: a common benign tumor consisting of cartilage and bone, commonly found near growth plates.

osteoclast: the bone cells responsible for the destruction and removal of old skeletal tissue.

osteofibrous dysplasia: a rare benign tumor of the long bones, most often the tibia and fibula of children.

osteogenesis imperfecta: a rare genetic condition that affects the development of the skeleton and connective tissues, leading to very fragile bones.

pars basilaris: one of the four bones that fuse in childhood creating the occipital bone. Pars basilaris makes up the posterior portion of the foramen magnum on the bottom of the skull.

pars lateralis: one the four bones that fuse in childhood creating the occipital bone. Pars lateralis is a paired bone and fuses with **pars basilaris** to make up the lateral borders of the foramen magnum on the bottom of the skull.

pediatrics: the branch of medicine concerned with children.

perinatal: referring to the period between twenty-eight weeks in utero to seven postnatal days.

periosteum: the osteogenic (bone-producing) sheath that encapsulates all bones of the skeleton except at the joints.

periostitis: the resulting skeletal changes from an inflammation of the periosteal tissue. Characterized by increased bone formation, periostitis can have a plaque-like appearance and has been associated with acute trauma and systemic stress.

postneonate: a term used to refer to a newborn from seven postnatal days to one year.

preeclampsia: a condition affecting pregnant women, manifesting initially in high blood pressure but potentially leading to organ failure and even death.

rickets: a condition resulting from vitamin D deficiency, leading to a softening of the skeletal system, among other symptoms.

scurvy: a condition resulting from vitamin C deficiency, leading to lesions on the skeletal system, among other symptoms.

sequestrum: a fragment of dead (necrotic) bone that is separated from normal bone.

sexual dimorphism: the differences between males and females in overall size, shape, or other external physical differences.

Slaughter of the Innocents: in Christian art, a depiction of the Gospel account of the slaying of all children under the age of two in Bethlehem upon order of King Herod, who intended to kill the newborn Jesus (Matthew 2, 13–23).

spina bifida cystica: a condition in which there is a cleft in the vertebral column where the spinal cord and the meningeal membrane protrude.

teratogens: substances that can cause congenital abnormalities by disrupting normal embryological or fetal development.

Treponema pallidum: the spirochete bacteria that causes treponemal disease (including syphilis) in humans.

Treponeme: anaerobic spirochete bacteria in the genus, *Treponema*.

triplody: a condition in which three sets of chromosomes are present in the cells instead of two.

trophectoderm: the tissue that comprises the outer cell mass of the **blastocyst**, giving rise to major components of the placenta and differentiating into the **trophoblast** cells.

trophoblast: the collective term for cells comprising the lining of the placenta, which include two primary types (cytotrophoblast and syncytiotrophoblast) important in regulating the transport of maternal nutrients, gases, and hormonal signals to the **conceptus**.

INDEX

Page numbers in *italics* indicate figures, tables, and illustrations. Archaeological and ethnographic sites are generally found under the current country or US state of location. Entries beginning with a number are indexed as if the number is spelled out (for example, "2PN embryo stage" is at "two").

A

abortion: as cause of maternal mortality, 95; Japan, *mizuko* in, 159–60, 228, 245, 281; legal efforts to restrict access to, 61, 66, 75; medicalization of pregnancy and, 67–68; Poland and Romania, abortion law in, 231; socially embedded person, interruption of process of producing, 228–29; spontaneous, 86, 114, 117, 118, 228. *See also* anti-abortion activism; Morocco, pregnancy and abortion in
achondroplasia, 122, 279
adoption and implantation of frozen embryos, 171–72, 173, 176, 177–79, 181–86
ASRM (American Society for Reproductive Medicine), 175
advertisements using fetal images and voice, 267
age-at-death for skeletal fetal remains, 37–40, *38*, 84, 88, 152–53
Alduc-Le Bagousse, A., 115
Ali, Kecia, 206
Alzheimer's disease, 176, 180
American Anthropological Association, 272–73n8
American College of Obstetricians and Gynecologists, 76n5
American Society for Reproductive Medicine (ASRM), 175
amino acid metabolism and fetal development, 22–23
AMLAC (Moroccan Association for the Fight against Clandestine Abortions), 218
amniocentesis, 68
Anderson, Bruce, 119–20
anemia, 43, 44, 96, 112, 279, 280
anencephaly, 121, 279
aneuploidy, *123*
animal bones, misidentification of fetal bones as, 88
anthropology of the fetus, 1–11, 276–78, xii–xiv; in archaeology and bioarchaeology, 6–8, 63, 73, 276–77 (*See also* bioarchaeology of fetuses; Gebel Ramlah Neolithic infant cemetery; paleopathology of fetuses; Poland, postmedieval, burial treatment of fetuses in); in biological anthropology, 3, 4–5 (*See also* borderless

Index

fetus; skeletal fetal remains); concepts of, 5 (*See also* concepts of the fetus in contemporary USA); defining the fetus, 4, 16, 35–36, *36,* 277; as field of study, 1–5; multidisciplinary approach to, 3, 276, xiii; public concerns and, 277–78; skeletal remains, 2–3 (*See also* skeletal remains); in sociocultural anthropology, 2–3, 8–11, 61–63, 276–77 (*See also* frozen embryos; heartbeat as fetal voice; Morocco, pregnancy and abortion in; Russian Orthodox anti-abortion activism)
anti-abortion activism: adoption and implantation of frozen embryos, 171–72, 173, 176, 177–79, 181–86; US heartbeat bills and legislative anti-abortion activism, 253, 254, 255–56, 259–60, 261–62, 266–69, 270, 272nn4–6; vulnerable politicized fetus in, 60, 64, 65–66. *See also* Russian Orthodox anti-abortion activism
Aramesh, Kiaresh, 204
archaeology and bioarchaeology of fetuses, 6–8, 63, 73, 276–77. *See also* bioarchaeology of fetuses; Gebel Ramlah Neolithic infant cemetery; paleopathology of fetuses; Poland, postmedieval, burial treatment of fetuses in
Ariès, Philippe, 158
Arizona, Homol'ovi III state, 120
Arkansas, heartbeat bill, 255, 272n4
Armelagos, George, 97
Austria: Krems-Wachtberg site, 92; Upper Paleolithic neonate burials, 134

B

baptism, 151–52, 159, 160, 228, 236–37, 245

Bargach, Jamila, 203, 206
Beaumont, Julia, 98
"belly talk," 60, 69–72
Benkirane, Abdelilah, 218–19
Bennett, Kenneth, 121
Betsinger, Tracy K., 1, 8, 11, 47, 73, 146, 163, 276, 277
bioarchaeology of fetuses, 6, 83–100; demographic analyses using, 95; differentiating burial contexts, *87,* 87–88; etiology of infant death and, 93; first and second trimester fetuses, 86, *87;* historical background, 85–87; infanticide, 94; maternal and fetal mortality and stress, evidence for, 95–98; misidentification as animal bones, 88; multiple fetal pregnancies and births (twins), 91–92; nonsurvivor bias, 87–88, 93; perinates buried with adults, 89–91, *90;* postmortem births or "coffin births," 92–93; premature and SGA babies as proxies for fetuses, 83, 84, 87–88; significance of, 83–84, 94–95; social identity and burial treatment, 98–100; terminological issues, 84; in utero fetuses, 83, 84, 86, 88–89. *See also* skeletal fetal remains
biological anthropology, 3, 4–5. *See also* borderless fetus; skeletal fetal remains
birth control: in Morocco, 212, 213; in Russia, 230, 240
birth trauma, 88, 93, 124–25
birthmarks *(twahima),* attributed to cravings *(twahm)* in Morocco, 208
Black, Sue, 35, *36,* 120
Blake, Kathleen Ann Satterlee, 4–5, 34, 49, 84, 88, 276
blastocyst implantation, *23,* 23–24, 279
Blondiaux, J., 115

Blossom Embryo Adoption program, 171–72, 173, 176, 177–79, *178*, 181–86, 191, 192
Boltanski, Luc, 228
bone density, 44
the borderless fetus, 4, 15–27; distinct biological entity, difficulty in defining fetus as, 15–16; epigenetic changes, environmentally-triggered, 18–19; genetic connectivity to past and future generations, 16, 17–19; linear timeline of fetal experience, moving beyond, 26–27; maternal experiential connectivity, 16, 19–22; placental synchronicity, 16, 22–26, *23*; spatiotemporal borderlessness of fetal experience, 16, *17*
Bourqia, Rahma, 208
Bowen, Donna Lee, 204
Bower, Dennis and Jolene, and Bower embryos, 181–86, 191, 192
Bradtmiller, Bruce, 89
Brazilian favelas, deferred emotional investment in infants in, 248n3
breastfeeding and isotope analysis, 98
Brickley, Megan, 46
brittle bone disease (osteogenesis imperfecta), 124, 282
Brothwell, Don, 120
Buddhism, 63, 245, 281
Buikstra, Jane, 48, 122, 148
burial treatment: age of infant at death and, 99; baptism and, 151–52, 159, 160; differentiating contexts, *87*, 87–88; of pregnant women, 99; social identity and, 98–100, 146–50. *See also* Gebel Ramlah Neolithic infant cemetery, Egypt; Poland, postmedieval, burial treatment of fetuses in
Bush, George W., 179

Bychenkova, Irina, 236

C
Caffey, John, 116, 124
Cagigao, E., 121
calcium levels, 44, 45
California Institute for Regenerative Medicine (CIRM), 179, 187
Canada: Belleville, Ontario, skeletal fetal remains from, 47; Elmbank Cemetery, Toronto, 121
cancer, 116, 176
Canguilhem, Georges, xiii
cardiovascular disease, 74
Casper, Monica, 67
Catholicism: on abortion, 228, 246; in Ecuadorian Andes, 228, 246; Holy Innocents in, 242, 248n5; in North America, 246; Russian Orthodox anti-abortion activism and, 233; unbaptized infants in, 152, 160
Centre for Trophoblast Research, University of Cambridge, 26
cerebral palsy, 117
chastity *(tselomudrie)*, concept of, in Russian Orthodox anti-abortion activism, 238
Chelsea Old Church site, London, 86
chimpanzees, 70, 84
chordoma, infantile, 116
Chraïbi, Chafik, 217, 218, 219, 221
Christianity: adoption of frozen embryos and, 171–72, 178; baptism in, 151–52, 159, 160, 228, 236–37, 245; burial treatment of fetuses in postmedieval Poland and, 150–52, 157, 159; social identity of fetus in, 148, 157. *See also* Catholicism; Russian Orthodox anti-abortion activism
chromosomal defects, 122–23, *123*
CIRM (California Institute for Regenerative Medicine), 179, 187

Index 287

cleft lip and palate, 121
cloverleaf deformity, 121
"coffin births" or postmortem births, 92–93
Combined Prehistoric Expedition (CPE), 132, 133, 136
commercialized representations of fetuses, 71–73, 267
community health indicated from skeletal fetal remains, 43–47, 48
concepts of the fetus in contemporary USA, 5, 59–75; cross-cultural and historical perspectives on, 61–64; culpabilities of pregnant women for health of the fetus, 60, 67, 74–75; health and disease issues, 74–75; identity of child and mother, 59–60; the personal fetus and "belly talk," 60, 68–72; rematerialized fetus, 61, 72–75; terms for "it," 59–61; the vulnerable fetus, 60, 64–68
Condit, Celeste, 261
congenital defects, 119–24, *120, 123,* 279. *See also specific types*
congestive heart failure, 117
Conklin, Beth A., 257
conocer and *saber* (affective relation and expert knowledge), 254, 256, 261, 263, 265–66, 268, 271
contraception: in Morocco, 212, 213; in Russia, 230, 240
Cope, Darcy, 121
copper coins, postmedieval Polish burials with, 153, *154,* 155, 156–57
coronary disease, 26
cortical hyperostosis, infantile, 115, 280
CPE (Combined Prehistoric Expedition), 132, 133, 136
cravings and birthmarks *(twahm* and *twahima)* in Morocco, 208
Crawford, Sally, 148, 152

CRISPR/Cas9, xiii–xiv
Cromer, Risa D., 8–9, 68, 171, 193, 277
culpabilities of pregnant women for health of the fetus, 60, 67, 74–75
cultural anthropology. *See* sociocultural anthropology and ethnology
Cyprus, Iron Age, infant and perinatal remains, 99
Czekaj-Zastawny, Agnieszka, 7, 132, 136, 141, 276

D

de Silva, P., 114
deafness, 117
Dedick, Andrew P., 124
deMause, Lloyd, 158
demographics: bioarchaeology of fetuses and, 95; burial treatment of fetuses in postmedieval Poland and, 157–58; paleodemographics, 48; Russian abortion as problem of, 230–31
Denmark, embryo donation for stem cell research in, 233
dental enamel defects, 97–98
depression, 44
Derevenski, Johanna Sofaer, 158
developmental origins of health and development (DOHaD), 74, 97
"deviant" burials in postmedieval Poland, 153, 159–61, 162
diabetes, 22, 26, 36, 44, 74, 176
"Diary of an Unborn Child," 242, 248n2
Dickenson, Sam, 92
disease. *See* health and disease; *specific diseases and health conditions*
DNA and genetics: connectivity of fetuses to past and future generations, 16, 17–19; CRISPR/Cas9, xiii–xiv; epigenetics, 18–19, 44–45,

74–75, 97; fetuses buried with adults, determining genetic relationship between, 91; homeobox (Hox) genes, 113; paternity testing, 214; syphilis diagnosis via, 119
DOHaD (developmental origins of health and development), 74, 97
Down syndrome, 122–23
Dubow, Sara, *Ourselves Unborn: A History of the Fetus in Modern America* (2011), 2
Dudar, J. Christopher, 121
Duden, Barbara, 62, 65
Duma, Pawel, 159
Dunn, Pat, 179, 180, 189
Dutch "hunger winter" (WWII), 25, 45
dysplasia, skeletal, 120–22, 279

E
East, A. L., 122
Ecuadorian Andes, 62, 228, 246, xii
Egypt: Dakhleh Oasis, 121, 124; Deir el Medina, 99; Elkab child cemetery, Nile Valley, 134; Hermopolis catacomb, 121. *See also* Gebel Ramlah Neolithic infant cemetery, Egypt
Einwögerer, Thomas, 92
Eisenberg et al., *What to Expect When You're Expecting* (1996), 71
Elliott, Gail E., 6, 83, 100–101, 276
embryotomy, 124, 279
encephalocele, 121
England: Andover Road, Winchester, 123; Anglo-Saxon burial sites, skeletal fetal remains in, 47; Birmingham (18th-19th century), fetal skeletal remains from, 46; Lechlade, early medieval remains from, 120; London, postmedieval, 116; medieval populations, 46, 120; Milton Cemetery, Portsmouth, 151; Poundbury Camp, Dorset, 47, 124; Yewden Roman villa site, Hambledon, Buckinghamshire, 94, 124
ensoulment in Islam, 202, 210
environmentally-triggered epigenetic changes, 18–19, 44–45, 74–75
epigenetics, 18–19, 44–45, 74–75, 97
ethnography. *See* sociocultural anthropology and ethnology
ethnogynecology in Morocco, 205–17, 222. *See also* Morocco, pregnancy and abortion in
experiential maternal connectivity, 16, 19–22
Eyers, Jill, 94

F
Faerman, Marina, 94
feminist theory and the fetus, 173, 218, 220, 229, 254, 256, xii
fertility estimates using perinatal and infant mortality data, 95
fetality, 74
fetology, 67
fetus, anthropology of. *See* anthropology of the fetus
Feucht, Erika, 99
Finlay, Nyree, 152
fontanelles, enlarged, 117, 280
Food and Drug Administration, 66
"For Life" *(Za Zhizn')* festival, Russia, 233
Fordyce, Lauren, 272–73n8
Forensic Fetal Collection, Smithsonian Museum of Natural History, 41, 121
fractures, perimortem *versus* postmortem, 124–25
France: Costebelle, 118–19; Lisieux, 115
Franklin, Sarah, 192
frontal bossing, 122
frozen embryos, 8–9, 171–92; adoption and implantation of, 171–72, 173, 176, 177–79,

178, 181–86; Bower embryo, ethnographic study of, 181–86; categorical ambiguity of, 172–73, 177, 181, 186, 191–92; cryopreservation process, 174; double reproductive value of, 192; ethical and political controversy regarding, 172, 176, 191; feminist theory and, 173; as financial, emotional, and legal burdens, 175–76, 184–86, 188–89, 191; methodological approaches to, 173; ranking and grading, 179, 184, 185, 189–90; saving and storage of, 174–77; in stem cell research, 171, 173, 176, 179–81, 186–91; Stoll embryo, case study of, 186–91, 192; waiting as condition of, 172–73, 181–82, 186, 191–92; at Western Fertility Clinic, 173, 174–75, 179
funerary treatment. *See* burial treatment
Furtado, Larissa, 120

G
Galen, 202
Gardeła, Leszek, 158–59
Gebel Ramlah Neolithic infant cemetery, Egypt, 7, 132–41; adult female burials in, 139; African societies, ethnographic evidence from, 135–36, 141; age and sex determinations, 139; changing mortuary practices at, 133–34; condition of remains, 137, 139, 140; dating, 136–37, 138; disposition of remains and grave goods, 138, 140–41; excavation at, 132, 133, 136–38, *137, 138*; Neolithic societies generally, 133; number and distribution of inhumations, *137*, 138–39; related sites, 134–35, 140; twin burial, 139

Geber, Jonny, 92
Geddes, Anne, 178
gender archaeology, 146
genetics. *See* DNA and genetics
Ghana, Ashanti people, 99, 135, 149
al-Ghazali, Abu Hamid, 204
glabella bossing, 120
Gleser, K., 115
González Martin, A., 116
Gottlieb, Alma, 148
grandmothers: on belly talk, 69; skeletal fetal remains indicating health of, 18–19, 44–45
Grasshopper Pueblo, 120
grave goods: at Gebel Ramlah Neolithic infant cemetery, Egypt, 138, 140–41; in postmedieval Poland, 153, *154, 155*, 158–59
graves. *See* burial treatment
Greece: Kylindra site, Astypalaia, 86; "Rich Athenian Lady" burial, 99
grief, archaeology of, 100
Griffith, J. P. Crozer, 116
Guatemala, osteogenesis imperfecta in, 124

H
Haeckel, Ernst, 235
Halcrow, Siân, 6, 47, 83, 92, 100, 276
Hamm, Dulcia, 267
Han, Sallie, 1, 5, 11, 59, 76, 276, 277
Haraway, Donna, xii
head binding, 116
health and disease: adult health outcomes and placental characteristics, 26; concepts of the fetus and, 74–75; culpabilities of pregnant women for health of the fetus, 60, 67, 74–75; DOHaD theory, 74, 97; grandmothers, fetal indications of health of, 18–

19, 44–45; population health, maternal and fetal health as measure of, 95; skeletal fetal remains indicating, 42–47, 48. *See also specific diseases and health conditions*
heartbeat as fetal voice, 10–11, 252–71; advertisements using fetal images and voice, 267; anthropology of, 269–71, 272–73n8; biological and social forms of being, slippage between, 256; *conocer* and *saber* (affective relation and expert knowledge), 254, 256, 261, 263, 265–66, 268, 271; emotional and biomechanical heart, terms for, 273n12; medical and diagnostic technology linked to political and cultural fetal claims, 256–58, 268–69, 270; methodological approaches to, 258–60; in Oaxacan reproductive health settings, 253–54, 258, 260–61, 262–66, *263*, 269, 270, 272n3, 272n7, 273n11; rhetorical propositions about fetal personhood and, 252–56, 270, 272n1; in US heartbeat bills and legislative anti-abortion activism, 253, 254, 255–56, 259–60, 261–62, 266–69, 270, 272nn4–6
hematoma, 124
Hillson, Simon, 86
Hinkes, M., 120
Hirst, Damien, "The Miraculous Journey" (2013), 72
holoprosencephaly, 121, 280
Holy Innocents, 242–44, *243*, 248n5, 283
homeobox (Hox) genes, 113
Hook, Ernest, 122
Hooton, Earnest, 85
Howes-Mischel, Rebecca, 7, 10, 70, 252, 271–72, 277
Hox (homeobox) genes, 113

humoral medical theory in Morocco, 205, 210, 223n1
hydrocephaly, 116, 280
hypertension, 26, 44, 95
hypoxia, 22, 280

I
Ibrahimi, Mustapha, 218, 220
identity, fetal: African societies, ethnographic evidence from, 135–36, 141; burial treatment and social identity, 146–50, 157–63; frozen embryo adoption and, 172, 177–79, 181–86; liminality of fetuses, 2, 152, 227–28, xii, xiii; as moral entity, in Russian Orthodox anti-abortion activism, 233–34; organism-centered view of, 16; perinates distinguished from older children, 146–48, 149; as protosocial, in Russian Orthodox anti-abortion activism, 229, 234–37, 245–46. *See also* borderless fetus; personhood of fetuses
in utero fetuses, 83, 84, 86, 88–89
in vitro fertilization (IVF) embryos: in Denmark, 233; redemption of (*See* frozen embryos); in Russia, 233–34
India, Rajasthan, infant burial in, 99
infant mortality. *See* mortality, infant
infanticide, 48, 94, 147
infantile cancer, 116
infantile chordoma, 116
infantile cortical hyperostosis, 115, 280
infantile myofibromatosis, 116
infections: bioarchaeology and, 93, 94, 95, 96; birth trauma and, 125; birth weight retarded by, 39; as cause of infant mortality, 93; in fetal paleopathology, 113–14, 117–19; maternal, 36, 94, 95, 96; osteomyelitis, neonatal, 117–18; placenta,

spreading via, 96, 113–14; skeletal fetal remains affected by, 36, 39, 42, 43, 44
influenza, 114
informed consent, concept of, 261
Ingold, Tim, 70; "Materials against Materiality" (2011), 73
iniencephaly, 121, 280
International Right to Life Federation, 267
intracranial hemorrhage, perinatal, 116
Ireland, postmedieval, *cillíní* sites in, 160, 162
Irish, Joel D., 7, 131, 132, 136, 276
iron deficiency, 96
Islam: circumcision and personhood in, 228; Qur'an, 203, 209; Ramadan, restricted food intake during, 25. *See also* Morocco, pregnancy and abortion in
Islamist Justice and Development Party (PJD), Morocco, 218–19
isotope analysis, 98
Israel, Ashkelon perinatal skeletal sample, 94
Italy: L'Aquila, 124; Rome, pre-Christian infant burials in, 99
Ives, Rachel, 46, 116
IVF. *See* in vitro fertilization (IVF) embryos
Ivory Coast: Anyi people, 135; Beng people, 135–36, 148
Ivry, Tsipy, 69

J
Jansson, Thomas, 24
Jantz, Richard, 86
Japan: "belly talk" in, 69–70; *mizuko* in, 159–60, 228, 245, 281
Jews and Judaism, 228, 244
jinn possession in Morocco, 208–10
Johnston, Francis, 85

K
Kabaciński, Jacek, 7, 131, 132, 136, 277

Kausmally, Tania, 116
Keith, Arthur, 122
Kentucky, Indian Knoll skeletal sample, 85
Khrushchev, Nikita, 231
Kirill (Russian Orthodox patriarch), 232
Klippel-Feil syndrome, 120
Kobusiewicz, Michał, 136
Kovacs, Christopher, 45
Kritzer, Robert, 63
Kurki, Helen, 96

L
Lagia, Anna, 48
Langerhans cell histiocytosis (LCH), 116
Law, Jane Marie, 2, 62
Layne, Linda, 158, 162
LCH (Langerhans cell histiocytosis), 116
leprosy, 114
Lewis, Mary E., 6, 35, *36,* 73, 88, 91, 95, 96, 112, 125–26, 149, 276
Life Center (Tsentr Zhizn'), 232, 235, 236, 241, 242, 244, 248n5
Life magazine fetal portraits (1965), 64–65
Lillehammer, Grete, 85, 157
liminality of fetuses, 2, 152, 227–28, xii, xiii
Live Science, 74
low birth weight. *See* small for gestational age (SGA) or low birth weight babies
Lubbock, Sir John, 133
Luehrmann, Sonja, 9–10, 64, 227, 247–48, 277
lytic cranial lesions, 116, 281

M
MacPhee, Marybeth, 210
malaria, 96, 114
Maliki school of Sunni Islam, 201, 203–4, 206
Malinowski, Bronislaw, 61, 62
Margulis, Lynn, 15

marmosets, 20–22
Mathews, Steve, 121
Matthews, Sandra, 64
Mays, Simon, 42, 47, 94
medicalized fetus, 60, 65, 66–68
medieval populations: Christian burial traditions, 151; in England, 46, 120; low life expectancy and high child mortality in, 157–58, 161–62; in Poland, 150; in Romania, 45, 47
Mendonça de Souza, Sheila, 116
meningocele, 116, 121, 124, 281
menstrual irregularity in Morocco, 208–10
Meskell, Lynn, 99
Mexico: Oaxacan reproductive health spaces, heartbeat as fetal voice in, 253–54, 258, 260–61, 262–66, *263*, 269, 270, 272n3, 272n7, 273n11; Olmec culture, La Venta, 62–63, 64, 73, 76n3; Oportunidades (public assistance program), 262
mice, 19
Michaels, Meredith, and Lynn Morgan, eds., *Fetal Subjects, Feminist Positions* (1999), 72, 227
Middle Ages. *See* medieval populations
Millennium Development Goals, 213
miscarriage/spontaneous abortion, 86, 114, 117, 118, 211, 228
Mitchell, Lisa, *Baby's First Picture: Fetal Ultrasound and the Politics of Fetal Subjects* (2001), 2
mizuko, in Japan, 159–60, 228, 245, 281
Mohammed VI (king of Morocco), 221
Mol, Annemarie, 258
Molleson, Theya, 47
Mooney, Max, 121

Morgan, Lynn, 148, 190, 228, 246, 257; *Fetal Subjects, Feminist Positions* (ed., with Meredith Michaels; 1999), 72, 227; *Icons of Life: A Cultural History of Human Embryos* (2009), 1, 63, xii; "Imagining the Unborn in the Ecuadorian Andes" (1997), 62, xii
Moroccan Association for the Fight against Clandestine Abortions (AMLAC), 218
Moroccan Right to Life Association (Jamiyya Maghrebia ul Haq al Hayat), 220
Morocco, pregnancy and abortion in, 9, 200–222; bioethics and health cultures in Islam, 202–4; biomedicalization, authority, and fetal knowledge, 217–19; birth control, 212, 213; contradiction and conjunction between ethnogynacological and biomedical models, 210–13; cravings and birthmarks *(twahm* and *twahima)*, 208; ensoulment in Islam, 202, 210; ethnogynecology of, 205–17, 222; hammam (public bath), use of, 209–10; herbal medications, 209–10, 212; humoral medical theory, 205, 210, 223n1; legal strictures on abortion and illicit sex, 201–2, 208; liberalization of abortion, arguments for, 217–19; "lost" periods due to menstrual irregularity or jinn possession, 208–10; maternal mortality and unsafe abortions, 213; midwives, traditional *(qablat)*, 212–13; permissibility of abortion in Islam, 203–4; politicization and personification of fetuses, 219–21; rape laws, 208; "the sleeping child" *(ragued)*, 205–7,

207, 214; unwed mothers, illegitimate children, and illicit sex in, 201, 204, 206, 209, 213–17, 220
mortality, infant: bioarchaeological evidence for, 95–98; causes of, 93, 112; deferred emotional investment in infants due to, 236–37, 248n3; fertility estimates using perinatal and infant mortality data, 95; neonatal *versus* post-neonatal, 96; prenatal pathology, 113–14; sex differentiated mortality rates, 89, 113; study of, 146
mortality, maternal: causes of birth-related deaths, 89, 95; fetal bioarchaeological evidence for, 95–98; in Morocco, 213
mortuary treatment. *See* burial treatment
mothers: concept of maternal-fetal relationship, 61; culpabilities of pregnant women for health of the fetus, 60, 67, 74–75; experiential maternal connectivity, 16, 19–22; Gebel Ramlah infant cemetery, adult female burials in, 139; identity of child and mother, 59–60; infections, maternal, 36, 94, 95, 96; Morocco, unwed mothers, illegitimate children, and illicit sex in, 201, 204, 206, 209, 213–17, 220; perinates buried with, 89–91, *90*; skeletal fetal remains, maternal health indicated from, 43–47, 48; smoking, maternal, 114. *See also* mortality, maternal
Mount Holyoke College, preserved embryos at, 190
Mudawana reforms (2004), Morocco, 221
multiple cranial suture fusion, 121
multiple fetal pregnancies and births (twins), 91–92, 139

mummies and mummification, 116, 124
Murphy, Eileen, 160
Musallam, Basim, 202, 204
myofibromatosis, infantile, 116

N
National Institutes of Health, 66
National Right to Life Committee, 65
Neanderthals, 70
Neolithic fetuses. *See* Gebel Ramlah Neolithic infant cemetery, Egypt
new bone formation, 114–15, 118
New England Journal of Medicine, 66
New York, 1950s, birth injuries in, 124
Newberry, Dan, 253
Newman, Jessica Marie, 9, 200, 223, 276–77
Nilsson, Lennart, 64–65, 257
nitrogen isotopes, 98
nonsurvivor bias, 40, 87–88, 93
North Dakota, heartbeat bill, 255, 272n4
North Sudan, Late Neolithic cemetery R12, Northern Dongola area, 134

O
Oaks, Laury, 67
Oaxacan reproductive health spaces, heartbeat as fetal voice in, 253–54, 258, 260–61, 262–66, *263,* 269, 270, 272n3, 272n7, 273n11
obstructed labor, 88, 89, 92, 93, 95, 121
Ochs, Elinor, 69
O'Donovan, Edmond, 92
Office of Population Affairs (OPA), 177
Ogden, John, 117–18
Ohio: heartbeat bill, 253, 260, 266–67, 268; Libben sample, 86
Oklahoma, heartbeat bill, 253

OPA (Office of Population Affairs), 177
Oportunidades (Mexican public assistance program), 262
orality, as language ideology, 76n7
Ortner, Donald, 46, 47
osteochondritis, 45, 118, 281
osteochondromas, 121, 282
osteofibrous dysplasia, 120, 282
osteogenesis imperfecta (brittle bone disease), 124, 282
osteomyelitis, neonatal, 117–18
Osterholtz. Anna, 46, 47
Otaño, L., 123
El Othmani, Sâad Eddine, 219
Owsley, Douglas, 86, 89
Oxenham, Marc, 91

P

Pacific Adoptions, 177, 178, 184
paleodemographics, 48
paleopathology of fetuses, 6, 112–25; congenital defects, 119–24, *120, 123*; Down syndrome and other chromosomal defects, 122–23, *123*; infections, 113–14, 117–19; lytic cranial lesions, 116; new bone formation, 114–15, 118; osteogenesis imperfecta (brittle bone disease), 124; osteomyelitis, neonatal, 117–18; potential and limitations of, 112–13; prenatal pathology, 113–14; rubella, 117; skeletal dysplasia, 120–22; skeletal lesions, 114–16; syphilis, congenital, 114, 115, 117, 118–19; trauma and birth injuries, 124–25
Pálfi, György, 118–19
pars basilarus, 116, 282
Patel, Tulsi, 99
paternity testing in Morocco, 214
pathology. *See* health and disease; paleopathology of fetuses; *specific diseases*
Patriarchal Commission on the Family and the Protection of Motherhood and Childhood (Russia), 232
Pearson, Michael Parker, 149
periostitis, 45, 114, 282
Perry, Megan, 151
personhood of fetuses: as concept of the fetus, 60, 68–72; heartbeat as fetal voice and rhetorical propositions about, 252–56, 270, 272n1; Morocco, politicization and personification of fetuses in, 219–21; personhood as socially ascribed status, 147–48; socially embedded personhood, abortion viewed as interrupting, 228–29, 246–47
Peru: Palpa urn burial, 121; skull lesions on six-month-old mummy from, 116
pet talk, 69
Petchesky, Rosalind Pollack, 65–66, 69, 261, 268–69; "Fetal Images: The Power of Visual Culture in the Politics of Reproduction" (1987), 257–58
PJD (Islamist Justice and Development Party), Morocco, 218–19
placenta: infections spreading via, 96, 113–14; malaria infection, placental, 96; medical re-imagination from barrier to sieve, 67; synchronicity, placental, 16, 22–26, *23*
Planned Parenthood of Southeastern Pennsylvania v. Casey (1992), 261, 273n9
Plasmodium vivax or *Plasmodium falciparum* infections, 96
pluripotency, 180
Poland: abortion law in, 231; Russian Orthodox anti-abortion activism and, 233
Poland, postmedieval, burial treatment of fetuses in, 8, 146–63; Christianity and, 150–52,

157, 159; coffins, use of, 153, 154, 156; copper coins, burials with, 153, *154*, 155, 156–57; demographic analysis of, 157–58, 161–62; "deviant" burials, 153, 159–61, 162; Drawsko site, Noteć River Valley, 150; excavation findings, *154–56*, 154–57; grave goods, 153, *154*, 155, 158–59; materials, and methods, 152–53; perinates *versus* older children, 146–48, 149, *155–56*, 156–59, 161, 162; social identity and burial treatment, 146–50, 157–63; survival and preservation of fetal remains, 47; terminology used for, 149–50
Polich, Laura, 76n7
postmortem births or "coffin births," 92–93
"potentiality" of children, as concept, 8, 157, 158, 159, 162
Powell, Theresa, 24
Powers, R., 120
preeclampsia, 44, 45, 49, 265, 273n10, 282
premature births: as cause of infant mortality, 93, 112; as proxies for fetuses in bioarchaeology, 83, 84, 87–88
prenatal diagnostic testing, 68
pro-life activism. *See* anti-abortion activism
Pustye Pesochnitsy (Empty sandboxes; 2012), 245
Putin, Vladimir, 232, 241

Q
Qur'an, 203, 209

R
Rakita, Gordon, 148
Ramadan, restricted food intake during, 25
rape laws in Morocco, 208
Rapp, Rayna, 68, xii
ras al hanout, 209

REDEEM Bionbank, Bay Unversity, 171, 173, 176, 179–81, 186–91, 192
redemption of frozen embryos. *See* frozen embryos
religion. *See specific religions and sects*
rematerialized fetus, 61, 72–75
respiratory distress syndrome, in premature infants, 93
rickets, 43–44, 117, 282
Right to Life, 272n5, xii
Robbins, Gwen, 95
Roe v. Wade (1973), 256
Roman Catholicism. *See* Catholicism
Romania: abortion law in, 231; medieval skeletal population from, 46, 47
Rose, Nikolas, 247
Rothman, Barbara Katz, 68
rubella, 117
Rudolf, Arnold, 117
Russia: birth control in, 230, 240; experience and practice of abortion in, 229–34; Lokomotiv Mesolithic cemetery, Irkutsk, Lake Baikal area, Siberia, 134; materialist embryonic development theory in, 235–36; peasant terms for miscarried/aborted fetuses, 228
Russian Orthodox anti-abortion activism, 9–10, 227–47; baptism in Russian Orthodoxy, 228, 236–37, 245; chastity *(tselomudrie)*, concept of, 238; experience and practice of abortion and, 229–34; Holy Innocents, use of, 242–44, *243*, 248n5; involvement of church in pro-life movement, 232–34; on IVF embryos, 233–34; liminality of the fetus and, 227–28; moral entity, fetus regarded as, 233–34; protosocial qualities of fetus in, 229, 234–37, 245–46;

psychological counseling sessions, abortion candidates required to attend, 231, 234–35; relating unborn children to born children, *239,* 239–40; socially embedded personhood, abortion viewed as interrupting, 228–29, 246–47; spectral fetus imagery, 240–42; visual media used by, 237–44, *239, 243*; Western/North American activism compared, 228, 229, 230, 231, 233, 234, 237, 241, 244, 245, 246; women with past abortions and, 244–46
Rutherford, Julienne, 2, 4, 15, 27, 44, 61, 276, xiii

S

saber and *conocer* (expert knowledge and affective relation), 254, 256, 261, 263, 265–66, 268, 271
Sagan, Dorion, 15
Samoa, treatment of infants in, 69
Sasson, Vanessa, 2, 62
Saudi Arabia, placental study in, 25
Saunders, Shelley, 47
Sayer, Duncan, 92
Scheper-Hughes, Nancy, 152, 248n3
Scheuer, Louise, 35, *36,* 120
Schieffelin, Bambi, 69
Schild, Roman, 133, 138
Schwartz, Jeffrey, 135
sclerosis, long-bone, 45
Scott, Amy B., 1, 8, 11, 47, 73, 146, 163, 276, 277
Scott, Eleanor, 147
scurvy, 43, 44, 45–46, 282
sex differentiated infant mortality rates, 89, 113
sex of skeletal fetal remains, determining, 40–42, *41,* 139, 153
SGA. *See* small for gestational age (SGA) or low birth weight babies
Shoener, Tim, 177, 184

Shopfner, Charles, 114–15
The Silent Scream (film, 1984), 65–66
Sjøvold, T., 121
skeletal dysplasia, 120–22
skeletal fetal remains, 2–3, 34–49; age-at-death, determining, 37–40, *38,* 84, 88, 152–53; community and maternal health indicated from, 43–47; defining the fetus in biological anthropology, 35–36, *36*; health and disease, evidence for, 42–43, 48; inclusion or exclusion from cemetery population, 47–48, 147; methodological approaches, 36–43; nonsurvivor bias, 40; sex, determining, 40–42, *41,* 139, 153; significance of, 34; survival and condition of, 47, 49, 137, 139, 140, 147; taphonomic influences, 47. *See also* Gebel Ramlah Neolithic infant cemetery, Egypt; Poland, postmedieval, burial treatment of fetuses in
Slaughter of the Innocents, 242–44, *243,* 248n5, 283
"the sleeping child" *(ragued)* in Morocco, 205–7, *207,* 214
small for gestational age (SGA) or low birth weight babies: as cause of infant mortality, 93, 112; as proxies for fetuses in bioarchaeology, 83, 84, 87–88, 89–91; rubella as cause of, 117
Smithsonian Forensic Fetal Collection, 41, 121
smoking, maternal, 114
social identity and burial treatment, 98–100
sociocultural anthropology and ethnology, 2–3, 8–11, 61–63, 276–77. *See also* frozen embryos; heartbeat as fetal voice; Morocco, pregnancy and abortion in; Russian Orthodox anti-abortion activism

sonogram. *See* ultrasound and sonogram
soul: Christianity, social identity of fetus in, 148, 157; Islam, ensoulment in, 202; in Russian Orthodox anti-abortion activism, 233–34, 244
sound and hearing: "belly talk," 60, 69–72; fetal heartbeat (*See* heartbeat as fetal voice)
South Africa, 117
South Dakota, Arikara sites, 86, 89
Southwest National Primate Research Center, 20
Spain: Cartagena infant cemetery (8th-2nd century BC), 135; El Molon, 120; Huelva, 119; Olèrdola, Barcelona, 92
spina bifida, 116, 120, 283
spontaneous abortion/miscarriage, 86, 114, 117, 118, 211, 228
stable isotope analysis, 98
Stalin, Joseph, 230, 231
Staphylococcus, penicillin-resistant, 118
Steel, Louise, 99
stem cell research: in Denmark, 233; frozen embryo redemption via, 171, 173, 176, 179–81, 186–91
Stephens, Nicole, 27
Steyn, M., 117
stillbirth, 117, 118, 158, 159, 160
Stoll, Angela, and Stoll embryo, 186–91, 192
Stormer, Nathan, 220, 222
Sunni Islam. *See* Morocco, pregnancy and abortion in
Sweden, St. Clement, Visby, Gotland, 121
syphilis, congenital, 45, 114, 115, 117, 118–19

T
talking to fetuses, 60, 69–72
tamarins, 20–22
Tate, Carolyn, 63, 76n3
Tayles, Nancy, 6, 83, 100, 276

Taylor, Janelle, *The Public Life of the Fetal Sonogram* (2008), 2
TB (tuberculosis), 44, 116
Tennessee: Elizabeth Mounds, 122; fetal homicide law, 75
Teotia, M. and S. P. S., 45
Thailand: Khok Phanom Di, *90*; Non Bon Jak, *87*; twin burials, 92
Thompson, Charis, 187
3D Babies, 72–73
time period definitions for embryo to neonate, 35–36, *36*
Tocheri, M. W., 157
Toga, Basari and Konkombe peoples, 135
Tophet or sacrifice place, 135
trauma, 124–25
Trends in Genetics, 74
Treponema pallidum, 118, 119, 283
triplody, 123, 283
trisomy 21 (Down syndrome), 122–23
Trotter Fetal Collection, Washington University, 41, 47
tuberculosis (TB), 44, 116
twins, 91–92, 139
2PN embryo stage, 171, 184, 189, 190

U
Ucko, Peter, 149
ultrasound and sonogram: first introduction of, 65; images of fetuses, 65, 66, 69, 76n5, 206, 261, 267, 268; legal mandates for abortion candidates, 66, 76n4, 261–62
United Kingdom. *See* England
Utah: multiple cranial suture fusion, 121; perinatal study, 120

V
Verdery, Katherine, 229, 242
viability, concept of, 66, 76n6, 256
Virginia sonogram/transvaginal ultrasound law, 66, 76n4
Visayan traditions (Philippines), 22

Vitamin C levels, 46
Vitamin D levels, 44, 45, 49
voice, fetal. *See* heartbeat as fetal voice
the vulnerable fetus, 60, 64–68

W
Wachtmann, Lynn, 266
Warriors of Life, 237, 238
WebMD, 74
Wendorf, Fred, 133, 138
West Africa: Anyi people, 135; Ashanti people, 99, 135, 149; Beng people, 135–36, 148
Western Fertility Clinic, 173, 174–75, 179
Wexler, Laura, 64
WHO (World Health Organization), 213
Wilke, John, 267
Willis, Anna, 91
Wimberger's sign, 118
World Health Organization (WHO), 213

Z
Zabolotskii, Boris, *Two Mothers (Dve mamy)*, 240–41

Fertility, Reproduction and Sexuality

GENERAL EDITORS:

Soraya Tremayne, Founding Director, Fertility and Reproduction Studies Group, and Research Associate, Institute of Social and Cultural Anthropology, University of Oxford.

Marcia C. Inhorn, William K. Lanman, Jr. Professor of Anthropology and International Affairs, Yale University.

Philip Kreager, Director, Fertility and Reproduction Studies Group, and Research Associate, Institute of Social and Cultural Anthropology and Institute of Human Sciences, University of Oxford.

Volume 1
Managing Reproductive Life: Cross-Cultural Themes in Fertility and Sexuality
Edited by Soraya Tremayne

Volume 2
Modern Babylon? Prostituting Children in Thailand
Heather Montgomery

Volume 3
Reproductive Agency, Medicine and the State: Cultural Transformations in Childbearing
Edited by Maya Unnithan-Kumar

Volume 4
A New Look at Thai AIDS: Perspectives from the Margin
Graham Fordham

Volume 5
Breast Feeding and Sexuality: Behaviour, Beliefs and Taboos among the Gogo Mothers in Tanzania
Mara Mabilia

Volume 6
Ageing without Children: European and Asian Perspectives on Elderly Access to Support Networks
Edited by Philip Kreager and Elisabeth Schröder-Butterfill

Volume 7
Nameless Relations: Anonymity, Melanesia and Reproductive Gift Exchange between British Ova Donors and Recipients
Monica Konrad

Volume 8
Population, Reproduction and Fertility in Melanesia
Edited by Stanley J. Ulijaszek

Volume 9
Conceiving Kinship: Assisted Conception, Procreation and Family in Southern Europez
Monica M. E. Bonaccorso

Volume 10
Where There Is No Midwife: Birth and Loss in Rural India
Sarah Pinto

Volume 11
Reproductive Disruptions: Gender, Technology, and Biopolitics in the New Millennium
Edited by Marcia C. Inhorn

Volume 12
Reconceiving the Second Sex: Men, Masculinity, and Reproduction
Edited by Marcia C. Inhorn, Tine Tjørnhøj-Thomsen, Helene Goldberg, and Maruska la Cour Mosegaard

Volume 13
Transgressive Sex: Subversion and Control in Erotic Encounters
Edited by Hastings Donnan and Fiona Macgowan

Volume 14
European Kinship in the Age of Biotechnology
Edited by Jeanette Edwards and Carles Salazar

Volume 15
Kinship and Beyond: The Genealogical Model Reconsidered
Edited by Sandra Bamford and James Leach

Volume 16
Islam and New Kinship: Reproductive Technology and the Shariah in Lebanon
Morgan Clarke

Volume 17
Childbirth, Midwifery and Concepts of Time
Edited by Christine McCourt

Volume 18
Assisting Reproduction, Testing Genes: Global Encounters with the New Biotechnologies
Edited by Daphna Birenbaum-Carmeli and Marcia C. Inhorn

Volume 19
Kin, Gene, Community: Reproductive Technologies among Jewish Israelis
Edited by Daphna Birenbaum-Carmeli and Yoram S. Carmeli

Volume 20
Abortion in Asia: Local Dilemmas, Global Politics
Edited by Andrea Whittaker

Volume 21
Unsafe Motherhood: Mayan Maternal Mortality and Subjectivity in Post-War Guatemala
Nicole S. Berry

Volume 22
Fatness and the Maternal Body: Women's Experiences of Corporeality and the Shaping of Social Policy
Edited by Maya Unnithan-Kumar and Soraya Tremayne

Volume 23
Islam and Assisted Reproductive Technologies: Sunni and Shia Perspectives
Edited by Marcia C. Inhorn and Soraya Tremayne

Volume 24
Militant Lactivism?: Infant Feeding and Maternal Accountability in the UK and France
Charlotte Faircloth

Volume 25
Pregnancy in Practice: Expectation and Experience in the Contemporary US
Sallie Han

Volume 26
Nighttime Breastfeeding: An American Cultural Dilemma
Cecília Tomori

Volume 27
Globalized Fatherhood
Edited by Marcia C. Inhorn, Wendy Chavkin, and José-Alberto Navarro

Volume 28
Cousin Marriages: Between Tradition, Genetic Risk and Cultural Change
Edited by Alison Shaw and Aviad Raz

Volume 29
Achieving Procreation: Childlessness and IVF in Turkey
Merve Demircioğlu Göknar

Volume 30
Thai *in Vitro*: Gender, Culture and Assisted Reproduction
Andrea Whittaker

Volume 31
Assisted Reproductive Technologies in the Third Phase: Global Encounters and Emerging Moral Worlds
Edited by Kate Hampshire and Bob Simpson

Volume 32
Parenthood between Generations: Transforming Reproductive Cultures
Edited by Siân Pooley and Kaveri Qureshi

Volume 33
Patient-Centred IVF: Bioethics and Care in a Dutch Clinic
Trudie Gerrits

Volume 34
Conceptions: Infertilities and Procreative Technologies in India
Aditya Bharadwaj

Volume 35
The Online World of Surrogacy
Zsuzsa Berend

Volume 36
Fertility, Conjuncture, Difference: Anthropological Approaches to the Heterogeneity of Modern Fertility Declines
Edited by Philip Kreager and Astrid Bochow

Volume 37
The Anthropology of the Fetus
Edited by Sallie Han, Tracy K. Betsinger, and Amy K. Scott

www.ingramcontent.com/pod-product-compliance
Lightning Source LLC
Chambersburg PA
CBHW070910030426
42336CB00014BA/2356